Medium-Energy Antiprotons and the Quark–Gluon Structure of Hadrons

ETTORE MAJORANA
INTERNATIONAL SCIENCE SERIES
Series Editor:
Antonino Zichichi
European Physical Society
Geneva, Switzerland

(PHYSICAL SCIENCES)

Recent volumes in the series:

A Continuation Order Plan is available for this series. A continuation order will bring
delivery of each new volume immediately upon publication. Volumes are billed only
upon actual shipment. For further information please contact the publisher.

Medium-Energy Antiprotons and the Quark–Gluon Structure of Hadrons

Edited by

R. Landua
CERN
Geneva, Switzerland

J.-M. Richard
Institut des Sciences Nucléaires
Grenoble, France

and

R. Klapisch
CERN
Geneva, Switzerland

Springer Science+Business Media, LLC

Library of Congress Cataloging-in-Publication Data

International School of Physics with Low-Energy Antiprotons on Medium
  -Energy Antiprotons and the Quark Gluon Structure of Hadrons (4th :
  1991 : Erice, Italy)
    Medium-energy antiprotons and the quark-gluon structure of hadrons
  / edited by R. Landua, J.-M. Richard, R. Klapisch.
        p.   cm. -- (Ettore Majorana international science series.
  Physical sciences ; v. 58)
    "Proceedings of the Fourth Course of the International School of
  Physics with Low-Energy Antiprotons on Medium-Energy Antiprotons and
  the Quark Gluon Structure of Hadrons, held January 25-31, 1990, in
  Erice, Sicily, Italy"--T.p. verso.
    Includes bibliographical references and index.
    ISBN 978-0-306-44087-8
    1. Antiprotons--Congresses. 2. Hadrons--Congresses. 3. Quark
  -gluon interactions--Congresses.   I. Landua, R.   II. Richard, J.-M.
  (Jean-Marc), 1947- III. Klapisch, R.   IV. Title.   V. Series.
  QC793.5.P72I57 1991
  539.7'212--dc20   91-41362
                                                                   CIP

Proceedings of the Fourth Course of the International School of
Physics with Low-Energy Antiprotons on Medium-Energy Antiprotons
and the Quark–Gluon Structure of Hadrons, held January 25–31, 1990,
in Erice, Sicily, Italy

ISBN 978-1-4615-9581-6   ISBN 978-1-4615-9579-3   (eBook)
DOI 10.1007/978-1-4615-9579-3

© Springer Science+Business Media New York 1991
Originally published by Plenum Press, New York in 1991
Softcover reprint of the hardcover 1st edition 1991

PREFACE

The fourth course of the International School on Physics with Low Energy Antiprotons was held in Erice, Sicily, at the Ettore Majorana Centre for Scientific Culture from 25 to 31 January, 1990. The previous courses covered topics related to fundamental symmetries, light and heavy quark spectroscopy, and antiproton-nucleus interactions. The purpose of this school is to review theoretical and experimental aspects of low energy antiproton physics concerning the quark-gluon structure of hadrons and the dynamics of the antiproton-nucleon interaction. Another important objective is the discussion of future directions of research with low- and medium-energy antiprotons in the context of future medium energy facilities at CERN and elsewhere.

These proceedings contain both the tutorial lectures and the various contributions presented during the school by the participants. The proceedings have been organised in three sections.

The first section is devoted to the theoretical lectures and contributions. The selection of the various subjects wants to emphasize the correlation between antiproton-nucleon physics and the underlying description in terms of quarks and gluons.

The second section contains an overview about 35 years of experiments with antiprotons. It gives an introduction to the particle physics aspects of the field by outlining the historical development of experiment and theory, and by describing the motivation and the results of three recent LEAR experiments in more detail.

The third section contains most of the contributions of the participants describing in more detail certain aspects of current or planned experiments at LEAR.

We thank the lecturers and the participants of the school for their efforts and their contributions. We hope that they found the school as interesting as we did, and we apologize for the six days of foggy weather which however provided for a complete presence of all participants during the whole school. We should like to thank Dr. Alberto Gabriele and the staff of the Ettore Majorana Centre who made the running of the School very easy and our stay extremely pleasant. We are particularly grateful to Mrs. Anne Marie Bugge for her very friendly and efficient help during the preparation of the school as well as for the editing of these proceedings.

R. Landua
J.M. Richard
R. Klapisch

CONTENTS

QCD AND THE ANTINUCLEON – NUCLEON INTERACTION

B. L. Ioffe

Institute of Theoretical and Experimental Physics
117259 Moscow

I   INTRODUCTION

The end of the sixties and the beginning of the seventies were marked by a tremendous progress in elementary particle physics: the strong interaction theory (quantum chromodynamics) was created and the universal theory of electroweak interaction was proposed. Although there are no doubts about basic points of quantum chromodynamics and electroweak interaction theory, both of them possess a number of important problems which are so far unclear.

In spite of full confidence that quantum chromodynamics (QCD) is the true theory of strong interaction, such that all the strong interaction processes, at least within the c.m.s. energy region ≤ 100 GeV can be described in QCD, a number of important problems of this theory, including those of rather low energy physics, are so far unclear (for the contemporary status of of QCD, see ref. 1). This is not surprising—in the history of science there are many such examples. Thus, the superconductivity theory has been formulated only in 1957, although there were almost no doubts that it is a consequence of quantum mechanics, the basic principles of which had been established thirty years before.

The experimental study of unsolved QCD problems does not require to increase the energy of already existing accelerators, but may instead be achieved by their improvement and by improving the experimental methods. Following this way, some of the unsolved problems of the electroweak interaction theory may be elucidated. The solution of these problems will be valuable not only for QCD and the electroweak theory, but also for the development of theories unifying all the interactions existing in Nature, since all of them are based on the same principles of gauge symmetry.

The statement that QCD is a true theory of strong interactions implies that all properties of hadrons, like their masses, the widths of hadronic resonances, the strong interaction cross sections, etc. follow from the QCD Lagrangian

Medium-Energy Antiprotons and the Quark–Gluon Structure of Hadrons
Edited by R. Landua *et al.*, Plenum Press, New York, 1991

$$L = i \sum_q \bar{\psi}_q^a (\nabla_\mu + i m_q) \psi_q^a - \frac{1}{4} G_{\mu\nu}^n G_{\mu\nu}^n$$

$$\nabla_\mu = \partial_\mu + ig \frac{\lambda^n}{2} A_\mu^n \qquad (1)$$

$$G_{\mu\nu}^n = \partial_\mu A_\nu^n - \partial_\nu A_\mu^n - g f^{nml} A_\mu^m A_\nu^l$$

Here ψ_q^a and A_μ^n are quark and gluon fields, $a = 1,2,3$; $n = 1,2 \dots 8$ are colour indices, $q = u$, d, s... are flavour indices, λ_n and f^{nml} are the Gell-Mann matrices and the structure constants of the SU(3) colour group.

We are now sure of many outstanding features of QCD. One of the most important is asymptotic freedom, the statement that the coupling constant in QCD is a running coupling constant

$$\alpha_s (Q^2) = \frac{g^2(Q^2)}{4\pi} = \frac{4\pi}{(11 - \frac{2}{3} n_f) \ln \frac{Q^2}{\Lambda^2}} \qquad (2)$$

decreasing with the logarithm of the square of the momentum transfer (Q^2) in the given process (in eq. 2 n_f is the number of flavours). The asymptotic freedom of QCD implies that QCD is a selfconsistent theory at small distances and the characteristics of the processes originating from small distances - the so-called hard processes - can be calculated in perturbation theory.

Many of the QCD properties at large distances are not well understood up to now - there appear unsolved QCD problems to which I turn.

II   UNSOLVED QCD PROBLEMS

1.   The first unsolved problem of QCD is the **confinement problem** (the confinement of colour objects - quarks and gluons), i.e. the proof of the fact that free quarks or gluons cannot be produced and therefore are not observed in Nature. Although this problem is mainly theoretical, some aspects related to it may be clarified experimentally.

But even if the quark confinement will be mathematically proven, it will not solve all principal problems of QCD. The reason is that the explanation of hadronic physics should not only incorporate the fact of quark confinement itself but also the confinement mechanism, e.g. how quarks or gluons, which are produced at small distances in hard processes, transform into observable hadrons. In other words, what is the mechanism of quark and gluon hadronization? It should be emphasized that all corroborations of QCD based on the study of hard processes possess one general feature: they are independent of the hadronization mechanism. Thus, for example, strong confirmations of QCD were obtained in the processes of hadronic jet production, but not for the production of individual hadrons. The jet production is determined by quark and gluon production at small distances and depends only weakly on the hadronization mechanism, while for the given hadron production process the hadronization mechanism is of importance. In the other confirmation of QCD — in studying deep-inelastic lepton-nucleon scattering — the theory only predicts the dependence of the cross section $\sigma(Q^2,x)$ on the momentum transfer squared at fixed ratio $x = Q^2/2\nu$, where ν is the energy transfer to the nucleon. But a complete theoretical description of the x-dependence of $\sigma(Q^2,x)$ at a fixed Q^2 is absent. The x-dependence is determined by the inverse of the hadronization process of nucleon fragmentation into quarks, which occurs at large distances and is not sufficiently understood theoretically. This is the reason why new experimental information on the properties of quark fragmentation into hadrons and of hadrons into quarks in various hard processes is absolutely necessary in order to check various theoretical models, although these informations will not help to solve the quark confinement problem.

2.   The second important unsolved problem in QCD (after confinement) is the problem of chiral symmetry and its violation. It is well-known that the masses of the u- and d-quarks entering the QCD Lagrangian are very small: $m_u = 4.2$ MeV, $m_d = 7.5$ MeV. The quarks figuring in the QCD Lagrangian are usually called current quarks, and their masses differ strongly from the masses of the constituent quarks $m_u^c \approx m_d^c \approx 350$ MeV, which appear in the nonrelativistic quark model of hadrons. Since m_u and m_d are much smaller than the characteristic hadron masses, they may be neglected with good accuracy and therefore the u- and the d-quark can be considered as massless. Because quarks in QCD interact via the exchange of gluons, the left- and the right-handed quark fields become independent in the massless quark approximation, and two new conservation laws appear for the left-handed (j_μ^L) and the right-handed (j_μ^R) quark currents.

The existence of two independent conservation laws corresponds to the appearance of an additional symmetry, called "chiral symmetry", in the QCD Lagrangian with massless quarks. The absence of transitions between left-handed (with spin opposite to the momentum direction) and right-handed (with spin parallel to the momentum) quarks is analogous to the known situation with massless neutrinos in weak interactions. The reason is that in the Dirac equation for fermions interacting with a vector field, the origin of the spin flip is only the mass term.

However, it can be shown that the chiral symmetry is violated for the physical states, i.e. only one class of solutions is realized in Nature. The chiral-symmetrical class of solutions is not observed. Such a phenomenon is called spontaneous chiral symmetry breaking. The fact that the chiral symmetry is violated follows immediately from the existence of baryons, whose mass is by no means small. Indeed, if the longitudinal quark polarizations do not change in an interaction, the same must be true for baryons consisting of three quarks. But the presence of a baryon mass leads to spin flip reactions when baryons are involved. The other fact demonstrating chiral symmetry breaking is the existence of quark condensate in the QCD vacuum $<o|\bar{\psi}\psi|o> = <o|\bar{\psi}_R\psi_L|o> + <o|\bar{\psi}_L\psi_R|o>$, corresponding to the transition in vacuum of left-handed fields ψ_L into right-handed ones ψ_R with a characteristic hadronic scale $<o|\bar{\psi}\psi|o> = -(240 \text{ MeV})^3$. These two phenomena - the presence of the quark condensate and the presence of baryon masses are tightly related — this is just the idea on which the calculation of the baryon masses in QCD [1] is based.

Evidently, the chiral symmetry breaking must be nonperturbative and proceed at large distances. But a great number of unsolved problems immediately arises here: which is the mechanism of spontaneous chiral symmetry breaking; is there only one mechanism responsible for confinement and chiral symmetry breaking, or are there different mechanisms; at which distances do each of them occur; which is the role of chiral symmetry violation in different processes, and so on.

3.   It is at present not clear which configurations of gluon and quark fields are most important in the QCD vacuum. Among many such configurations intensively studied in the last few years a special role appertains to instantons [2,3]. Instantons are theoretically well-defined configurations of the vacuum gluon field. They have specific properties under colour gauge transformations which result in definite physical consequences. So, I dwell for some time on a discussion of instantons and on their use in low-energy hadron physics.

Instantons are the classical solutions of the QCD equations in euclidean space for gluonic fields which realize the minimum of action and satisfy the selfduality condition

$$G_{\mu\nu}^{n} = \tilde{G}_{\mu\nu}^{n} , \quad \tilde{G}_{\mu\nu}^{n} = \frac{1}{2} \varepsilon_{\mu\nu\lambda\sigma} G_{\lambda\sigma}^{n} \qquad (3)$$

Formally, the instanton is a vector (triplet) representation of the SU(2) subgroup of the SU(3)$_c$ colour group.

The instanton solution is remarkable in the following aspect. It can be shown that in QCD there appears a new quantum number n — the so-called topological charge or winding number

$$n = \frac{\alpha_s}{8\pi} \int d^4x \, G_{\mu\nu}^{n} \, \tilde{G}_{\mu\nu}^{n} \qquad (4)$$

(n is integer). As a consequence, in QCD there exists an infinite set of vacuum states each of which corresponds to a definite eigenvalue n. The true QCD vacuum is a coherent superposition of vacua with different winding numbers. If the instanton solution is substituted into (4), the result is n = 1. Hence it follows the physical meaning of instantons as classical trajectories in the gluonic field space on which the tunnel transition between vacua proceeds. The winding numbers of these vacua differ by unity.

It can be proven that in the region of small distances ($\alpha_s/2\pi \ll 1$) the instanton action is minimal among all the non-perturbative solutions, and in the first quasi-classical approximation   given by

$$S = \frac{2\pi}{\alpha_s} \qquad (5)$$

Therefore, the amplitude of the tunnel transition between two vacua is equal to

$$\exp(-S) = \exp\left(-\frac{2\pi}{\alpha_s}\right) \qquad (6)$$

The running coupling constant α_s should be substituted into Eq.(6) at the value of the characteristic energy scale Q in the given process $\alpha_s = \alpha_s(Q^2)$ [3]. After substituting Eq. (2) into Eq. (6), we have

$$\exp(-S) \sim \left(\frac{Q}{\Lambda}\right)^{-\left[11 - \left(\frac{2}{3}\right)n_f\right]} \qquad (7)$$

As seen in eq. (7), the instanton contribution decreases very rapidly with increasing Q. Another important property of instantons is that they violate chiral invariance: in the presence of the instanton there is a transition of the left-hand quark into the right-hand one, $\psi_L \to \psi_R$.

You see that qualitatively instantons have all the desired properties which are necessary for a description of nonperturbative QCD effects. Now I wish to demonstrate the use of instantons for the qualitative explanation [4] of the Okubo-Zweig-Iizuka (OZI) rule [5]. It is well known that in the nonet of vector mesons ($J^{PC}= 1^{--}$) the ω-meson contains with a rather high accuracy only u- and d-quarks (ω ≈ ūu + đd), while the φ-meson contains only s-quarks (φ ≈ s̄s), and the admixture of u and d quarks in φ-meson is small — about a few or maybe ten percent. The same situation (with even better accuracy) takes place in the tensor meson nonet ($J^{PC} =2^{++}$): the f_2-meson is built of (ūu + đd), and the f_2' of s̄s (OZI rule). In the pseudoscalar nonet the situation is quite opposite: η ≈ (ūu + đd - 2s̄s) is an octet in SU(3) symmetry, and η' ≈ (ūu + đd + s̄s) is a singlet. In quark models the suppression of transitions s̄s → ūu + đd in the vector and tensor channels is "explained" by the OZI rule, which states that the pin-like quark diagrams are suppressed. Of course, in QCD we are not satisfied with such an explanation and the question arises why in QCD the nondiagonal matrix element

$$< ūu + đd \mid s̄s >$$

is small in V- and T-channels, but is equal (times some Clebsch-Gordan coefficient) to the diagonal one in the P-channel

$$< ūu + đd \mid s̄s >_P \quad ≈ \quad < ūu + đd \mid ūu + đd >_P$$

It is also necessary to explain why the situation changes drastically when we go over from light to charmed quarks ($\eta_c ≈ c̄c$). The standard explanation in the framework of perturbative QCD is that the mixing between s̄s and ūu + đd in the V-channel requires a three gluon exchange, while in the P-channel only a two-gluon one (Fig. 1).

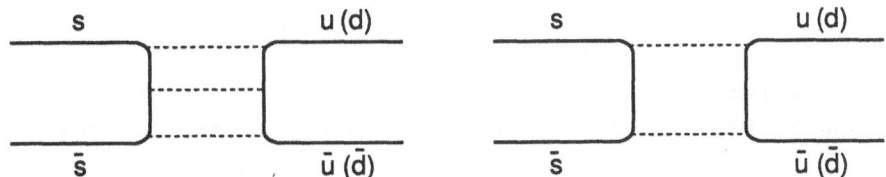

Fig.1. The mixing between |s̄s> and |ūu + đd> states a) in V and b) in P channels

But such an explanation is not satisfactory for the following reasons:

(i) the ratio of the matrix elements is

$$\frac{< s̄s \mid ūu + đd >_P}{< ūu + đd \mid ūu + đd >_P} \sim \alpha_s^2 \tag{8}$$

instead of ~ 1 which is required by experiment;

(ii) in the tensor channel the two-gluon exchange is allowed and in perturbative QCD we expect

$$\frac{< s̄s \mid ūu + đd >_T}{< s̄s \mid ūu + đd >_P} \sim 1 \tag{9}$$

instead of the experimental value ~1/30;

5

(iii) the ratio

$$\frac{< \bar{s}s \mid \bar{u}u + \bar{d}d >_V}{< \bar{s}s \mid \bar{u}u + \bar{d}d >_P} \sim \alpha_s \sim 0.3 \tag{10}$$

is probably too large. For this reason I reject the perturbative QCD explanation.

At the same time the nonperturbative instanton mechanism qualitatively explains all the facts connected with the OZI rule. Let us neglect all perturbative effects and consider quarks as moving in the instanton gluonic field. Then the nondiagonal matrix element, $<\bar{s}s \mid \bar{u}u+\bar{d}d>$ comes from the tunnel transition $\bar{u}u + \bar{d}d \rightarrow \bar{s}s$ in the instanton field. It can be proven that such a transition is strictly forbidden (in the dilute instanton gas approximation) for the case of V- and T-channels and allowed for the case of P- or A-channels. For the case of the vector channel the proof is very simple. If perturbative effects are disregarded, then the nondiagonal matrix element of interest is proportional to

$$< 0 \mid \bar{s} \gamma_\mu s \mid 0 >_{inst} \left[<0 \mid \bar{u} \gamma_\mu u \mid 0 > + <0 \mid \bar{d} \gamma_\mu d \mid 0 > \right]_{inst} \tag{11}$$

and is represented by the diagram of Fig. 2.

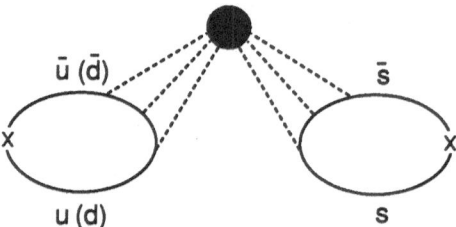

Fig. 2 The diagram representing the transition $\bar{u}u + \bar{d}d \rightarrow \bar{s}s$ in the instanton field. The black
point is the centre of the instanton, the crosses are the sources of quark fields.

Since the instanton gluonic field $A_\mu{}^n$ is a vector in the $SU(2)_c$ subgroup of the $SU(3)_c$ colour group, the whole Lagrangian of quarks and instantons is invariant under transformation $G_c = CI_c$, where C is the charge conjugation and I_c is a 180° rotation around the y-axis in the $SU(2)_c$ subgroup. But $\bar{s}\gamma_\mu s$ and $\bar{u}\gamma_\mu u + \bar{d}\gamma_\mu d$ are odd under this transformation, i.e. (11) is equal to zero.

If we suppose that instantons are responsible for the large $\bar{s}s \rightarrow \bar{u}u + \bar{d}d$ mixing in the pseudoscalar nonet, then the strong energy (or mass) dependence of instanton effects (see eq. 7) explains why the mixing disappears for charmed states. Unfortunately, the dilute instanton gas approximation used above is not selfconsistent: each instanton is characterized by its size r, but the integral over all possible values of r diverges. Therefore, instantons can only give a qualitative explanation and we should instead speak of instanton-like vacuum configurations (see e.g. refs. 6 and 7).

4.   I turn to the next unsolved problem in QCD — the problem of exotic states. In QCD one should expect the appearance of a great number of states (resonances), which are absent in the simple

quark model: these are glueballs — hadrons consisting only of gluons (or possibly with some admixture of quark–antiquark pairs); hybrids (meiktons) — mesons consisting of a quark–antiquark pair in a colour octet state and a gluon; the four-quark states — mesons consisting of two quark–antiquark pairs each in a colour octet state; six-quark states not reduced to two baryons, and so on. Nowadays, there are candidates for some of these states but so far there are no definite conclusions. As for glueballs, the main difficulty to find them experimentally seems to be the fact that glueballs, at least the low-lying ones, contain a rather large admixture of quark-antiquark pairs and as a result there are no explicit experimental signatures for their identification. Clear experimental signatures are also absent permitting to firmly identify four-quark states built of light (u,d) quarks or light meiktons (hybrids). One of the possible ways to experimentally observe exotic states (except for glueballs) is the search for such systems consisting of heavy quarks (strange or charmed). It should be expected that exotic states containing heavy quarks will be rather narrow and their clear signatures can be used to identify them experimentally.

5.   The last unsolved QCD problem which I wish to discuss is the problem of hadronic matter at high temperatures and densities. In some aspects this problem is related to the problem of description of many-body systems (nuclei) in QCD. We wish to have the answers to the questions:

(i) Is there a phase transition of usual hadronic matter into a quark-gluon plasma state and which are the properties of such a transition (1st order, 2nd order, etc.)?
(ii) Are there two phase transitions: one confinement-deconfinement, the other restoration of chiral symmetry?

There are several theoretical approaches to finding the answers to these questions. They are based on:

a) Lattice calculations (see e.g. ref. 8). In these calculations there are clear indications to phase transitions in QCD, but unfortunately most of them were obtained in pure gluodynamics (without quarks); the results in QCD including quarks are not so clear.
b) Chiral theory of massless pions at low temperatures attempting to extrapolate this theory to the temperature of the phase transition (refs. 9, 10) and to match it to high temperature calculations in perturbative QCD (ref. 11). In this approach, some indications to a phase transition were found — the restoration of chiral symmetry at temperatures about 150-200 MeV.

III   THE ANTINUCLEON-NUCLEON INTERACTION. EXPERIMENTS WITH ANTIPROTON BEAMS WHICH MAY ELUCIDATE UNSOLVED QCD PROBLEMS

From my point of view the best suitable instrument for the experimental elucidation of the unsolved QCD problems is an accelerator with an intense monochromatic antiproton beam of variable energy (from ~1 GeV up to 70 - 80   GeV). The required parameters of the beam are

flux $j_{\bar{p}} > 10^7 \, \bar{p}/s$

monochromaticity $\Delta p/p \sim 10^{-4} - 10^{-5}$.

In the calculations below I assume $j_{\bar{p}} = 10^7 \bar{p}/s$ and the luminosity in experiments with a gas jet target $L = 10^{31} \, cm^{-2} \, s^{-1}$. With such a beam one may:

1) study in detail the properties of heavy quarkonia (mainly of charmonium, but there are some chances to study also upsilonium, which would be extremely interesting); find new states, study the quarkonium production mechanism and understand the role of chiral symmetry in these processes; study quarkonia decays;

2) systematically search for exotic states expected in QCD (in particular, hybrid, 4-quark and 6-quark states), and in the case of their discovery - for which there are good chances -study their production and decay mechanisms;

3) study hard exclusive processes ($\bar{p}p \to e^+e^-$, $\bar{p}p \to \gamma\gamma$, $\bar{p}p \to \bar{B}B$ etc.) aiming at elucidating the role of confinement and chiral symmetry breaking in these processes;

4) study the processes of charm and beauty production in $\bar{p}p$–collisions;

5) study the problem of colour transparency and nuclear shadowing as well as the processes with large energy release in nuclear matter in experiments on antiproton-nuclei collisions.

The most promising point of this list is the study of quarkonia. As has been shown during the last ten years, the study of heavy quarkonia was very important in establishing quantum chromodynamics and in elucidating its principal features. There are all reasons to believe that the role of heavy quarkonia is far from being completed and many facts could be understood by studying them. This is so because when going from heavy quarkonia to lighter ones, we gradually move from the small distance region (where all is clear from a theoretical viewpoint) to the large distance region (where all the above mentioned and unsolved problems appear).

An antiproton accelerator has essential advantages over other methods of studying the problems in view, especially of studying charmonium and exotic states. The reason is that $\bar{p}p$ –annihilation can produce practically any such state as a direct channel resonance (in the so-called formation experiments). As in the case of resonances with large mass (larger than 2–3 GeV), their widths must be rather small, so by choosing the appropriate initial energy, the separate study of each resonance, its production mechanism (i.e. angular and spin dependence) and its decay is possible. Since in QCD the production of charmonium or exotic $\bar{c}c$- or $\bar{s}s$-quark states in $\bar{p}p$–annihilation occurs via gluon exchange, i.e. in two stages, it becomes possible to elucidate the chiral symmetry breaking mechanism by analyzing spin and angular correlations. Thus, the antiproton accelerator is a unique machine for studying unsolved problems of quantum chromodynamics. This idea is not new, it was forwarded, for instance, as a result of the 1st Workshop on Antimatter Physics at Low Energies at Fermilab [12]. The above mentioned experiments will be considered in detail below. A part of them was discussed in ref. 12, a part is new.

1.   Study of Charmonium

An antiproton beam with the above parameters will open a new era in studying the charmonium states. So far charmonium was basically studied in e^+e^- collisions where only states ψ_n with the photon numbers $J^{PC}=1^{--}$ are directly produced. All other states could only be obtained as decay products of ψ_n , giving only limited experimental possibilities of their study. In e^+e^- collisions, owing to the large energy spread of e^+e^- beams, the width of a resonance cannot be measured directly. The interaction of charmonium with usual hadrons cannot be studied at all in e^+e^- experiments. Some charmonium data were also obtained from experiments studying charmonium production on proton accelerators (and partially at the ISR), but they were very scarce because of poor statistics.

The use of an antiproton beam will allow the production of all the charmonium states (including C-even) as direct channel resonances and their study on the basis of high statistics. There are very good chances for the discovery of the theoretically expected, but not yet experimentally observed states $^1P_1$ (J^{PC} = 1 $^{+-}$), $^3D_2$ (J^{PC} = 2 $^{--}$), and $^1D_2$ (J^{PC} = 2 $^{-+}$). All of these resonances may be observed by their decay modes into J/ψ (see the discussion of possibilities of their observation in $\bar{p}p$ annihilation in ref. 13).

Of special interest would be the observation of the $^1P_1$ state (m_{th} = 3.51 GeV) which is extremely hard to see in e^+e^- experiments. The $^1P_1$ state could be observed by the isospin violating decay $^1P_1 \to J/\psi + \pi^\circ$, by hadronic decays $^1P_1 \to \varphi\eta$ or $^1P_1 \to \varphi\eta'$ (which must occur with rather high probabilities), and by $^1P_1 \to \gamma\eta, \gamma\eta'$ decays.

Because of the high monochromaticity of the beam (at $\Delta p/p = 10^{-4} - 10^{-5}$ the uncertainty of the charmonium mass will be 10-100 keV), it will be possible to directly measure the level widths and to produce the charmonium states directly at the centre of the peak. The measurement of the angular distribution of the decay products will make it possible to determine the production amplitudes and thereby to study the interaction mechanism of charmonium with usual hadrons. The production cross section of the state χ as a resonance in the direct channel of $\bar{p}p$ annihilation is described by the Breit-Wigner formula

$$\sigma(\bar{p}p \to \chi) = \frac{\pi (2j+1)}{m_\chi^2 - 4m^2} \frac{\Gamma_{tot} \Gamma_{(\chi \to \bar{p}p)}}{(E - m_\chi)^2 + \frac{1}{4}\Gamma_{tot}^2} \tag{12}$$

where m_χ and m are the masses of the χ and the proton, j is the χ spin, Γ_{tot} and $\Gamma(\chi \to \bar{p}p)$ are the total and partial widths of the χ, E is the c.m.s. energy. At the peak

$$\sigma(\bar{p}p \to \chi)_{peak} = \frac{4\pi (2j+1)}{m_\chi^2 - 4m^2} B_2 (\chi \to \bar{p}p), \quad B_2 (\chi \to \bar{p}p) = \frac{\Gamma_{(\chi \to \bar{p}p)}}{\Gamma_{tot}} \tag{13}$$

Consider as an example the production of the C-even state χ_2 (m_χ = 3555 MeV, J^{PC} = 2^{++}, Γ_{tot} = 2.9 ± 1.8 ± 1.1 MeV). According to the data of ref. 14, the $\chi_2 \to \bar{p}p$ decay probability is Br($\chi_2 \to \bar{p}p$) ≈ 10^{-4}. After substituting this value into eq. (13) we find that the cross section of annihilation $\bar{p}p \to \chi_2$ at the peak is equal to 0.3 μb. In experiments with an internal gas jet target at the above-mentioned beam parameters one may expect a luminosity L = 10^{31}cm^{-2}s^{-1}, and the rate would be ≈ 3 χ_2 per second [15].

A fixed target experiment using a liquid hydrogen target would also be possible in this case. The thickness of the hydrogen target would be fixed by the requirement that ionization losses do not move the antiproton momentum outside the resonance peak: for $\Delta m(\chi_2)$ = 1 MeV (in the lab. system ΔE = 3 MeV), the number of χ_2 produced in the experiment would be 0.5 χ_2 per second. (In this case the requirement of the monochromaticity of the beam is comparatively loose — it is sufficient to have $\Delta p/p \approx (3 - 5) \cdot 10^{-4}$). The χ_2 may be observed via the decays $\chi_2 \to \gamma$ J/ψ, $J/\psi \to \mu^+\mu^-$ and $J/\psi \to e^+e^-$. Accounting for the decay branching ratios, the number of events in the experiment (at 100% efficiency) will be ~ 250/hour using a gas jet target at the above luminosity.

The measurement of the γ and μ(e) angular distributions in the processes $\bar{p}p \to \chi_2 \to \gamma$ $J/\psi, J/\psi \to \mu^+\mu^-$ (e^+e^-) permits the choice between various theoretical models of hadronic production of C-even charmonium states being presently discussed, which would elucidate the role of chiral symmetry in this process. For example, if the process occurs according to the gluon fusion model and if chiral symmetry is conserved (gluons may interact with one another and with light quarks), then the angular distribution is proportional to [16]

$$(1 + \cos^2 \theta) (1 + \cos^2 \vartheta) \tag{14}$$

where θ and ϑ are the photon and lepton emission angles relative to the beam direction in the χ_2 rest system. Up to now these angular distributions were measured only by the R704 group at ISR [17]. The statistics was very poor and no definite conclusions can be obtained from these data. Since muons or electrons carry off almost all antiproton energy the background in this experiment must be small.

The production of the other charmonium P-states χ_o and χ_1 may be investigated in an analogous way. Of special interest is the experimental study of the process $\bar{p}p \to \chi_1$ ($J^{PC} = 1^{++}$). The production mechanism of this state in hadronic collisions is completely unknown, since the usual mechanism of gluon-gluon fusion is forbidden (Landau-Yang theorem). There are several possibilities for χ_1 production:

the perturbative mechanism (where g_T and g_L are transverse and longitudinal gluons)

$$(g_L + g_T)_{\text{off mass shell}} \to \chi_1$$

$$(g_T + g_T)_{\text{off mass shell}} \to \chi_1 \tag{15}$$

and the nonperturbative mechanism

$$g + \text{"inst"} \to \chi_1 \tag{16}$$

where "inst" implies some nonperturbative configuration of the gluon field. The perturbative gluon is necessary in order to share the total energy, since the "inst" contribution may very rapidly decrease with energy. The measurement of the angular distribution of the χ_1 decay products can help to solve the problem of χ_1 production and may be very important, especially if it is found that longitudinal gluons (or instantons) contribute to this process. It would also be interesting to check the prediction following from chiral symmetry conservation that the production of χ_o must be suppressed.

Another example is the production of the η_c (m = 2961 MeV, Γ = 11 ± 4 MeV, Br ($\eta_c \to \bar{p}p$) = 0.12 ± 0.06%) [18]. The cross-section of η_c production at the peak is equal to $\sigma(\bar{p}p \to \eta_c) \approx 1\mu b$. The rate for the production with a gas jet target is ~ 10 η_c /s. The η_c may be observed via decays $\eta_c \to 2\gamma$ and for a (theoretically estimated [19]) Br($\eta_c \to 2\gamma$) = 4·10^{-4}, the number of events will be ~ 10/hour. It is quite important in this experiment to measure the partial decay width $\eta_c \to 2\gamma$. The knowledge of this quantity is a crucial test of various theoretical approaches [19]. The background in this experiment arises from the process $\bar{p}p \to \gamma\gamma$. Its estimate (see below) shows that the effect and the background are of the same order of magnitude. Therefore, it will probably be possible to subtract the background by performing the experiment inside and outside of the η_c peak.

There are also good chances to observe and investigate hadronic decay modes of C-even charmonium states, like η_c, $\chi \to K^*\bar{K}^*$, $\varphi\varphi$, $\eta'\pi\pi$, $\eta\pi\pi$ etc. Cross sections times branching ratios for such processes are typically ~10–50 nb and event rates about 100 events/hour. The background problem is more complicated here and requires a special investigation, but exploiting the fact that all detected particles must have large momenta and using the phase analysis it will perhaps be possible to get rid of the background. All the experiments discussed above can also be performed on extracted antiproton beams with momenta p = 4–5 GeV/c.

In a $\bar{p}p$–annihilation experiment with high statistics it is possible to study the well-known charmonium states, like J/ψ, ψ' and observe their rare decay modes. Due to the small J/ψ and ψ' decay widths such experiments should be made with an internal gas jet target. The cross-section of $\bar{p}p \to J/\psi$ at the peak is $\sigma(\bar{p}p \to J/\psi) = 5\,\mu b$. In this case, in order not to leave the peak, the monochromaticity must be at least $\Delta p/p = 2 \cdot 10^{-5}$. The rate is then $50\, J/\psi\, s^{-1}$ or about $5 \cdot 10^8\, J/\psi$ per year , exceeding the total world statistics of observed J/ψ by two orders of magnitude [20].

All the charmonium states discussed above may also be produced in a τ-c factory [21]: J/ψ, ψ' directly in e^+e^- annihilation; χ, η_c states via decays $\psi' \to \gamma\chi$, $J/\psi \to \gamma\eta_c$. The rate of production at the proposed τ-c factory (luminosity $L = 10^{33}\, cm^{-2}\, s^{-1}$) is by an order of magnitude higher for J/ψ, and χ, and of the same order for η_c. But by experiments in a τ-c factory the main goal of the experiments discussed above cannot be achieved, namely to get information about the mechanism of charmonium production in hadronic collisions.

The observation of upsilonium states (especially C-even) in $\bar{p}p$ annihilation, $\bar{p}p \to \chi_b$, would be a very exciting possibility. For such experiments it is necessary to have an antiproton beam energy of ≈ 60 GeV. Let us estimate the rate of events in the process $\bar{p}p \to \chi_{2b} \to \gamma\Upsilon$, $\Upsilon \to \mu^+\mu^-$ (e^+e^-). The width $\Gamma(\chi_{2b} \to \bar{p}p)$ can be estimated from the known value of $\Gamma(\chi_{2c} \to \bar{p}p)$ using the asymptotic quark counting rule relations

$$\Gamma(\chi_{2b} \to \bar{p}p) = \Gamma(\chi_{2c} \to \bar{p}p) \left(\frac{m_{\chi_{2c}}}{m_{\chi_{2b}}} \right)^7 \approx 0.2\, eV \qquad (17)$$

It must be taken into account that in the mass region m = 3-9 GeV the asymptotic relations may not be correct, and due to the large power of the exponent in eq. 17 the error may be as large as one order of magnitude. If we accept for the total χ_{2b} width, the value $\Gamma_{tot}(\chi_{2b}) = 100$ keV obtained in ref. 22, then $Br(\chi_{2b} \to \bar{p}p) = 2 \cdot 10^{-6}$ and

$$\sigma(\bar{p}p \to \chi_{2b})_{peak} = 0.5\, nb \qquad (18)$$

The widths of all upsilonium states below the $\bar{b}b$ threshold are expected to be small — of the order of a few hundred keV or less [22]. Thus, their observation in formation experiments requires a high monochromaticity of the beam ($\Delta p/p \approx 10^{-5}$) and an experimental setup with an internal gas jet target. For the luminosity assumed above the event rate in the process $\bar{p}p \to \chi_{2b} \to \gamma\Upsilon$, $\Upsilon \to \mu^+\mu^-$, e^+e^- will be about 200 events/month.

An unclear problem in such an experiment is the problem of the background. I wish to emphasize again the importance of such an experiment: up to now we have almost no experimental information about the mechanism of hadronic production of upsilonium states and such an information is necessary for the realization of the programme outlined above.

2. The Search for Exotic States

The antiproton beam provides possibilities for searching narrow mesonic states by formation experiments in $\bar{p}p$ annihilation. One of the advantages of this method is that it allows the study of exotic resonances built of strange and charmed quarks where the level of theoretical understanding is higher, and in such a way — starting from heavy quark masses — comes to an understanding of the systems made of light quarks and gluons.

Speaking of the exotic states containing s and c quarks, the most promising are the searches for four-quark mesons ($\bar{s}s\bar{s}s$ and $\bar{c}c\bar{s}s$) and hybrids (meiktons) $\bar{s}sg$ and $\bar{c}cg$. It is expected that all these states are rather narrow, their widths being estimated as $\Gamma_{tot} \leq 50$ MeV. This circumstance enhances the chances for their observation in $\bar{p}p$–annihilation (if, of course, the partial width of their decays into $\bar{p}p$ is not very small).

Let us start the discussion with the consideration of 4-quark $\bar{s}s\bar{s}s$ mesons (κ_{4s}) consisting of two strange quarks and two strange antiquarks. (Each $\bar{s}s$ pair is coloured, so that the transition into a pair of $\bar{s}s$ mesons is suppressed). The calculations performed in the potential model [23] neglecting spin-spin and spin-orbital interactions show that in this approximation there are 4 lowest degenerate states κ_{4s} with common mass $M(\kappa_{4s}) = 2.39$ GeV and quantum numbers $J^{PC} = 0^{++}$, S = 0 (3x3); $J^{PC} = 0^{++}$, S = 0 (6x6); $J^{PC} = 1^+$, S = 1 (3x3) (in parentheses the colour eigenstates of diquarks ss and antidiquarks $\bar{s}\bar{s}$ are shown). The spin-spin interaction results in a splitting (≈ 100 MeV) of the degenerate levels and in some lowering of the mean energy of the four states ($\Delta M \approx 100$ MeV).

The partial width of the decay $\kappa_{4s} \to \bar{p}p$ can be roughly estimated in the following way. We calculate first, using the quark counting relations, the width of the $\eta_c \to \bar{p}p$ decay for an η_c mass of 2.3 GeV. Using the experimental value for the width of the decay $\eta_c \to \bar{p}p$ we find

$$\Gamma(\eta_c \to \bar{p}p)_{m = 2.3 \text{ GeV}} = \Gamma(\eta_c \to \bar{p}p)\left(\frac{m_{\eta_c}}{2.3}\right)^8 = 150 \text{ keV} \qquad (19)$$

Then the estimate of the partial width of the 4-quark state may be obtained from a comparison with the width of the two-quark-state [23]:

$$\frac{\Gamma(\kappa_{4s} \to \bar{p}p)}{\Gamma(\eta_c \to \bar{p}p)} \sim \alpha_s^2 (2m_s)\left(\frac{k_s}{m_s}\right)^6 \sim 10^{-2} - 10^{-3} \qquad (20)$$

where k_s is the momentum in the 4-quark system and $k_s/m_s \approx 1/2$. It follows from eqs. 19 and 20 that $\Gamma(\kappa_{4s} \to \bar{p}p) \approx 100$ eV - 1 keV. (This estimate holds if the process $\kappa_{4s} \to \bar{p}p$ proceeds via two perturbative gluon fusions as well as if it goes in a nonperturbative way via gluon + "inst" [24]). Using this estimate of the partial width and supposing that the total width $\Gamma_{tot}(\kappa_{4s}) \approx 50$–100 MeV, we find the following estimates of the cross sections for $\bar{p}p$–annihilation into 4-quark $\bar{s}s\bar{s}s$ states

$$\sigma(\bar{p}p \to \kappa_{4s}) \sim 10 - 100 \text{ nb} \qquad (21)$$

The main decay mode of the κ_{4s} state with $J^{PC} = 2^{++}$ must be the decay $\kappa_{4s} \to \varphi\varphi$ [24], the expected event rate being about 10^2–10^3/hour. The 4-quark κ_{4s} resonance with quantum numbers $J^{PC} = 1^+$, which is probably about 100 MeV lighter than the resonance with $J^{PC} = 2^{++}$, may be observed via its main decay modes into $\varphi\eta$ and $\varphi\eta'$ on about the same statistical level. Of course, one should not forget that the above estimates are very rough.

Now there are some candidates for exotic states in the mass region ≥ 2 GeV, but the experimental information is very scarce and the theoretical interpretation is unclear. In the lower mass region ~1.5 GeV there are also several candidates for exotic states, in particular, the resonances found at IHEP with the masses m = 1.59 GeV [25] and m = 1.49 GeV [26]. Their nature is also unclear and probably their observation in $\bar{p}p$–annihilation (in production experiments) may help to solve the problem.

The analogous estimates for 4-quark mesons $c\bar{c}s\bar{s}$ built of charmed and strange quarks are $M(c\bar{c}s\bar{s}) \approx 4.5$ GeV, $\Gamma(\kappa_{4cs} \to \bar{p}p) \sim 1\text{--}10$ eV, $\sigma(\bar{p}p \to \kappa_{4s}) \approx 0.1 - 1$ nb. The main decay mode of the resonance with $J^{PC} = 2^{++}$ is $\kappa_{4cs} \to J/\psi + \varphi$.

For meiktons (hybrids) only mass calculations are available, showing that $m(\bar{s}sg, J^P=1^+) = 1.75$ GeV [27] and $m(\bar{c}cg) > 4$ GeV [28], and there are not even rough estimates of the partial $\bar{p}p$ decay widths. On general grounds one may expect that these partial decay widths are not small (in any case they are much larger than the partial $\bar{p}p$ decay widths of 4-quark states). Therefore, the search for $\bar{c}cg$ states and $\bar{s}sg$ states in $\bar{p}p$ formation experiments is not hopeless.

It is very interesting to search in $\bar{p}p$–annihilation for the narrow resonances recently observed at CERN [29] and JINR [30] with masses in the vicinity of 3.1 GeV and widths $\Gamma < 20$ MeV, whose decay branching ratios into states involving baryon-antibaryon pairs is rather high. The nature of these states is unclear (6-quark states?) but their comparatively large production cross-sections in NN collisions ($\geq$ 1 μb), their narrow widths and the large probability of decays into baryon–antibaryon pairs — all this tells us that if such resonances do exist, the chances to observe them and to investigate their properties in $\bar{p}p$–annihilation are very high.

3.   Hard Processes

In the discussion of the possibilities of an antiproton machine for the study of hard processes it is useful to dwell on the following items:

a) The check of perturbative QCD by measuring electromagnetic form factors in the region of time-like momentum transfers $Q^2 = s$ in the reaction $\bar{p}p \to e^+e^-$ (or $\mu^+\mu^-$). In the space-like region the proton form factors are known up to $Q^2 = -30$ GeV^2, but in the time-like region there are, up to now, only data in the threshold region $Q^2 = s = 4\text{--}5$ GeV^2. The experimental data on the proton form factors in the time-like region at large Q^2 would be very important for the check of different approaches describing exclusive processes in QCD at high momentum transfer. In particular they may give an answer to the question at what momentum transfer the asymptotic region starts and what the quark wave function of proton is (see the review in ref. 31). At $Q^2 = 4\text{--}5$ GeV^2, $\sigma(\bar{p}p \to e^+e^-) \approx 1$ nb. A tremendous advance would be achieved if it were possible to measure $\sigma(\bar{p}p \to e^+e^-)$ up to values of order 1 pb, i.e. to go up to $Q^2 \sim 10 - 15$ GeV^2.

b) The study of reaction $\bar{p}p \to \gamma\gamma$. As was repeatedly emphasized by Brodsky (see, e.g. refs. 32 and 33), the measurements of the cross-section $\bar{p}p \to \gamma\gamma$ (or the inverse process $\gamma\gamma \to \bar{p}p$) are extremely important for checking the theory of exclusive processes in QCD. It is expected [32, 33] that in this process the asymptotic region starts at $s \approx 3$ GeV^2, i.e. much earlier than in the nucleon formfactor. Up to now there are only the JADE group data [34] on the inverse process $\gamma\gamma \to \bar{p}p$ in the threshold region $\sqrt{s} = 2.0 - 2.5$ GeV. According to these data $\sigma(\gamma\gamma \to \bar{p}p)_{\sqrt{s}=2.3\,\text{GeV}} = 4 \pm 1$ nb. This corresponds to a cross section of the $\bar{p}p \rightarrow \gamma\gamma$ reaction $\sigma(\bar{p}p \rightarrow \gamma\gamma)_{\sqrt{s} = 2.3\,GeV} = 7$ nb. The asymptotic theory of exclusive processes predicts that $\sigma(\bar{p}p \rightarrow \gamma\gamma) \sim s^{-5}$. Accepting this energy dependence and supposing that it will be possible to measure the cross-sections of the $\bar{p}p \rightarrow \gamma\gamma$ process up to the values of ~10 pb, we come to the conclusion that in experiments with $\bar{p}$ beams such a reaction may be studied up to $s \approx 20\,GeV^2$. The measurement of the γ angular distribution would be very important to check the theory in more detail.

c) The investigation of the Drell-Yan process — production of lepton pairs in the reaction

$$\bar{p} + N \rightarrow \mu^+ \mu^- + \text{all} \quad (\text{or } e^+ e^- + \text{all}) \tag{22}$$

The use of antiprotons instead of protons in the Drell-Yan process has the advantage that the antiquark distribution, which gives the main contribution to this reaction, is well known in the antiproton but badly known in proton. Therefore, in the cases where a projectile is a proton, the observation of some subtle effects in Drell-Yan processes may be obscured by the poor knowledge of the antiquark distributions. In this respect, the results of the measurements of the above reaction (22) are more transparent. The process (22) was experimentally studied in the ISR and SPS-colliders at high energies and for large lepton pair masses. The region of smaller energies was not sufficiently studied. Its examination may help to understand the preasymptotic QCD regime — the role of higher twist terms, the accuracy of the factorization hypothesis, etc. In the heavy nucleus experiments it is interesting to check the prediction on the increase of longitudinal distances with energy. Such an effect must result in violation of factorization which is restored again with the increase of lepton pair mass.

d) Investigation of the reaction $\bar{p}p \rightarrow \bar{Y}Y$ (where Y is a hyperon) at high momentum transfer. In these cases the measurements of the asymmetry in the hyperon and antihyperon decays give information about their respective polarizations, permitting to check the predictions of QCD in more details than from the measurements of cross-sections alone.

4. Production of Charm and Beauty

At ~ 50 GeV antiproton energies it is possible to observe the production of $\bar{D}D$ meson pairs by measuring their leptonic decays. There are also good chances to observe charmed baryon pair production in inclusive and exclusive processes like $\bar{p}p \rightarrow \bar{\Lambda}_c \Lambda_c$. One may expect that the cross section of the latter process at antiproton momenta about 20–40 GeV/c will be of order 1 μb.

Of special interest would be the observation of beauty particles in the threshold region $p_{\bar{p}} \approx 80$ GeV/c

$$\bar{p}p \rightarrow \bar{B}B \quad \text{or} \quad \bar{p}p \rightarrow \bar{\Lambda}_b \Lambda_b \quad (+ \text{maybe } n\pi)$$

The optimistic estimate of the cross-section of this process is $\sigma \sim 1$ nb. Perhaps more accurate calculations can be performed.

5. Interactions of Antiprotons with Nucleus

In the experiments on $\bar{p}$ + nucleus interactions the problems of colour transparency and nuclear shadowing may be elucidated. An interesting example of such an experiment was recently proposed by

Brodsky and Mueller [35]. They showed that the cross section of J/ψ interactions with nucleons can be found by studying the A-dependence of J/ψ production in p̄–nucleus collisions in the region near the threshold of J/ψ production. In such an experiment, the J/ψ state is formed inside the nucleus and the deviation of cross section from $\sigma(\bar{p}A \rightarrow J/\psi + ..) \sim A$ is connected with the J/ψ interaction with nucleons inside the nucleus. The situation in this experiment is quite different from that in the J/ψ photoproduction on nuclei, where the J/ψ formation length is much larger than the nuclear size and objects which interact inside the nucleus are free c̄c quarks.

p̄+ nucleus collisions at momenta ~ 10 GeV/c differ from p + nucleus collisions by the fact that in the former a much larger part of the energy is dispersed inside the nucleus due to the larger cross section $\sigma(\bar{p}N)$ compared to $\sigma(pN)$. Therefore, the study of these processes is interesting from the point of view of obtaining high local densities of energy inside the nucleus and of approaching the state of the quark-gluon plasma.

IV. CONCLUDING REMARKS

The list presented above of possible experiments which can be performed with an antiproton beam is by no means complete. For example, it leaves aside experiments with polarized targets or (and) beam, which, as last year's experience shows, give very important information. I have no doubts that a lot of other good experiments can be suggested, maybe even much better than those discussed above. The goal of the exposition was to demonstrate that even on the above-stated base an extended and deep programme of physical investigations for an antiproton accelerator may be formulated.

In conclusion, I wish to repeat my point of view that experiments with antiprotons with good monochromaticity is now the best way for studying physical problems of strong interactions — quantum chromodynamics.

REFERENCES

[1]   B.L. Ioffe, Status of QCD, Proc. XXII Int. Conf. on High Energy Physics, Leipzig, II:176 (1984); B.L. Ioffe, Proc. XX Winter School LINP, Leningrad, p.113 (1985).

[2]   A.A. Belavin, A.M. Polyakov, A.S. Schwartz, Yu.S. Tyupkin, Phys.Lett. 59B:85 (1975).

[3]   G. 't Hooft, Phys.Rev. D14:3432 (1976).

[4]   B V. Geshkenbein, B.L. Ioffe, Nucl. Phys. B166: 340 (1980) .

[5]   S. Okubo, Phys. Lett. 5:165 (1963). G. Zweig, "Symmetries in Elementary Particle Physics", Academic Press, N. Y., p. 162 (1965).

[6]   D.I. Dyakonov, V.Yu. Petrov, Nucl. Phys. B245:259 (1984), and B272: 457 (1986) .

[7]   E.V. Shuryak, Phys. Rep. 115:151 (1984).

[8]   E.V.E. Kovacs, D.K. Sinclair, J.B. Kogut, Phys.Rev.Lett. 58:751 (1987); A.Ukawa, Report at VII Int. Conf. on Ultra-Relativistic Nucleus, Nucleus Collisions and Quark Matter, Lenox, Mass. (1988), preprint CERN-TH 5266/88 (1988); R.V. Gavai et al., preprint CERN-TH 5530/89 (1989).

[9]   J.Gasser, H Leutwyler, Phys.Lett. B188:477 (1987).

[10]   P. Gerber, H. Leutwyler, Bern Univ. preprint BUTP-88/30 (1988).

[11]   V.L. Eletsky, B,L. Ioffe, Yad. Fiz. 48:602 (1988).

[12]   Proc. First Workshop on Antimatter Physics at Low Energy, FNAL, Chicago, April 1986.

[13]   R.L. Jaffe, ibid, p.1.

[14]   C. Baglin et al., Phys. Rev. 172B:455 (1986).

[15]   P. A. Rapidis, Ref. 12, p. 83 .

[16]   B. L. Ioffe, Phys. Rev. Lett, 39:1589 (1977).

[17]   C. Baglin et al., preprint CERN-EP/87-94 (1987).

[18]   "Review of Particle Properties", Phys.Lett. B204, April (1988).

[19]   V.A. Khose, M.A. Shifman, Uspekhi Fiz. Nauk. 140:3 (1983).

[20]   M.S. Chanowitz, Ref. 12, p. 393.

[21]   R.H. Schindler, preprint SLAC- PUB-4995, 4996 (1989).

[22]   M.B. Voloshin, Yu.M. Zaitzev, Uspekhi Fiz. Nauk. 152:361 (1987).

[23]   A.M. Badalyan, B.L. Ioffe, A.V. Smilga, Nucl. Phys. B281:85 (1987).

[24]   A.M. Badalyan, B.L. Ioffe, Yad . Fiz . 43:1340 (1986) .

[25]   D. Alde et al., Nucl.Phys. B269:485 (1986).

[26]   S.I. Bityukov et al., Pis'ma Zh. Exp. Theor. Fiz. 42:384 (1985); Yad. Fiz. 38:1205 (1983).

[27]   I.I. Balitsky, D.I. Dyakonov, A.V. Yung, Z. Phys. C33: 265 (1986) .

[28]   J.Covaerts, L.J. Reinders, P. Francken, X. Couze, J. Weyers, preprint Inst. de Phys. Theor., Univ. Catholique de Louvain, BUTP-86/14 (1986) .

[29]   M. Bourquin et al., Phys. Lett. B172:113 (1986).

[30]   A.N. Aleev et al., JINR Rapid Communications N-19-86 (1986).

[31]   V.L. Chernyak, A.R. Zhitnitsky, Phys. Rep. 112, No. 3, 4 (1984).

[32]   S.J. Brodsky, VII Int. Workshop on Photon-Photon Collisions, College de France, Paris, April 1986, preprint SLAC-PUB-4022 (1986) .

[33]   S.J. Brodsky, Ref. 12, p.131.

[34]   R. Brandelik et al., Phys. Lett. 108B: 67 (1982).

[35]   S.J. Brodsky, A.H. Mueller, Phys. Lett. B206:685 (1988).

ASPECTS OF MESON SPECTROSCOPY WITH $\overline{N}N$ ANNIHILATION

Carl B. Dover

Department of Physics
Brookhaven National Laboratory
Upton, New York 11973

ABSTRACT

We focus on the potentialities of nucleon–antinucleon $(\overline{N}N)$ annihilation as a means of producing new mesonic states. The case for the existence of quasinuclear $\overline{N}N$ bound states is discussed in detail. Strong evidence for a $2^{++}(0^+)$ state of this type has been obtained at LEAR in annihilation from the p–wave $(L = 1)$ $\overline{N}N$ system, in support of earlier sightings of this object in $L = 0$ annihilation at Brookhaven. In the next generation of LEAR experiments, the emphasis shifts to the search for mesons containing dynamical excitations of the gluonic field, namely glueballs and hybrids $(Q\overline{Q}g)$. We discuss some features of the masses, decay branching ratios and production mechanisms for these states, and suggest particular $\overline{N}N$ annihilation channels which are optimal for their discovery.

1. INTRODUCTION

Through the annihilation process, the $\overline{N}N$ system provides a rich source of mesons. Some mesonic states, for instance the E(1420), were first seen in the debris of $\overline{N}N$ annihilation products[1]. Although the final state is composed mostly of pions (multiplicity of about 5 at threshold), the main features of annihilation at low energies can be accounted for via a sequential process $\overline{N}N \rightarrow M_1M_2$, followed by the strong decay of M_1 and M_2 into two or three pions[2]. The quasi–two–body modes M_1M_2 stand out prominently in Dalitz plots of the invariant mass distributions of such two and three meson sub–systems; for instance, the vector mesons (ρ, ω) occur frequently as annihilation products. To search for new mesons X, the two–body mode $\overline{N}N \rightarrow \pi X$ is of special interest. Accordingly, we focus attention on the selection rules for such reactions, as well as the very important kinematical factors which intervene for production of X from initial $L = 0, 1$ states.

These lectures are organized as follows: We first review (in Sect. 2) the general features of the $\overline{N}N$ interaction. The key point here is the existence of very strong tensor forces for isospin $I = 0$ $\overline{N}N$ configurations. This coherent tensor interaction reg-

Medium-Energy Antiprotons and the Quark–Gluon Structure of Hadrons
Edited by R. Landua *et al.*, Plenum Press, New York, 1991

17

isters its effect in cross section differences for $\bar{p}p$ and $\bar{n}p$ at low energies[3], dramatic behavior in certain spin observables[4], and a band of natural parity $\left(0^{++}, 1^{--}, 2^{++} \ldots\right)$ quasinuclear (QN) bound states with $I = 0$.[5]

In Sect. 3, we argue that at least one member of the QN band has in fact been seen by the ASTERIX collaboration at LEAR, namely the AX(1565) meson[6]. We present calculations of the mass and width, as well as the production and decay branching ratios of such objects. The preferential production of the AX(1565) from $L = 1$ rather than $L = 0$ $\overline{N}N$ states is naturally explained, using techniques developed[7] for a general discussion of reactions of the type $\overline{N}N \to \gamma X, \pi X$. The AX(1565) is most likely the same object as seen earlier at Brookhaven in reactions $\bar{p}n \to 2\pi^-\pi^+$[8] and $\bar{p}n \to 3\pi^-2\pi^+$.[9] In the $\overline{N}N$ potential model, the coherent tensor interaction provides the strong attraction necessary to achieve the rather substantial binding energy observed[10]. This is an essentially non–perturbative effect; in QCD, the tensor force obtained from one–gluon exchange is much weaker. We also consider alternative explanations of the nature of the AX(1565), which must lie outside of the nonet of 2^{++} tensor mesons. Interpretations in terms of a glueball, four quark $\left(Q^2\overline{Q}^2\right)$ or hybrid $(Q\overline{Q}\,g)$ state are found to be unsatisfactory. Crucial experimental tests of the quasinuclear hypothesis are suggested (search for the $\eta\eta$ decay mode, etc.).

In Sec. 4, we review the theoretical predictions for the masses and preferred decay modes of the $Q\overline{Q}\,g \equiv X_g$ hybrid mesons. It is argued that the reaction $\overline{N}N \to \pi X_g$ (or γX_g, $\pi\pi X_g$ in favorable cases) will proceed with measurable branching ratios. The case of *exotic* X_g (ex. 1^{-+}) is worthy of special attention; the candidate X_g state seen in the $\pi^0\eta$ decay channel[11] should be seen in the $\bar{p}p \to \pi^0\pi^0\eta$ reaction, for instance. We present arguments for identifying the E(1420) meson seen in the $\bar{p}p \to \pi^+\pi^-E$ reaction as the 0^{-+} (0^+) member of the $Q\overline{Q}\,g$ multiplet obtained by coupling a $Q\overline{Q}$ pair with the quantum numbers of the ω meson to a transverse electric (TE) gluon normal mode.

The $\overline{N}N$ annihilation modes involving ϕ mesons are of particular interest. Here the signal of new physics may be found in the order of magnitude violations of the Zweig rule which are found in certain channels, for instance $\overline{N}N \to \pi\phi$. It has been suggested[12], and disputed[13,14], that such Zweig rule violations signal a soft strange quark $(Q^3s\bar{s})$ component in the nucleon wave function. An alternate interpretation in terms of broad four quark $(Q\overline{Q}s\bar{s})$ resonances has also been suggested[15].

A global conclusion of these lectures is that $\overline{N}N$ annihilation offers exciting prospects for the discovery of new mesonic states, including the elusive exotics. It forms an essential complement to πp, pp, K^-p, e^+e^- and $\gamma\gamma$ channels, for instance, as a means of studying meson spectroscopy. The physics program of the new generation of LEAR detectors (Crystal Barrel, Obelix and Jetset) should be given high priority and vigorously pursued.

2. QUASINUCLEAR $\overline{N}N$ BOUND STATES

Models for the $\overline{N}N$ potential at low energies, as well as the nature of the bound state spectrum, have been discussed by numerous authors[16-25]. The components of the NN and $\overline{N}N$ potentials due to t–channel meson exchange are related by the

Table 1. Signs of Meson Exchange Potentials for the NN System

	$I = 0$				$I = 1$			
	V_0	V_σ	V_{LS}	V_T	V_0	V_σ	V_{LS}	V_T
π	0	$-$	0	$-$	0	$+$	0	$+$
η	0	$+$	0	$+$	0	$+$	0	$+$
ρ	$-$	$-$	$+$	$+$	$+$	$+$	$-$	$-$
ω	$+$	$+$	$-$	$-$	$+$	$+$	$-$	$-$
δ	$+$	0	$+$	0	$-$	0	$-$	0
σ	$-$	0	$-$	0	$-$	0	$-$	0

G–parity transformation

$$V_{NN} = \sum_i V_i$$

$$V_{\overline{N}N} = \sum_i G_i V_i \tag{1}$$

where $i = \{\pi, \eta\}, \{\rho, \omega\}, \{\delta, \sigma\}$ correspond to exchanges of pseudoscalar (0^-), vector (1^-) and scalar (0^+) mesons, respectively. The real potential $V_{\overline{N}N}$ must be supplemented by a complex annihilation potential V_A to obtain a realistic optical potential V_{opt} to describe low and medium energy elastic scattering, charge exchange, annihilation cross sections, and spin observables. The potential $V_{\overline{N}N}$ is strongly attractive in certain channels, and supports a spectrum of bound states; these acquire a strong decay width Γ through the action of Im V_A. Generally, these quasinuclear (QN) bound states of the $\overline{N}N$ system will have typical hadronic widths of the order of $\Gamma \approx$ 100–200 MeV, although in some cases narrower states are possible. These QN states are not to be confused with the putative long–lived "baryonium" states of $Q^2\overline{Q}^2$ character, whose narrow width had its origin in a topological selection rule[26].

The level order of the lowest–lying states of the $\overline{N}N$ spectrum is essentially model independent (not the absolute mass values!), and can be predicted[18] on the basis of the *strong coherences* of the spin–isospin dependent components of the meson exchange potential $\sum_i G_i V_i$. Each component V_i can be decomposed into central, spin–spin, spin–orbit and tensor pieces:

$$V_i = \begin{pmatrix} 1 \\ \tau_1 \cdot \tau_2 \end{pmatrix} (V_0 + V_\sigma \sigma_1 \cdot \sigma_2 + V_{LS} L \cdot S + V_T S_{12}) \tag{2}$$

where $S_{12} = 3\sigma_1 \cdot \hat{r}\sigma_2 \cdot \hat{r} - \sigma_1 \cdot \sigma_2$ is the tensor operator. The *signs* of the various terms V_0, V_σ, V_{LS} and V_T are determined quite generally from the form of the interaction Lagrangian which couples mesons of different spin–parity to nucleons. The pattern of signs for pseudoscalar (π, η), vector (ρ, ω) and scalar (δ, ϵ) exchange is displayed in Tables 1 and 2 for NN and $\overline{N}N$ systems. A *coherence property*[18] becomes immediately evident: for NN, V_σ and V_{LS} terms all have the same sign for $I = 1$, while for $\overline{N}N$ in $I = 0$ states, V_0 and V_T components are coherent. This striking contrast between NN and $\overline{N}N$ spin–isospin dependences is an effect of the *G*–parity transformation.

Table 2. Signs of Meson Exchange Potentials for the $\overline{N}N$ System

	$I = 0$				$I = 1$			
	V_0	V_σ	V_{LS}	V_T	V_0	V_σ	V_{LS}	V_T
π	0	+	0	+	0	−	0	−
η	0	+	0	+	0	+	0	+
ρ	−	−	+	+	+	+	−	−
ω	−	−	+	+	−	−	+	+
δ	−	0	−	0	+	0	+	0
σ	−	0	−	0	−	0	−	0

The influence of coherences is dramatic, both for NN scattering and for the spectrum of $\overline{N}N$ bound states. As an example, we display in Figure 1 the energy dependence of the $^{33}P_0$ and $^{13}D_2$ phase shifts δ for NN scattering. In both channels, the dominant tensor component of the π exchange potential is attractive, and generates a substantial $\delta > 0$. For $^{33}P_0$, the shorter range but coherent *repulsive* spin-orbit potentials due to scalar and vector exchange counteract the pion tensor force at higher energies, and produce the observed zero in δ. For $^{13}D_2$, on the other hand, the coherence is absent, and the exact δ remains close to the result for π exchange only.

In the $\overline{N}N$ system, the effects of spin dependent potentials are also dramatic, but now the tensor rather than spin–orbit components are most important. In Fig. 2, we illustrate how the predicted $\bar{p}p$ polarization is altered when various spin dependent terms are set to zero. $P(\theta)$ is seen to arise almost entirely from the coherent tensor forces, while $L \cdot S$ and $\sigma_1 \cdot \sigma_2$ terms play a minor role. The tensor effects also dominate for other $\bar{p}p$ spin observables[4].

The spectrum of quasinuclear bound states is influenced in a qualitative way by the tensor force. For $I = 0$, natural parity $(L = J \pm 1)$ states, the $\overline{N}N$ eigenfunctions[5] are close to one of the configurations $|\alpha_J\rangle$ or $|\beta_J\rangle$ which diagonalize S_{12}, namely

$$|\alpha_J\rangle = \left[(J+1)^{1/2}|L = J - 1\rangle + J^{1/2}|L = J + 1\rangle\right] / (2J + 1)^{1/2}$$
$$|\beta_J\rangle = \left[-J^{1/2}|L = J - 1\rangle + (J+1)^{1/2}|L = J + 1\rangle\right] / (2J + 1)^{1/2} \qquad (3)$$

In this basis, the $\overline{N}N$ potential has the form

$$V_{\alpha\alpha} = V_0 + V_\sigma - V_{LS} + 2V_T + \frac{\hbar^2}{m_N} \frac{J(J+1)}{r^2}$$
$$V_{\beta\beta} = V_0 + V_\sigma - 2V_{LS} - 4V_T + \frac{\hbar^2}{m_N} \frac{J^2 + J + 2}{r^2}$$
$$V_{\alpha\beta} = [J(J+1)]^{1/2} \left[-V_{LS} + 2\hbar^2/m_N r^2\right] \qquad (4)$$

For $I = 0$, the effective tensor potential is repulsive in channel α and attractive in channel β; the effect of the coupling term $V_{\alpha\beta}$ is small[5]. In Fig. 3, we plot $V_{\beta\beta}(r)$

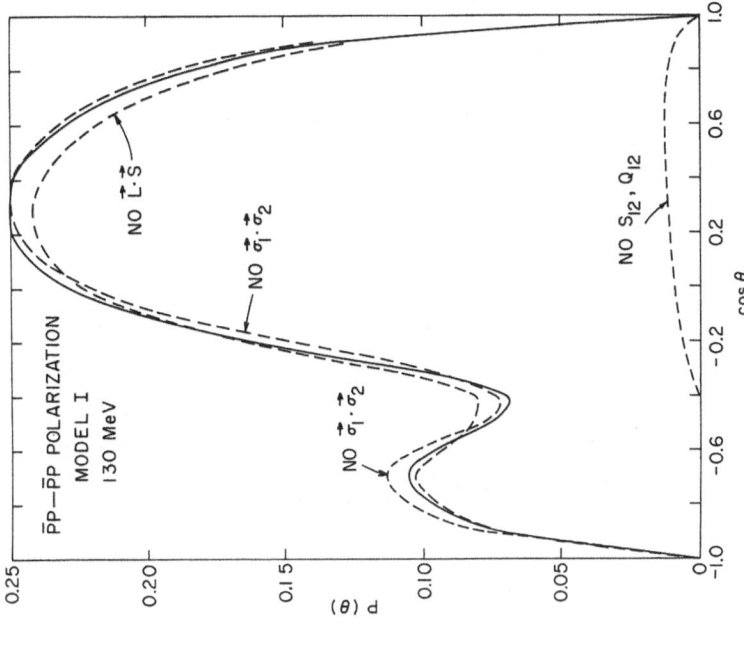

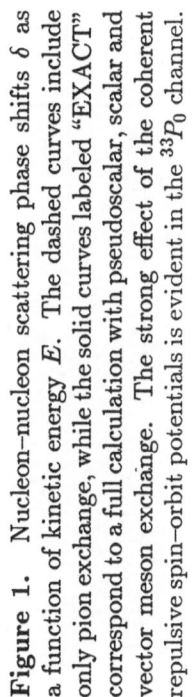

Figure 1. Nucleon–nucleon scattering phase shifts δ as a function of kinetic energy E. The dashed curves include only pion exchange, while the solid curves labeled "EXACT" correspond to a full calculation with pseudoscalar, scalar and vector meson exchange. The strong effect of the coherent repulsive spin–orbit potentials is evident in the $^{33}P_0$ channel.

Figure 2. Angular distribution of elastic $\bar{p}p$ polarization $P(\theta)$. The solid curve is the result of an optical model calculation[4], and the dashed curves show the effect of turning off various spin–dependent components of the meson exchange potential. The effect of removing the tensor (S_{12}) term is dramatic.

21

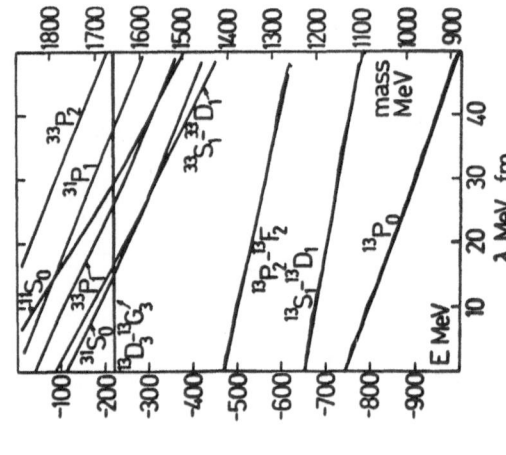

Figure 4. Binding energies (left–hand scale) and masses (right–hand scale) for $\overline{N}N$ bound states, as a function of the strength λ of the annihilation potential; from Niskanen and Green, Ref. (22).

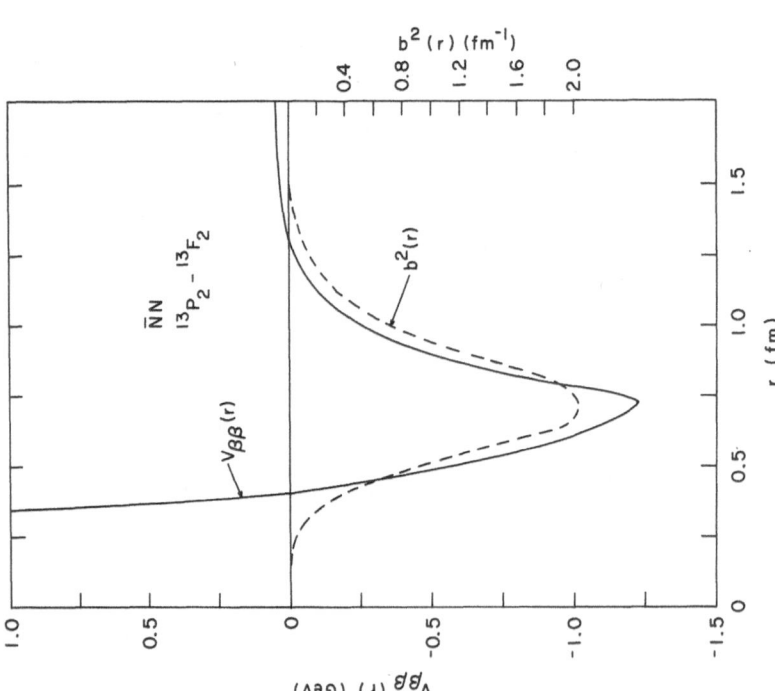

Figure 3. The potential $V_{\beta\beta}(r)$ of Eq. (4) as a function of r for the $2^{++}(0^+)$ state of the $\overline{N}N$ system, together with $b^2(r)$, the square of the bound state wave function; from Ref. (5).

for the $^{13}P_2 - ^{13}F_2$ state of the $\overline{N}N$ system. The square of the radial wave function $b(r)$ for the bound state in this channel is also shown. The r.m.s. radius of the 2^{++} state is of normal hadronic size (0.8 fm); the centrifugal term in Eq. (4) prevents the probability $b^2(r)$ from leaking into the short distance regime $r < 0.5$ fm. In a realistic annihilation potential, such a configuration is expected to have a normal hadronic decay width, of order 100 – 200 MeV.

The presence of strong tensor forces favors the formation of an isoscalar band of $\overline{N}N$ states with quantum numbers

$$J^{\pi C}(I^G) = 0^{++}(0^+), 1^{--}(0^-), 2^{++}(0^+), 3^{--}(0^-)\ldots \tag{5}$$

The structure of the low–lying $\overline{N}N$ spectrum is thus considerably simplified. For $I = 1$, there is no coherence of tensor forces, so bound states tend to cluster closer to the $\overline{N}N$ threshold. This pattern is evident in Fig. 4, which shows an $\overline{N}N$ spectrum[22] as a function of a parameter λ, the strength of a complex annihilation potential. Other calculations[16,17,18,21] yield a qualitatively similar picture. The absolute binding energies are very sensitive to the prescription used to regularize the $\overline{N}N$ potential at short distances, but the level order of the lowest lying states is the same in most of the existing calculations.

We emphasize again that the dynamical mechanism for producing relatively high spin (3^{--} in Fig. 4) $\overline{N}N$ bound states is the attractive coherence for $I = 0$ of tensor potentials from pseudoscalar (π, η) and vector (ρ, ω) meson exchange. The off–diagonal tensor force induces virtually complete mixing of orbital states with $L = J \pm 1$, and is particularly important in bringing $J \geq 2$ states down to energies below the $\overline{N}N$ threshold.

The spectrum of $\overline{N}N$ states is quite distinct from that for a four quark $\left(Q^2\overline{Q}^2\right)$ system, as calculated by Jaffe[27] or Badalyan and Kitoroage[28], among others. A $Q^2\overline{Q}^2$ spectrum, with Q^2 in a color $\{\overline{3}\}$ and $\overline{Q}^2$ in $\{3\}$, is shown in Fig. 5. We can have the diquark in a spin–isospin state α with $I = S = 0$ or β with $I = S = 1$. This gives trajectories $A = \alpha\bar{\alpha}$, $B = \alpha\bar{\beta} \pm \bar{\alpha}\beta$ and $C = \beta\bar{\beta}$. The trajectory A, which couples most strongly to $\overline{N}N$, has the same quantum numbers as the lowest–lying band of $\overline{N}N$ states given in Eq. (5). Trajectory B displays G–parity doublets, while C has degenerate $I = 0, 1, 2$ states. These features are not shared by the spectrum of $\overline{N}N$ quasinuclear states.

The considerations of this section are speculative. In the next section, we argue that at least some of the predicted $\overline{N}N$ bound states may be realized in nature.

3. EXPERIMENTAL EVIDENCE FOR $\overline{N}N$ QUASINUCLEAR BOUND STATES

The ASTERIX group at LEAR studied the $\bar{p}p \to \pi^+\pi^-\pi^0$ reaction from an initial $L = 1$ atomic state[6]. They saw a peak in the $\pi^+\pi^-$ mass spectrum in the vicinity of 1565 MeV, which they denoted as the AX(1565). This data is shown in Fig. 6. From an analysis of decay angular distributions and the non–observation of a peak in the $I = 1$ $\pi^\pm\pi^0$ mass spectrum, an assignment of $2^{++}(0^+)$ quantum numbers was given. The branching ratio is

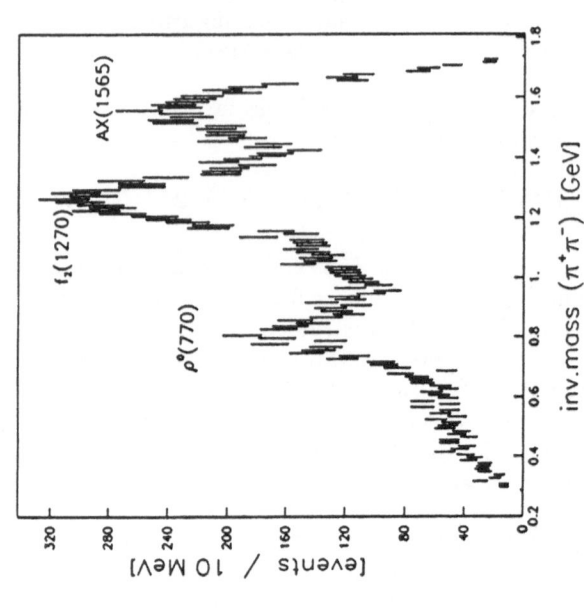

Figure 6. The $\pi^+\pi^-$ mass spectrum for $\bar{p}p \rightarrow \pi^+\pi^-\pi^0$ annihilation from atomic P-states; from May *et al.*, Ref. (6).

Figure 5. Masses of $Q^2\bar{Q}^2$ states, from Jaffe, Ref. (27).

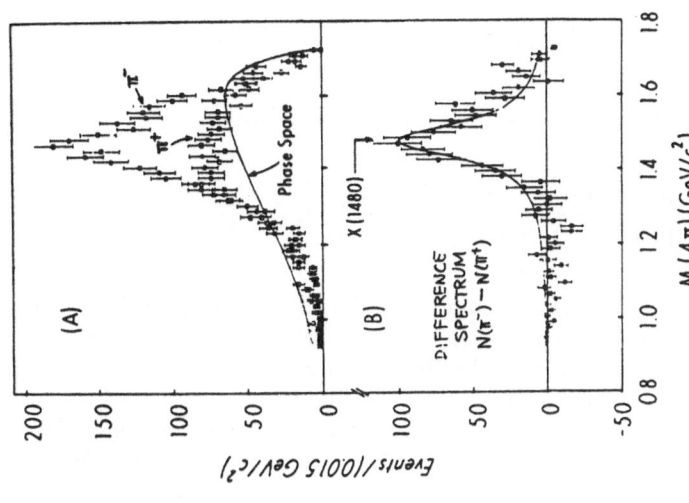
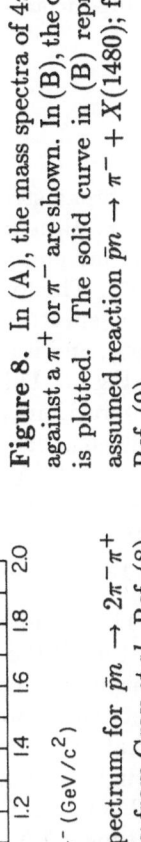

Figure 8. In (A), the mass spectra of 4π systems recoiling against a π^+ or π^- are shown. In (B), the difference spectrum is plotted. The solid curve in (B) represents a fit to an assumed reaction $\bar{p}n \rightarrow \pi^- + X(1480)$; from Bridges *et al.*, Ref. (9).

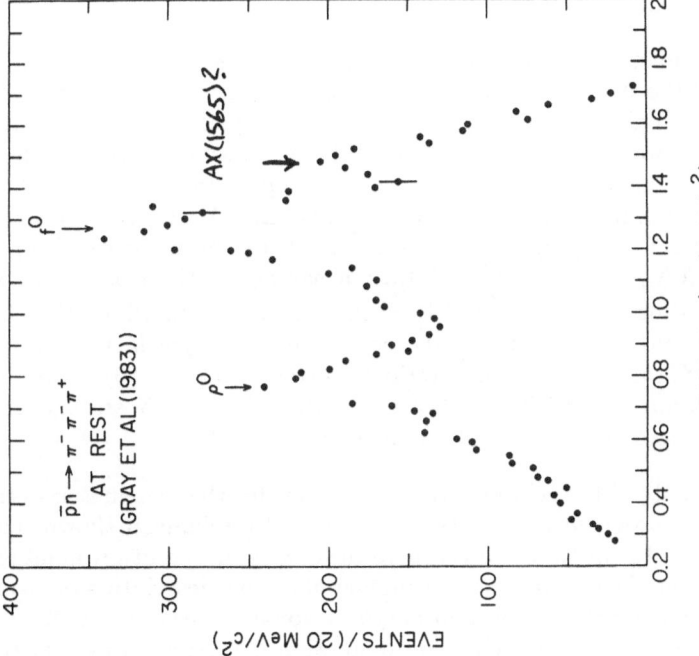

Figure 7. The $\pi^+\pi^-$ mass spectrum for $\bar{p}n \rightarrow 2\pi^-\pi^+$ annihilation from atomic S–states; from Gray *et al.*, Ref. (8).

$$B\left(\bar{p}p\,(L=1)\to\pi^0+\mathrm{AX}(1565)\right)\cdot B\left(\mathrm{AX}(1565)\to\pi\pi\right)=(5.6\pm0.9)\times10^{-3}\quad(6)$$

There is good evidence that the AX(1565) was seen earlier by Gray $et\ al.$[8] in bubble chamber experiments (essentially pure $L=0$) in the $\bar{p}n\to2\pi^-\pi^+$ and $\bar{p}p\to3\pi^0$ channels. Some of this data is plotted in Fig. 7. The quoted[8] branching ratio is

$$B\left(\bar{p}n\,(L=0)\to\pi^-+\mathrm{AX}(1565)\right)\cdot B\left(\mathrm{AX}(1565)\to\pi\pi\right)\approx(3.9\pm0.6)\times10^{-3}\quad(7)$$

Correcting for an isospin factor. Eqs. (6) and (7) imply the ratio

$$\frac{B\left(\bar{p}p\,(L=0)\to\pi^0+\mathrm{AX}(1565)\right)}{B\left(\bar{p}p\,(L=1)\to\pi^0+\mathrm{AX}(1565)\right)}\approx0.35^{+0.13}_{-0.10}\quad(8)$$

Another broad structure in the same mass region has been seen in the $\bar{p}n\to3\pi^-2\pi^+$ reaction at Brookhaven by Bridges $et\ al.$[9] and at LEAR by Ahmad $et\ al.$[29]. The quantum number assignment preferred by Bridges $et\ al.$[9] is $2^{++}(0^+)$, identical to the AX(1565). The four pion invariant mass plots are shown in Figs. 8 and 9. The peak in the 4π spectra from the $\bar{p}d\to3\pi^-2\pi^+p$ reaction lies somewhat lower than the AX(1565), and its apparent mass decreases with increasing spectator momentum, as seen in Fig. 9. This effect has been explained by Kolybasov $et\ al.$[30] in terms of final state interactions of the pions with the proton. A plausible hypothesis is that the AX(1565) has been seen by Bridges $et\ al.$[9] in its $\rho\rho$ decay mode; if this is tenable, it is clear that $B\left(\mathrm{AX}(1565)\to\rho\rho\right)>B\left(\mathrm{AX}(1565)\to\pi\pi\right)$. Furthermore, Gray $et\ al.$[8] indicate that the $\overline{K}K$ decay mode is strongly suppressed, so that the structure at about 1530 MeV in Fig. 7 cannot be identified with the $f'(1525)$, i.e., the well known $2^{++}(0^+)$ $s\bar{s}$ meson.

The arguments for interpreting the AX(1565) as the $^{13}P_2-{}^{13}F_2$ quasinuclear bound state of the $\overline{N}N$ system have been developed in detail by Dover $et\ al.$[31]. The mechanism which gives strong binding for this particualr $I=0$ configuration is the coherent tensor potential. Calculations employing a complex $\overline{N}N$ optical potential fit to scattering data predicted a quasinuclear (QN) state in the $2^{++}(0^+)$ channel with roughly the right mass and width: for instance, Vinh Mau[21] gives $M\approx1500$ MeV, $\Gamma\approx65$ MeV. Long ago, Dover $et\ al.$[7] predicted branching ratios for reactions of the type $\left(\overline{N}N\right)_{\mathrm{atom}}\to\pi+\left(\overline{N}N\right)_{\mathrm{QN}}$; a sample of these is shown in Fig. 10. When these estimates are adjusted to obtain the observed width $\Gamma\approx170$ MeV of the AX(1565) and to incorporate the correct phase space factor for the reaction $\overline{N}N\to\pi+\mathrm{AX}(1565)$, a branching ratio consistent with Eq. (6) is obtained[31]. The preferential production of AX(1565) from an initial $L=1$ $\overline{N}N$ state, as indicated by Eq. (8), is understood as essentially a kinematical effect.

An estimate of the relative branching ratios for the two–meson decays $\overline{N}N\to M_1M_2$ has been provided in Ref. 31, assuming the mechanism shown schematically in Fig. 11. The $Q\overline{Q}$ vertices are taken to have vacuum quantum numbers (the $^3P_0$ model). Such a model has been successfully applied to a description of $\bar{p}p$ annihilation at rest and in flight[32,33]. Meson and baryon resonance decay are well accounted for using an effective $^3P_0$ vertex. This approximation is motivated by flux tube models of QCD in the strong coupling limit[34,35]. The strength of the $^3P_0$ vertex has been determined from the $Q\overline{Q}$ and gluon condensates by Weber[36].

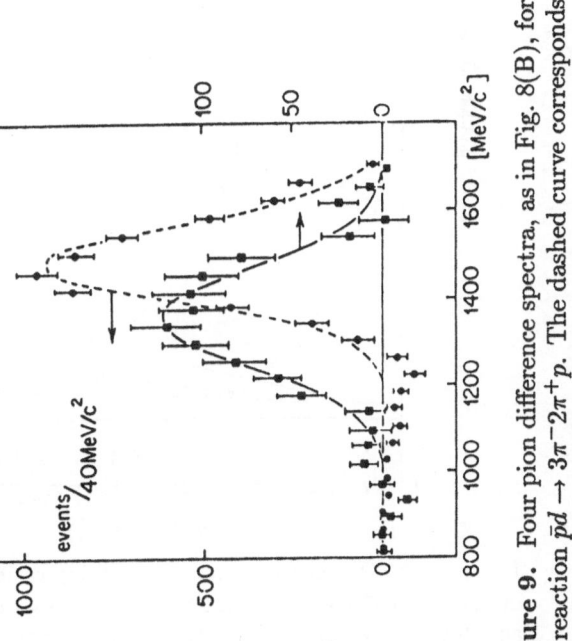

Figure 9. Four pion difference spectra, as in Fig. 8(B), for the reaction $\bar{p}d \rightarrow 3\pi^-2\pi^+p$. The dashed curve corresponds to recoil protons in the low momentum "spectator" region, while the solid curve represents the data for "non-spectator" protons of higher momentum; from Ahmad *et al.*, Ref. (29).

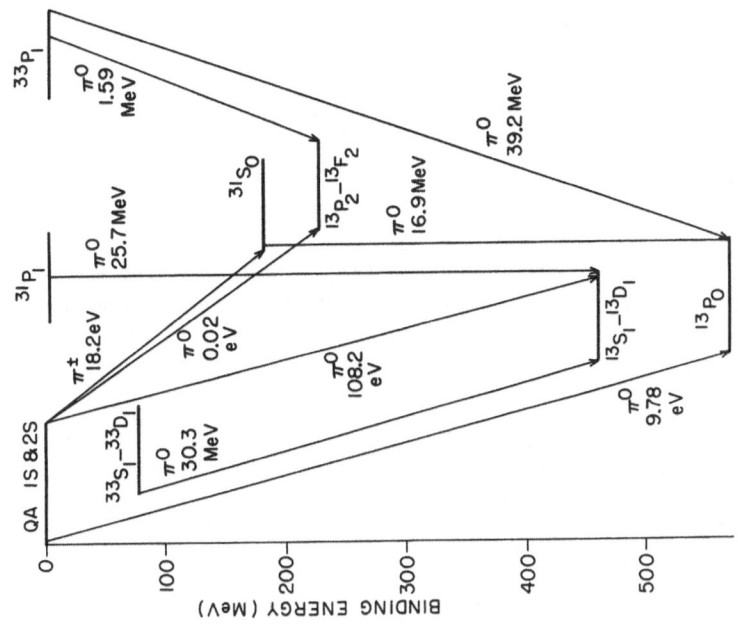

Figure 10. Widths Γ for π^0 emission from quasi-atomic (QA) $\bar{N}N$ bound states to deeply bound quasi-nuclear (QN) states. Selected transitions between QN states are also shown.

27

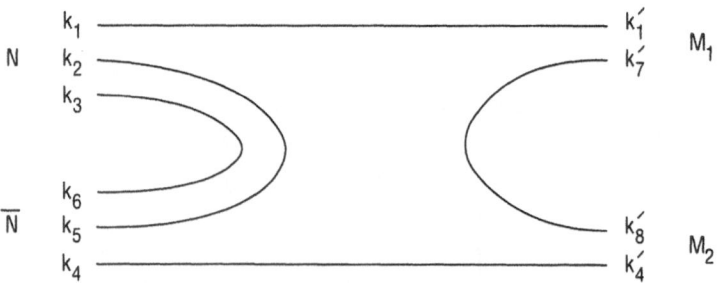

Figure 11. Mechanism for $\overline{N}N$ annihilation into two mesons M_1, M_2 via quark–antiquark ($Q\overline{Q}$) creation and destruction.

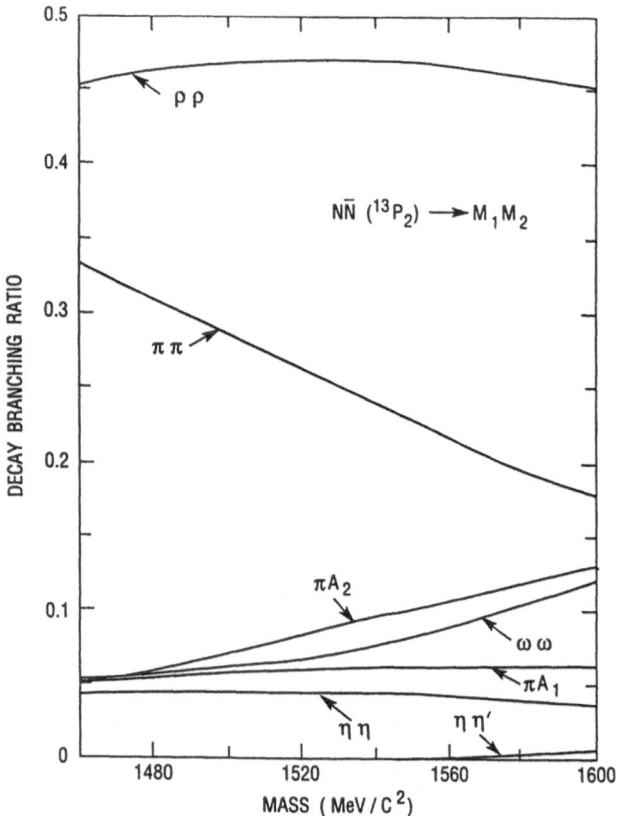

Figure 12. Predicted relative branching ratios for the decay of the $2^{++}(0^+)$ $\overline{N}N$ bound state into two meson channels, as a function of mass; from Dover *et al.*, Ref. (31).

28

The transition amplitude T corresponding to Fig. 3 is given by[33]

$$T = \int d^3 k_1 \ldots d^3 k_6 d^3 k'_1 d^3 k'_4 d^3 k'_7 d^3 k'_8 \psi^\dagger_{M_1 M_2} \mathcal{O} \psi_{\overline{N}N} \qquad (9)$$

where $\mathcal{O}$ is the operator

$$\mathcal{O} = \lambda \delta \left(k'_1 - k_1 \right) \delta \left(k'_4 - k_4 \right) \delta \left(k_3 + k_6 \right) \delta \left(k_2 + k_5 \right) \delta \left(k'_7 + k'_8 \right)$$

$$\sigma^{36}_{-\mu} y_{1\mu} \left(k_3 - k_6 \right) \sigma^{25}_{-\nu} y_{1\nu} \left(k_2 - k_5 \right) \sigma^{7'8'}_{-\gamma} y_{1\gamma} \left(k'_7 - k'_8 \right) (-)^{\mu+\nu+\gamma} / 3\sqrt{3} \qquad (10)$$

where the factor $(-)^\mu / \sqrt{3} = (1\mu 1 - \mu | 00)$ ensures that each vertex (with z–component of spin $-\mu$) is coupled to 0^{++} quantum numbers. We now sum over allowed spin-flavor assignments for the Q and $\overline{Q}$ lines in Fig. 11, and assume $SU(6)$ wave functions for N, $\overline{N}$, M_1 and M_2. For each transition $\overline{N}N \to M_1 M_2$ with final c.m. momentum q, we obtain

$$\Gamma = C q F(q) W_{12} \qquad (11)$$

where $F(q)$ is a kinematical form factor proportional to $|T|^2$ and W_{12} is a spin–flavor weight. These have been tabulated by Maruyama et al.[33].

The predicted[31] branching ratios for the decay of a $^{13}P_2$ quasinuclear $\overline{N}N$ state are displayed in Fig. 12. We note that decay is likely to occur at a measurable rate to several final states. The $\rho\rho$ mode is largest, which is consistent with the large signal in this channel seen by Bridges et al.[9], assuming that the X(1480) is in fact the same object as the AX(1565). A $\pi\pi$ branching ratio of order 0.2, together with the calculated production rate[7,31] for $\overline{N}N (L = 1) \to \pi + $ AX(1565), is consistent with Eq. (6). The $\overline{K}K$ decay mode of the $^{13}P_2$ state (not shown in Fig. 12) is predicted to be strongly suppressed, in agreement with the absence of a signal in the experiment of Gray et al.[8]. Such a suppression is also seen in the recent data of Doser et al.[37], who give

$$B\left(\bar{p}p(L = 1) \to K^+ K^- \right) / B\left(\bar{p}p(L = 1) \to \pi^+\pi^- \right) \approx 0.06 \pm 0.012 \qquad (12)$$

From Fig. 12, we note that $B(\text{AX}(1565) \to \eta\eta) \approx 0.05$. This would be a clear exprimental signature, which could be looked for in the $\bar{p}p \to \pi^0 \eta\eta$ channel with the Crystal Barrel detector at LEAR.

The picture of $\overline{N}N$ quasinuclear states presented here is somewhat similar to the interpretation[38] of the S^* and δ scalar mesons as $I = 0, 1$ S–wave "$\overline{K}K$ molecules" rather than $^3P_0$ $Q\overline{Q}$ states. One could also attempt to interpret the AX(1565) in terms of a $Q^2\overline{Q}^2$ states[39,40]. Note, however, that in the Bag Model calculation of Jaffe[27], as seen in Fig. 5, the lowest–lying $2^{++}(0^+)$ state at 1550 MeV is accompanied by degenerate $I = 1, 2$ partners, for which there is no evidence in the $\bar{p}p \to \pi^+\pi^-\pi^0$ data.

4. PRODUCTION Of HYBRID MESONS In $\overline{N}N$ ANNIHILATION

We consider in this section the utility of the $\overline{N}N$ annihilation process as a means of producing hybrid mesons, namely $Q\overline{Q}g$ composites in which the gluon field g is excited to a transverse electric (TE) or transverse magnetic (TM) normal mode. Among these are found "exotic" mesons, which have $J^{\pi C}$ quantum numbers forbidden for a non–relativistic $Q\overline{Q}$ system, for example 0^{--}, 0^{+-} or 1^{-+}.

4.1 The Spectrum of $Q\overline{Q}\,g$ Hybrids

There are estimates of the masses of hybrids in a variety of models, for instance the Bag Model[41–43], the flux tube model[44], the technique of QCD sum rules[45–47], the constituent gluon model[48–50], and lattice QCD[51]. In the latter approach, a static $Q\overline{Q}$ potential is calculated in the presence of a gluon field with non–vacuum quantum numbers. A recent review of the phenomenology of hybrid mesons is due to Godfrey[52].

In the Bag Model, the lowest lying $Q\overline{Q}\,g$ states are obtained by coupling an S–wave $Q\overline{Q}$ pair (0^{-+} ($^1S_0$) or 1^{--} ($^3S_1$)) to a TE (1^{+-}) or TM (1^{--}) mode of the gluon field. In a spherical cavity, the TM mode lies about 500 MeV above the TE mode, so the lowest–lying $Q\overline{Q}\,g$ hybrids are 1^{--} ($^1S_0 \otimes TE$) and 0^{-+}, 1^{-+}, 2^{-+} ($^3S_1 \otimes TE$). The 1^{-+} exotic member comes in four flavor combinations ρ_g, ω_g, ϕ_g, and K_g^*, having the quark structure $(u\bar{u} - d\bar{d})/\sqrt{2}$, $(u\bar{u} + d\bar{d})/\sqrt{2}$, $s\bar{s}$ and $\bar{s}d$, respectively. According to Barnes et al.[41], the masses are approximately 1.4, 1.55, 1.52 and 1.75 GeV/c^2, respectively. The mass scale depends on the ratio of TE and TM self energies, and the values given by Chanowitz and Sharpe[42] are somewhat larger. The near degeneracy of ω_g and K_g^* is a signature of hybrid structure in the Bag Model.

In the flux tube model[44], the ρ_g and ω_g lie at about 1.9 GeV/c^2, and 0^{+-} ($0^-, 1^+$) and 2^{+-} ($0^-, 1^+$) exotics are predicted at the same mass, a degeneracy not present in the Bag Model. In this case, it would not be possible to produce exotics in $\overline{N}N$ annihilation at rest. In the sum rule approach[45], the ρ_g is predicted to lie in the range 1.6 – 2.1 GeV/c^2, with a most probable mass at the upper end of this range.

4.2 Decays of $Q\overline{Q}\,g$ Hybrid Mesons

Decay branching ratios of hybrids have been estimated in the flux tube model[44], using QCD sum rules[39], and from the constituent gluon picture[53,54]. In all of these calculations, *strong selection rules* restrict the possible decay modes of hybrids. In the $SU(6)$ approximation, for $Q\overline{Q}\,g$ hybrids constructed with a $1^{+-}\,TE$ gluon, we have

$$
\begin{aligned}
(Q\overline{Q})_{\ell=0} \otimes TE &\to sp\,(\ell_f = 0) \\
&\not\to sp\,(\ell_f = 2) \\
&\not\to ss\,(\ell_f = 1)
\end{aligned}
\tag{13}
$$

where s denotes a $(Q\overline{Q})_{\ell=0}$ meson and p is a $(Q\overline{Q})_{\ell=1}$ meson, and ℓ_f is the meson–meson relative orbital angular momentum. The decay mechanism which leads to these selection rules is depicted in Fig. 13. In the allowed decays of $Q\overline{Q}\,g$, the spin of the gluon field appears as an internal orbital excitation of a $Q\overline{Q}$ meson rather than as a relative P–wave between two s mesons. These selection rules emerge in both the constituent gluon and flux tube models.

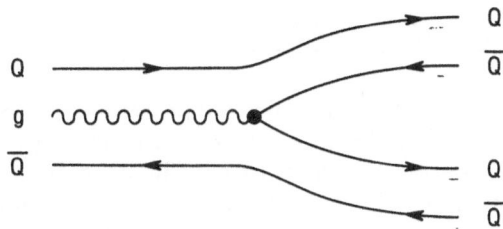

Figure 13. Mechanism for the decay of a $Q\overline{Q}\,g$ hybrid meson into two $Q\overline{Q}$ mesons.

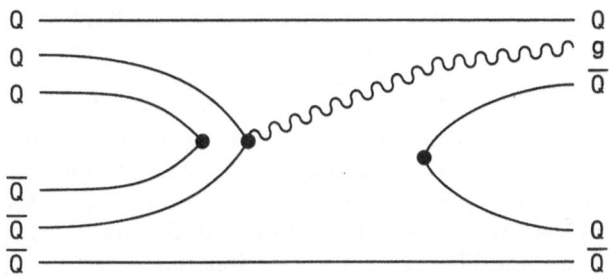

Figure 14. Mechanism for the production of $Q\overline{Q}\,g$ hybrid mesons in $\overline{N}N$ annihilation.

4.3. Candidates for $Q\overline{Q}g$ Mesons

Alde *et al.*[11] have presented evidence for the existence of a $1^{-+}(1^-)$ meson (the ρ_g) with a mass of 1406 ± 20 MeV/c^2 and a width of 180 ± 30 MeV. The putative ρ_g was seen in its $\pi^0\eta$ decay mode in the $\pi^- p \to \pi^0 \eta n$ reaction at 100 GeV. The observed mass lies close to the Bag Model prediction, but according to Eq. (13), $\rho_g \to \pi^0 \eta\,(\ell_f = 1)$ should be essentially a forbidden decay mode, except for $SU(3)$ breaking contributions. This is a puzzle. There are a number of other candidates for mesons which fall outside of the usual $Q\overline{Q}$ nonets. A recent comprehensive review of the experimental situation is due to Landsberg[55]. Further studies of the $\pi^- p \to \rho_g^- p$ reaction are underway at Brookhaven and KEK.

4.4 Hybrid Production in $\overline{N}N$ Annihilation

We expect that the $\overline{N}N$ annihilation process is a copious source of gluons through the successive destruction of $Q\overline{Q}$ pairs from the initial state. Nonperturbative $Q\overline{Q}$ annihilation can lead to glue in its ground state (with vacuum quantum numbers) or, in some cases, an excited gluonic configuration. Such excitations could combine with each other to form glueballs, or fuse with a $Q\overline{Q}$ pair to form a hybrid meson. A possible mechanism for $Q\overline{Q}g$ formation in $\overline{N}N$ annihilation is shown in Fig. 14. Calculations are underway[56] to estimate the relative branching ratios for annihilation reactions $\overline{N}N \to (Q\overline{Q}) + (Q\overline{Q}g)$, for instance $\bar{p}p \to \pi + \rho_g$.

The principal advantage of $\overline{N}N$ annihilation at rest is that one can tag the formation of an $L = 1$ atomic state[37] using a gaseous hydrogen target by observing a $3d \to 2p$ gamma ray in coincidence with the mesonic decay products. In liquid hydrogen, on the other hand, the annihilation occurs dominantly from the $L = 0$ state. This ability to filter the initial L is crucial. An example is provided by Eq. (8): the production of AX(1565) is predicted and observed to be enhanced from the $L = 1$ state, thus strengthening the interpretation of AX(1565) as a quasinuclear $\overline{N}N$ bound state. By comparing $\bar{p}n$ to $\bar{p}p$, one can filter the initial isospin. One further selects G or C by choosing certain exclusive final states. To illustrate the power of quantum number filtration, we consider the three reactions $\bar{p}p \to \pi^0 X^0$, $\bar{p}p \to \pi^\pm X^\mp$ and $\bar{p}n \to \pi^- X^0$. For $\bar{p}p \to \pi^0 X^0$, C parity conservation considerably diminishes the number of allowed transitions. In $\bar{p}p \to \pi^\pm X^\mp$, we select only isovector and isotensor mesons $X^\mp$, whereas in $\bar{p}n \to \pi^- X^0$, we prepare an $I = 1$ initial state. Among the $J^{\pi C}$ exotic mesons with $J = 0, 1$, only the ρ_g can be produced in all three of the above reactions for both $L = 0, 1$.

Consider the production of ρ_g. If ρ_g decays predominantly in sp modes like πB or πD, as predicted by Eq. (13), then a promising sequence of reactions is

$$\bar{p}p \to \pi^\pm \rho_g^\mp, \quad \rho_g^\mp \to \pi^\mp B^0, \quad B^0 \to \pi^0 \omega, \qquad \omega \to \pi^0 \gamma \qquad (14)$$

as proposed for study by the Crystal Barrel collaboration at LEAR. The $\omega \to \pi^0 \gamma$ radiative decay is chosen to minimize combinatorial background. If the observation of the ρ_g by Alde *et al.*[11] is correct, on the other hand, then the neutral channels $\pi^0 \pi^0 \eta$ and $\pi^0 \pi^0 f$ are worthy of study. For $L = 0$, these correspond to the decay chains

$$\bar{p}p\,({}^{11}S_0) \to \pi^0\rho_g^0\,(\ell_f = 1)\,, \quad \rho_g^0 \to \pi^0\eta\,(\ell_f = 1)\,, \quad \eta \to 2\gamma$$
$$\bar{p}p\,({}^{11}S_0) \to \pi^0\rho_g^0\,(\ell_f = 1) \quad \rho_g^0 \to \pi^0 f\,(\ell_f = 2)\,, \quad f \to 2\pi^0 \tag{15}$$

These are optimum channels for study by the Crystal Barrel detector. The $\pi^0\pi^0\eta$ system would be detected as six photons.

As another example for ρ_g, consider $\bar{p}p \to 2\pi^+2\pi^-$. The "signal" is $\bar{p}p\,({}^{11}S_0,\,{}^{33}S_1)$ $\to \pi^\pm\rho_g^\mp\,(\ell_f = 1)$, $\rho_g^\mp \to \pi^\mp\rho^0\,(\ell_f = 1)$, and the background due to ordinary two-body channels arises from $\bar{p}p\,({}^{11}S_0,\,{}^{33}S_1) \to \pi^\pm A_2^\mp\,(\ell_f = 2)$, $A_2^\mp \to \pi^\mp\rho^0\,(\ell_f = 2)$, for instance. As with the analysis of Alde et $al.$[11] for the $\pi^0\eta$ system, one could look for an interference of $\ell_f = 0$ and $\ell_f = 2$ $\pi\rho$ amplitudes.

The 1^{-+} isoscalar partner of the ρ_g, the ω_g, could be searched for in the neutral channels $\bar{p}p \to 4\pi^0$, $\pi^0\eta\eta$, $\pi^0\eta\eta'$. In $\bar{p}p \to 3\pi^0$, there is no background from $\pi\rho$, and $\pi^0 f$ is known to be small, so $\bar{p}p \to \pi^0\omega_g$ might be observable. Earlier experiments[8,57] show very puzzling differences in the Dalitz plots for $\bar{p}p \to 3\pi^0$, $\bar{p}p \to \pi^+\pi^-\pi^0$ and $\bar{p}n \to 2\pi^-\pi^+$, which are still unexplained.

4.5. The E(1420) as a $Q\overline{Q}\,g$ Hybrid?

The ASTERIX group has recently presented results[58] for the annihilation channel $\bar{p}p \to \pi^+\pi^-K^\pm\pi^\mp K^0$ with a gaseous target. In the invariant mass spectrum of the $K^\pm\pi^\mp K^0$ system, evidence is seen for the D(1285) and E(1420) mesons. In earlier experiments[1] in liquid hydrogen (mostly $L = 0$ annihilation), the E(1420) was seen with a larger branching ratio, indicating a formation rate which depends strongly on L. The dependence of the E(1420) branching ratio on the fraction of $L = 1$ annihilation [$61 \pm 6\%$ for ASTERIX in this channel] is shown in Fig. 15. Baillon et $al.$[1] saw no charged E's, so $I = 0$. Also, $E \to K_s K_s \pi^0$ was seen, but $E \to K_s K_L \pi^0$ was not, so $C = 1$ $G = 1$ for the E. The possible quantum number assignments for the E are then $J^{\pi C}(I^G) = 0^{-+}\,(0^+)$ or $1^{++}\,(0^+)$. On the basis of the observed suppression of E production from $L = 1$ initial states (see Fig. 15), Duch et $al.$[58] prefer $E\,(1420) = 0^{-+}\,(0^+)$. The argument is a simple kinematical one: if the E is a $0^{-+}\,(0^+)$ pseudoscalar meson, the allowed transitions from $L = 0, 1$ are

$$
\begin{aligned}
{}^{11}S_0 &\to [(\pi\pi)_{\ell=0} \otimes E]_{\ell_f=0} \\
{}^{33}S_1 &\to [(\pi\pi)_{\ell=1} \otimes E]_{\ell_f=1} \\
{}^{13}P_1 &\to [(\pi\pi)_{\ell=0} \otimes E]_{\ell_f=1} \\
{}^{31}P_1 &\to [(\pi\pi)_{\ell=1} \otimes E]_{\ell_f=0}
\end{aligned}
\tag{16}
$$

Since relatively little phase space is available, the $\ell = \ell_f = 0$ case is favored kinematically, implying dominant production of the E(1420) from the ${}^{11}S_0$ initial state. Similarly, for a $1^{++}\,(0^+)$ axial vector meson like the D(1285), we have

$$
{}^{11}S_0 \to [(\pi\pi)_{\ell=0} \otimes D]_{\ell_f=1}
$$

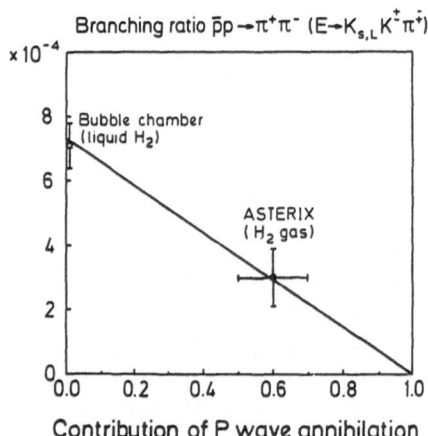

Figure 15. Branching ratio for $\bar{p}p \to \pi^+\pi^- E$ as a function of P wave annihilation fraction; from Duch *et al.*, Ref. (58).

$$^{33}S_1 \to [(\pi\pi)_{\ell=1} \otimes D]_{\ell_f=0}$$
$$^{13}P_1 \to [(\pi\pi)_{\ell=0} \otimes D]_{\ell_f=0}$$
$$^{31}P_1 \to [(\pi\pi)_{\ell=1} \otimes D]_{\ell_f=1} \tag{17}$$

The same phase space argument favors production of the D(1825) from the $^{13}P_1$ initial state. This is what is observed experimentally: the D(1285) is not seen by Baillon et al.[1] ($L = 0$) but is quite distinct in the ASTERIX data[58](60% $L = 1$).

In liquid hydrogen, the branching ratios for $\bar{p}p \to \pi^+\pi^-E(1420)$ are given[1] as $(2.0 \pm 0.2) \times 10^{-3}$ for the $^1S_0$ state and $(0.6 \pm 0.06) \times 10^{-3}$ from the $^3S_1$ state. Removing the $(2S+1)$ factor for an assumed statistical population of $^1S_0$ and $^3S_1$ initial states, we find

$$\frac{\Gamma\left(^{33}S_1 \to \pi^+\pi^-E\right)}{\Gamma\left(^{11}S_0 \to \pi^+\pi^-E\right)} \approx \frac{1}{10} \tag{18}$$

This seems to support the kinematical argument based on Eq. (16), since transitions from $^{33}S_1$ are suppressed because $\ell = \ell_f = 1$. However, then it is not clear that the $L = 1$ production of E is as strongly suppressed as indicated by the solid line in Fig. 1. Indeed, transitions $^{13}P_1, ^{31}P_1 \to \pi^+\pi^-E$ should be less suppressed than $^{33}S_1 \to \pi^+\pi^-E$, since only one relative P-wave is required (rather than two) in the final state. Note that a branching ratio for $L = 1 \to \pi^+\pi^-E$ up to 2×10^{-4} can easily be accommodated by keeping the straight line within the errors in Fig. 15.

The allowed two body decay modes of a $0^{-+}\left(0^+\right)$ meson are

$$0^{-+}\left(0^+\right) \to \pi\delta\left(\ell_f = 0\right),\ \eta\sigma\left(\ell_f = 0\right),\ \overline{K}K^* + c.c\left(\ell_f = 1\right),$$
$$\rho\rho\left(\ell_f = 1\right),\ \omega\omega\left(\ell_f = 1\right),\ \pi A_2\left(\ell_f = 2\right) \tag{19}$$

where $\sigma = (\pi\pi)_{\ell=0}$. For E(1420), the decay analysis of Duch et al.[58] prefers $E \to \pi\delta$ as the dominant $K^\pm\pi^\mp K^0$ decay mode, but $\overline{K}K^* + c.c$ is not ruled out at the 20 – 30% level.

The E(1420) has the quantum numbers of the η and η'. It lies too low in mass to be a radial excitation. Thus it is a prime candidate for a non-$Q\overline{Q}$ meson. It is not unreasonable to interpret the E(1420) as a $Q\overline{Q}g$ hybrid meson of the type $[\omega \otimes TE]_{0^{-+}}$, the pseudoscalar partner of the exotic 1^{-+} meson ω_g discussed earlier. From Eq. (13), we then have the approximate selection rules

$$E(1420) \to \pi\delta,\ \eta\sigma$$
$$E(1420) \not\to \overline{K}K^*,\ \rho\rho,\ \omega\omega,\ \pi A_2 \tag{20}$$

The mass of the E is close to the value predicted for a $0^{-+}\left(0^+\right) Q\overline{Q}g$ meson in the Bag Model[41]. Eq. (20) would explain naturally the preference for $\pi\delta$ over $\overline{K}K^*$ in the decay of the E. A puzzle concerns the decay $E \to \eta\sigma \to \eta\pi\pi$, which was not seen by Ando et al.[59]. It is possible that this is caused by a destructive interference between the two quasi-two-body modes $\pi\delta$ and $\eta\sigma$, which lead to the same $\pi\pi\eta$ final state. The consistency of the hypothesis that E(1420) is a $Q\overline{Q}g$ meson is being studied[56] in detail, based on the production and decay mechanisms of Figs. 13 and 14.

35

5. CONCLUSIONS

Studies of $\overline{N}N$ annihilation offer exciting prospects for the discovery of mesonic states beyond the usual $Q\overline{Q}$ nonets. We have presented plausible interpretations of the new objects (for instance, the AX(1565) and the E(1420)) already seen in the debris of $\overline{N}N$ annihilation. The $\overline{N}N$ system forms an essential complement to pp, e^+e^-, K^-p, $\gamma\gamma$ and other channels as a means of shedding new light on the meson spectrum. The LEAR facility is unique for these studies and its physics program should be vigorously pursued in the 1990's, employing the full range of new detectors coming on line.

ACKNOWLEDGMENTS

This work was supported by the Department of Energy under contract No. DE-AC02-76CH00016.

REFERENCES

1. P. Baillon *et al.*, Nuovo Cim. 50A:393 (1967).
2. J. Vandermeulen, Z. Phys. C37:563 (1988).
3. G.S. Mutchler *et al.*, Phys. Rev. D38:742 (1988).
4. C.B. Dover and J.M. Richard, Phys. Rev. C25:1952 (1982).
5. C.B. Dover and J.M. Richard, Phys. Rev. D17:1770 (1978).
6. B. May *et al.*, Phys. Lett. B225:450 (1989).
7. C.B. Dover, J.M. Richard, and M.C. Zabek, Ann. Phys. (N.Y.) 130:70 (1980).
8. L. Gray *et al.*, Phys. Rev. D27:307 (1983).
9. D. Bridges *et al.*, Phys. Rev. Lett. 56:211 and 215 (1986).
10. C.B. Dover, Phys. Rev. Lett. 57:1207 (1986).
11. D. Alde *et al.*, Phys. Lett. B205:397 (1988).
12. J. Ellis, E. Gabathuler, and M. Karliner, Phys. Lett. B217:173 (1989).
13. H.J. Lipkin, Phys. Lett. B225:287 (1989).
14. M. Rho, G.E. Brown, and B.-Y. Park, Phys. Rev. C39:1173 (1989).
15. C.B. Dover and P.M. Fishbane, Phys. Rev. Lett. 62:2917 (1989).
16. L.N. Bogdanova, O.D. Dalkarov, and I.S. Shapiro, Ann. Phys. 84:261 (1974); I.S. Shapiro, Phys. Rep. C35:129 (1978).
17. I.S. Shapiro, in: *Physics at LEAR with Low Energy Antiprotons*, Nuclear Science Research Conference Series, Vol. 14, Eds. C. Amsler *et al.*, Harwood Academic Publ., Chur (1988) p. 377.
18. W.W. Buck, C.B. Dover, and J.M. Richard, Ann. Phys. (N.Y.) 121:47 (1979); C.B. Dover and J.M. Richard, Ann. Phys. (N.Y.) 121:70 (1979).
19. C.B. Dover and J.M. Richard, Phys. Rev. C21:1466 (1980).
20. J. Coté *et al.*, Phys. Rev. Lett. 48:1319 (1982).
21. R. Vinh Mau, in: *Medium Energy Nucleon and Antinucleon Scattering*, Lecture Notes in Physics, Vol. 243, Ed. H.V. von Geramb, Springer–Verlag, Berlin (1985), pp. 3–23; see especially Table 7.

22. J.A. Niskanen and A.M. Green, Nucl. Phys. A431:593 (1984).

23. A.M. Green and J.A. Niskanen, Prog. in Particle and Nuclear Phys., 18:93 (1987).

24. M. Lacombe et al., Phys. Rev. C29:1800 (1984).

25. M. Kohno and W.Weise, Nucl. Phys. A454:429 (1986).

26. G.C. Rossi and G. Veneziano, Phys. Rep. 63:149 (1980).

27. R.L. Jaffe, Phys. Rev. D15:267 (1977); Phys. Rev. D17:1445 (1978).

28. A.M. Badalyan and D.I. Kitoroage, Yad. Fiz. 47:1343 (1988) [Sov. J. Nucl. Phys. 47:855 (1988)].

29. S. Ahmad et al., in: *Physics at LEAR with Low Energy Antiprotons*, Nuclear Science Research Conference Series, Vol. 14, Eds. C. Amsler et al., Harwood Academic Publishers, Chur (1988) p. 447.

30. V.M. Kolybasov, I.S. Shapiro, and Yu. N. Sokolskikh, Phys. Lett. B222:135 (1989).

31. C.B. Dover, T. Gutsche, and A. Faessler, preprint (1990).

32. C.B. Dover, P.M. Fishbane, and S. Furui, Phys. Rev. Lett. 57:1538 (1986).

33. M. Maruyama, S. Furui, and A. Faessler, Nucl. Phys. A472:643 (1987).

34. N. Isgur and J. Paton, Phys. Rev. D31:2910 (1985).

35. H.G. Dosch and D. Gromes, Phys. Rev. D33:1378 (1986) and Z. Phys. C34:139 (1987).

36. H.J. Weber, Phys. Lett. B218:267 (1989).

37. M. Doser et al., Nucl. Phys. A486:493 (1988).

38. J. Weinstein and N. Isgur, Phys. Rev. D27:588 (1983); Phys. Rev. D41:2236 (1990).

39. K.F. Liu and B.A. Li, Phys. Rev. Lett. 58:2288 (1987).

40. S.K. Bose and E.C.G. Sudarshan, Phys. Rev. Lett. 62:1445 (1989).

41. T. Barnes, F.E. Close, and F. deViron, Nucl. Phys. B224:241 (1983).

42. M. Chanowitz and S.R. Sharpe, Nucl. Phys. B222:211 (1983).

43. M. Flensburg, C. Peterson, and L. Sköld, Z. Phys. C22:293 (1984).

44. N. Isgur, R. Kokoski, and J. Paton, Phys. Rev. Lett. 54:869 (1985).

45. J.I. Latorre, P. Pascual, and S. Narison, Phys. Lett. B147:169 (1984); Z. Phys. C34:347 (1987).

46. J. Govaerts et al., Phys. Lett. B128:241 (1983); Nucl. Phys. B248:1 (1984).

47. I.I. Balitsky, D. Dyakanov, and A.V. Yung, Z. Phys. C33:265 (1986).

48. T. Barnes, Z. Phys. C10:275 (1981).

49. J. Cornwall and A. Soni, Phys. Lett. B120:431 (1983).

50. D. Horn and J. Mandula, Phys. Rev. D17:898 (1978).

51. N.A. Campbell, L.A. Griffiths, C. Michael, and P.E.L. Rakow, Phys. Lett. B142:281 (1984);
N.A. Campbell, A. Huntly, and C. Michael, Nucl. Phys. B306:51 (1988).

52. S. Godfrey, in: *Glueballs, Hybrids and Exotic Hadrons*, AIP Conf. Proc. No. 185 (Particles and Fields Series 36), Ed. Suh–Urk Chung, New York (1989), p. 373.

53. A. LeYaouanc *et al.*, Z. Phys. C28:309 (1985).

54. F. Iddir *et al.*, Phys. Lett. B205:564 (1988).

55. L.G. Landsberg, IHEP preprint 89–54, Serpukhov (1989); extended version of an invited talk at the XXIV Rencontres de Moriond, Les Arcs, France, March, 1989.

56. C.B. Dover, T. Gutsche, and A. Faessler, in preparation.

57. T.E. Kalogeropoulos *et al.*, Phys. Rev. D24:1759 (1981).

58. K.D. Duch *et al.*, CERN preprint (1989).

59. A. Ando *et al.*, Phys. Rev. Lett. 57:1296 (1986).

CHIRAL SYMMETRY AS AN EXPERIMENTAL SCIENCE

John F. Donoghue[†]

Theory Division, CERN
1211 Geneva 23, Switzerland

1. INTRODUCTION

The common impression of chiral symmetry is that it is:

a) incomprehensible,

b) a theorist's theory,

c) used mainly in the study of superstrings and/or

d) just another model.

It it likely true that the concepts of chiral effective Lagrangians are understood by more superstring theorists than those physicists using QCD. However, this is a shame as the foundation of these ideas is in the phenomenology of the strong interactions. In reality, chiral symmetry is

a) a true/direct consequence of QCD,

b) a rigorous way to calculate in the very low energy region

c) a rich phenomenological theory and

d) subtle and fun.

In these lectures, I wish to present an introduction to the subject, hopefully one where the essential ideas are not buried by too much formalism. Some of the

---

† Permanent address: Department of Physics and Astronomy, Univ. of Massachusetts, Amherst, MA 01003, U.S.A.

Medium-Energy Antiprotons and the Quark–Gluon Structure of Hadrons
Edited by R. Landua *et al.*, Plenum Press, New York, 1991

ideas and language are not familiar, as they have not become part of the standard graduate studies training. These then require some examination. Overall the goal is to describe how chiral symmetry is used in phenomenological applications, and to point to possible future theoretical and experimental directions.

The study of QCD can be divided into three regions of energy. We are all aware of the applications of perturbative QCD to the high energy region. As one comes down in energy, we enter a region where perturbative QCD is no longer applicable. Here we do not know how to calculate (apart from lattice computer work), and we are reduced to making models (quark models, pole models, Skyrme model, ...) which we hope capture some of the physics of QCD. However, these are not controlled approximations in that there is no well-defined way to calculate a next approximation or to put error bars on the predictions. However, at very low energies (say, $E \lesssim 1/2$ GeV) we again enter a region where a controlled approximation is possible, using symmetry methods. It is this very low-energy region which we are concerned with here.

An analogy with the calculational procedure of perturbative QCD will shed light on the similar procedure in the very low-energy region. At high energies, hard scattering processes are known in an expansion in $\alpha_s \sim 1/\ell n E$. However, there remains some dependence on (presently) non-calculable "soft" physics such as structure functions, fragmentation functions, $p_\perp$ distributions, etc., as well as the intrinsic parameter Λ_{QCD}. These must be determined phenomenologically. The content of high energy Λ_{QCD} is then relationships between scattering and decay processes, parametrized by the empirical values of Λ_{QCD}, structure functions, etc. At low energies, we will see that very soft processes are highly constrained by the symmetry. There remains some dependence on (presently) non-calculable hard physics, contained in F_π and the various low energy constants. These must be determined phenomenologically. The content of very low energy QCD is then relationships between scattering and decay processes, parametrized by the empirical values of F_π, lower energy constants, etc.

2. WHAT IS CHIRAL SYMMETRY?

First, let us describe the essential physics without mathematics. Pretend that the u, d, s quarks are massless. For each quark, there is a left-handed helicity state (spin antiparallel to the momentum) and a right-handed helicity state. QCD interactions are the same for left and right helicity and do not flip helicity. Under these conditions left-handed massless particles will always stay left handed and right handed will stay right handed. We have two separate worlds, a LH world and a RH world. Since each flavour is massless and has the same QCD coupling, there exists a separate flavour $SU(3)$ invariance in each world. We can make an $SU(3)$ rotation on the left-handed quarks without influencing the right-handed ones. The overall invariance is then called $SU(3)_L \times SU(3)_R$. This is the chiral symmetry.

Now consider adding a common mass. It is clear that we cannot maintain the separate L and R invariances. If one has a massive left-handed particle, one can always boost to the rest frame and then boost to a frame moving with other directions, such that the LH particle is now right handed. Kinematically, at the very least, the LH and RH worlds become related, and one no longer has separate $SU(3)$ invariances. However, one still has an invariance under common $L + R$ rotations, with a symmetry called $SU(3)_{L+R} = SU(3)_V$ (V is for "vectorial"). However, if the mass is small, the original $SU(3)_L \times SU(3)_R$ symmetry could also be an approximate symmetry, and the mass could be treated as a perturbation.

The real world is obtained by allowing m_u, m_d and m_s to all be different. This breaks the $SU(3)_V$ symmetry. However, to the extent that Δm is small, the $SU(3)_V$ symmetry could be considered as an approximate symmetry, and the mass differences could be treated as perturbations.

To see this same result more precisely, consider the field

$$\psi = \begin{pmatrix} u \\ d \\ s \end{pmatrix} \tag{1}$$

and the projection operators

$$
\begin{aligned}
\Gamma_L &= \frac{1}{2}(1 + \gamma_5) \\
\Gamma_R &= \frac{1}{2}(1 - \gamma_5) \\
\Gamma_L^2 &= \Gamma_L; \quad \Gamma_R^2 = \Gamma_R; \quad \Gamma_L \Gamma_R = 0 \\
I &= \Gamma_L + \Gamma_R
\end{aligned}
\tag{2}
$$

such that

$$
\begin{aligned}
\psi_L &= \Gamma_L \psi \\
\psi_R &= \Gamma_R \psi \\
\psi &= \psi_L + \psi_R
\end{aligned}
\tag{3}
$$

For massless particles these project out the helicity of the particle. For $m \neq 0$, Γ_L and Γ_R are still projection operators but do not yield exactly the helicity. For this reason, a new name is introduced and ψ_L and ψ_R are said to have left and right "chirality" (chiral being related to the Greek for "hand"). The Dirac Lagrangian can be rewritten using these fields

$$\mathcal{L} = \bar{\psi}(i\not{D} - m)\psi$$
$$= \bar{\psi}_L i\not{D}\psi_L + \bar{\psi}_R i\not{D}\psi_R + \bar{\psi}_L m\psi_R + \bar{\psi}_R m\psi_L \tag{4}$$

Here we see that if $m = 0$, there is a decoupling of left and right, as advertized above, and that the QCD Lagrangian is invariant under

$$\psi_L \rightarrow L\psi_L$$
$$\psi_R \rightarrow R\psi_R \tag{5}$$

with L and R being $SU(3)$ transformations, with in general $L \neq R$. The presence of a mass removes the separate invariances.

Given this symmetry, Noether's theorem says that we have a set of 16 conserved currents and charges. The eight vector currents and charges are the usual flavour $SU(3)_V$ ones: it is the 8 axial charges which are "new". The normal expectation is that symmetries require particles to appear in multiplets. If there exists some single particle state with

$$H|P\rangle = E_P|P\rangle \tag{6}$$

then a symmetry transformation will yield a state with the same energy, *i.e.* for an axial charge Q_5^j

$$H(e^{iQ_5^j}|P\rangle) = e^{iQ_5^j}H|P\rangle = E_P(e^{iQ_5^j}|P\rangle) \tag{7}$$

since conservation of the charge requires $[H, Q_5^j] = 0$. However, this does not seem to be a property of the real world; as the proton (for example) does not live in a multiplet with partners of the opposite parity. This occurs because the symmetry is "dynamically broken" or "dynamically hidden".

3. SYMMETRY BECOMES DYNAMICS

The hardest thing to understand about chiral symmetry is why it makes predictions in the first place. How can a symmetry predict a dynamical scattering amplitude, such as that for $\pi\pi$ scattering? The answer involves an understanding of "dynamical symmetry breaking" [1]. This terminology is misleading as it reinforces the mistaken impression that "if the symmetry is broken, it does not exist anymore and I can ignore the subject". The phrase would be better called "dynamical symmetry hiding" in that the symmetry, although still valid, appears to be hidden when one looks at the spectrum. Symmetry predictions still exist.

The generic case of hidden symmetry occurs when the Lagrangian is invariant under the symmetry, but when there exists a continuous family of ground state solutions, which are related to each other by the symmetry but which are not individually invariant under the symmetry. The classic example is that of magnetic domains. The magnet Hamiltonian is rotationally invariant, but the possible ground states consist of configurations with all the spins lined up in the same direction. The overall direction can be arbitrary (this is the continuous family of solutions), but each ground state is not invariant. Any of these solutions could be "the" ground state, and one is chosen. (We assume here that it fills all of space). Within this ground state one does not see the full rotational invariance; it is hidden.

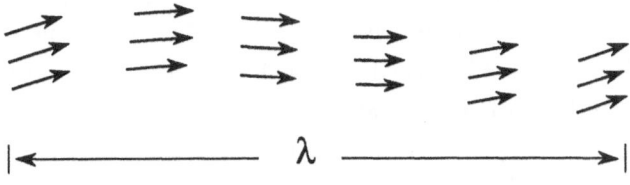

Fig. 1. A spin wave in a magnet system

An immediate consequence of this situation is that there will be massless particles in the theory. This is Goldstone's theorem, and it is not hard to see conceptually. Let us define the ground state to have $E = 0$. Then there clearly must be other states with $E = 0$. These are the partners of the vacuum state, all of which have the same energy because of the original rotational invariance of the Lagrangian (Hamiltonian). In quantum mechanics, all states are quantized and, apart from the vacuum itself, must be described by particles. To have $E = 0$, these particles must have $m = 0$. In the magnet example, the excitations are spin waves where the spins vary in a wave-like pattern, Fig. 1. This spin wave in general carries energy. However, as the wavelength gets large, the wave consists of just rotating the spins as a whole over a large region. As $\lambda \to \infty$, it is a rotation of the whole system of spins. However, by the rotational invariance, the rotation of the whole system does not require any energy. The energy therefore falls with λ for large λ, $E \sim c/\lambda \sim cp$, *i.e.* it corresponds to the massless particle. Here we see physically that a symmetry transformation corresponds to the excitation of a massless, zero energy Goldstone boson.

The situation with the chiral symmetry of QCD is similar. In the limit of massless quarks, there would be different possible vacua, all equivalent. It is not as easy to picture these as it was in the case of the spin vector of the magnet. My own mental picture imagines the vacuum to be a complex soup of virtual quark pairs. In the symmetry limit, there could be any combination of LH and RH quarks, as all their QCD interactions are the same. The continuous family of vacua then corresponds to differing LH/RH compositions of the vacuum. In any case, the lack of invariance of the true chosen vacuum can be seen by considering matrix elements like

$$\langle 0|\bar{\psi}\psi|0\rangle = \langle 0|\bar{\psi}_L\psi_R + \bar{\psi}_R\psi_L|0\rangle \tag{8}$$

Since the operator is not invariant under the symmetry, the matrix element must vanish if the vacuum is invariant. Theoretical studies of this operator indicate that it is non-zero if the symmmetry is hidden. Given this limit the π, K, η would be the massless Goldstone bosons.

We now come to the question of the nature of the predictions. Usual symmetries like isospin require that when one makes a symmetry transformation on a state (*i.e.* the proton), one obtains another state in the same multiplet (*i.e.* the neutron). The couplings of the neutron and proton are then related by the symmetry. For a hidden symmetry, one does not have the usual multiplet structure. The partner of the proton under an axial transformation is the state composed of a proton plus a zero energy pion. These have the same energy, and there was nothing in the general quantum mechanical argument of Eq. (7) that required that we allow only single particle states. More generally, if $|\beta\rangle$ is an arbitrary state, then the symmetry relates

$$|\beta\rangle \iff |\beta + \pi(p_\mu = 0)\rangle \tag{9}$$

and their couplings are related. This is manifest in the "soft pion theorem" of chiral symmetry [2]

$$lim_{p_\mu \to 0}\langle\beta\pi^i(p)|\theta|\alpha\rangle = -\frac{i}{F_\pi}\langle\beta|[Q_s^i, \theta]|\alpha\rangle \tag{10}$$

where β and α are arbitrary states, θ is some operator and $F_\pi = 93$ MeV is the pion decay constant.

The end result is that symmetry has led to non-trivial dynamics. It relates processes with differing numbers of pions, such as

$$
\begin{aligned}
K &\to 3\pi \iff K \to 2\pi \\
K &\to \pi\pi e\nu \iff K \to \pi e\nu \\
\pi^0 &\to \gamma\gamma \iff \gamma \to 3\pi \\
K &\to \pi\pi e\nu \iff \pi\pi \to \pi\pi
\end{aligned} \tag{11}
$$

Sometimes the predictions are very surprising, as in Weinberg's absolute prediction of $\pi\pi \to \pi\pi$ amplitudes. If one thinks of what a mess it would be to try to calculate these processes directly from QCD, the simplicity and power of the chiral predictions are truly remarkable.

4. HISTORY AND SOCIOLOGY

Much work has been done on the subject of chiral symmetry. This section contains some comments that may help one to understand the general patterns in the literature of the subject.

The subject of chiral symmetries started in the 1960s. There was an initial discovery period, where the general ideas were invented and explored. The issues were the idea of hidden symmetry, the patterns of symmetry breaking, the representation dependence, the relation to field theory, etc. There was also a period of phenomenology [2]. The idea was still a hypothesis, and one tested it by applying it to physics and seeing if it accurately represented reality. Kaon decay relations, the Goldberger-Treiman relations and Weinberg's $\pi\pi$ scattering amplitudes were explored at this time and experimental studies also addressed their issues. The theoretical techniques generally involved the soft-pion theorem plus current algebra, although effective Lagrangian methods were also developed. The issues and notations of this period are often obscure to present day readers. One can obtain a good flavour of this period from S. Coleman's Erice lectures on the subject [1].

The 1970s were a quiet period for the subject, both theoretically and experimentally, as most effort was devoted to the development of gauge theory and the standard model. However, there gradually developed a general awareness that the standard model implied the validity of chiral symmetry and that the idea was no longer a hypothesis but was obligatory.

One might refer to the 1980s as the modern theoretical period, as the theorists rediscovered chiral symmetry as an active field. Many of us date this new period from a characteristically clear paper of Weinberg [3] which is recommended reading for anyone interested in the subject. In this period, there are a variety of purposes for which chiral symmetry is used, ranging over QCD, CP violation, WW scattering, strings, etc. The language here uses effective Lagrangians, and the state of the art involves "next-to-leading order" calculations. The theoretical machinery has been explored again at a new level. In phenomenological applications, one has had to reanalyze old experiments, as this style of low energy experiments is no longer fashionable.

There is some hope that the 1990s could become the modern experimental period. There is a wide variety of high intensity, low energy machines planned such as $\pi, K, \eta, \phi, B, \ldots$ factories or $e, \bar{p}, \gamma, \ldots$ machines. Experimentalists (including many coming from nuclear physics) want to study QCD. The framework now exists, and theorists are interested. It requires some work to be aware of the issues and language of chiral symmetry and very low energy QCD, but the potential now exists for fruitful interplay between theory and experiment.

There is also an important cautionary note that must be added concerning the varieties of chiral symmetry. There are many different ways in which one can use the ideas of chiral symmetry, and one must carefully distinguish among them. A special place is held by what is now called chiral perturbation theory, which is the prime focus of these lectures. This uses chiral symmetry in its fullest generality, so that the predictions are those of the symmetry alone without additional dynamical assumptions. It is phenomenological in character. This use is the only one which is a rigorous controlled approximation. However, it is often somewhat limited in scope because of the need to determine the low-energy constants phenomenologically. A second form is what I call models of chiral symmetry. These are often motivated by the desire to remove some of the limitations of strict chiral perturbation theory. For example, dispersion relations may be used to extend the predictions to higher energy, or models for the low energy constants may be used when these are not known from experiment. These attempts are only as good as the physics which one puts into the model, and one can be led astray. Thus, these give up the rigour in hopes of being more widely useful. More comments are made on these in Section 7. Finally, there are some uses of effective chiral Lagrangians outside the range of validity of the energy expansion, *i.e.* when the energy is large. In this case, there is no perturbative expansion, and one can have 100% corrections, and we have no apriori clue as to the validity of predictions. A prime example of this class is the Skyrme model. This model uses a chiral Lagrangian with two terms, [the meaning of this will be clearer below], and the condition for forming a soliton requires that the two terms contribute equally to the energy function. This means that higher order terms could also contribute equally, and that many results can be changed drastically by the inclusion of more terms in the chiral Lagrangians. The Skyrme model is quite interesting, but it must be remembered that it is just a model. The uninitiated reader will often assume that all papers that use the words "chiral symmetry" are on the same footing, and then will be confused when not all agree. If you are aware of these distinctions, it is not hard to classify the framework of each paper and to sort out the differences.

5. An Example of Effective Lagrangian

The language of the field now uses non-linear effective Lagrangians as the basis of application of chiral symmetry [4,5]. Before presenting this language in general, it is useful to have a "hands-on" example, where all of the manipulations can be easily and explicitly seen. For this purpose, the linear sigma model is useful. It is defined by a Lagrangian involving a scalar field σ, the π's, and a " nucleon" doublet

$$\psi = \begin{pmatrix} p \\ n \end{pmatrix} \qquad (12)$$

with the form

$$\mathcal{L} = \bar{\psi}i\partial\!\!\!/\psi + 1/2[(\partial_\mu\sigma)^2] + (\partial_\mu\vec{\pi})^2] - g\bar{\psi}(\sigma + i\vec{\tau}\cdot\vec{\pi}\gamma_5)\psi + \frac{\mu^2}{2}(\sigma^2 + \vec{\pi}^2) - \frac{\lambda}{4}(\sigma^2 + \vec{\pi}^2)^2 \quad (13)$$

This has a $SU(2)_L \times SU(2)_R$ invariance where the fields transform as

$$\begin{aligned} \psi_L &\rightarrow \psi'_L = L\psi_L \\ \psi_R &\rightarrow \psi'_R = R\psi_R \\ \sigma + i\vec{\tau}\cdot\vec{\pi} &\rightarrow \sigma' + i\vec{\tau}\cdot\vec{\pi}' = L(\sigma + i\vec{\tau}\cdot\vec{\pi})R^+ \end{aligned} \quad (14)$$

with L, R being $SU(2)$ transformation matrices. This is a conventional renormalizable field theory. Apart from the Yukawa couplings, it is just like the Higgs sector of the standard model. The usual way to solve it is to minimize the energy to find the ground state at $\langle\sigma\rangle = v$, with $v = \sqrt{\mu^2/\lambda}$, then expand about this state using $\sigma = v + \tilde{\sigma}$. When one does this, the Lagrangian reads

$$\mathcal{L} = \bar{\psi}(i\partial\!\!\!/ - gv)\psi + 1/2[(\partial_\mu\tilde{\sigma})^2 - 2\mu^2\tilde{\sigma}^2] + 1/2(\partial_\mu\vec{\pi})^2 - g\bar{\psi}(\tilde{\sigma} + i\vec{\tau}\cdot\vec{\pi}\gamma_5)\psi$$

$$-\lambda v\tilde{\sigma}(\tilde{\sigma}^2 + \vec{\pi}^2) - \frac{\lambda}{4}(\tilde{\sigma}^2 + \vec{\pi}^2)^2 \quad (15)$$

This now describes massive σ and "nucleons", with

$$\begin{aligned} m_P &= gv \\ m_\sigma &= \sqrt{2}\mu \end{aligned} \quad (16)$$

and massless pions. The pion decay constant turns out to be $F_\pi = v$.

This Lagrangian has the chiral $SU(2)$ invariance and hence is consistent with the low-energy prediction of chiral symmetry. For example, to obtain Weinberg results for $\pi\pi$ scattering, one can have both the direct π^4 coupling plus the effect of $\tilde{\sigma}$ exchange from the $\tilde{\sigma}\pi^2$ coupling. In the case of $\pi^+\pi^0$ scattering one obtains

$$\begin{aligned} \mathcal{M}(\pi^+\pi^0 \rightarrow \pi^+\pi^0) &= -2i\lambda + (-2i\lambda v^2)^2\frac{i}{q^2 - m_\sigma^2} \\ &= -2i\lambda[1 + \frac{2}{q^2 - 2\lambda v^2}] \\ &= \frac{iq^2}{v^2} + ... = \frac{iq^2}{F_\pi^2} + \theta(q^4) \end{aligned} \quad (17)$$

which is the Weinberg result. Note the cancellation required in order to get a low energy result which is proportional to q^2.

One can better expose the content of the theory by using a change of variable. Let us define new scalar field "S", pion fields π', and "nucleon" fields B, using

$$(\sigma + i\vec{\tau} \cdot \vec{\pi}) \equiv (v + S)U$$

$$U = exp(i\frac{\vec{\tau} \cdot \vec{\pi}'}{v})$$

$$\equiv \xi\xi \qquad (18)$$

$$\xi^{+}\psi_L \equiv B_L$$

$$\xi\psi_R \equiv B_R$$

Again this is similar to the Higgs change of variables in the standard model. One does not change the content of the theory by renaming the fields. In the new basis the Lagrangian has the form

$$\mathcal{L} + \bar{B}(i\slashed{D} - gv)B + 1/2((\partial_\mu S)^2 - 2\mu^2 S^2)$$

$$-\lambda v S^3 - \frac{\lambda}{4}S^4 + \frac{v^2}{4}Tr(\partial_\mu U \partial^\mu U^+)[1 + \frac{S}{v}]^2 \qquad (19)$$

with

$$D_\mu = \partial_\mu + V_\mu + A_\mu \gamma_5$$

$$V_\mu = \frac{1}{2}(\xi^+\partial_\mu\xi + \xi\partial_\mu\xi^+) \qquad (20)$$

$$A_\mu = \frac{i}{2}(\xi^+\partial_\mu\xi - \xi\partial_\mu\xi^+)$$

The essential point is that, apart from the πNN couplings contained in D_μ, the interactions of pions have been compacted into a single term, the last one given above. This already has two derivatives so that it always contributes to matrix elements with a power of q^2. The nucleons and scalar S are heavy so that we are able to drop them at very low energy. [This makes only corrections at order q^4]. The result is then a very simple form

$$\mathcal{L} = \frac{v^2}{4}Tr(\partial_\mu U \partial^\mu U^+) + ... \qquad (21)$$

This is the effective low energy Lagrangian for the theory. It is very non-linear in that it contains all numbers of pion fields. However, it retains the $SU(2)_L \times SU(2)_R$ chiral symmetry, which from Eqs. (14), (18) can be seen to be

$$U \to U' = LUR^+ \qquad (22)$$

To obtain results for $\pi\pi$ scattering, one expands this Lagrangian (dropping primes)

$$\mathcal{L} = 1/2(\partial_\mu\pi)^2 + \frac{2}{3v^2}[(\vec{\pi}\partial_\mu\vec{\pi})^2 - \vec{\pi}^2(\partial_\mu\vec{\pi})^2] + ... \qquad (23)$$

and one may directly calculate the scattering amplitude, obtaining

$$\mathcal{M}(\pi^+ \pi \rightarrow \pi^+ \pi^0) = \frac{iq^2}{v^2} \tag{24}$$

as before. Likewise any other scattering process involving more pions will agree to order q^2 when calculated in the two approaches.

The essential point of this exercise has been to show that one can find a simple, but non-linear, effective Lagrangian that encapsulates all of the low energy predictions of a more complicated theory. The important feature is that the symmetry properties have been retained in the transition to the effective theory.

6. CHIRAL PERTURBATION THEORY

Despite my desire to keep these lectures as non-formal as possible, I have to introduce the language which theorists use in discussing chiral predictions. This involves a quick tour through the formalism of chiral perturbation theory. The method uses effective Lagrangians, a specific example of which was given in the previous section. The general idea is that if a prediction is to follow from a symmetry alone, one may use an effective theory which shares the same full symmetry properties as the original theory. The symmetry relations will be the same for each theory.

The general procedure of chiral perturbation theory is as follows:

1) Write out the most general effective Lagrangian consistent with the symmetry properties of the interaction.

2) Calculate all possible diagrams.

3) Renormalize all of the parameters in the effective Lagrangian from experiment.

4) Apply at low and moderate energies.

This sounds like a daunting program. In practice, it is made manageable by the expansion in energy. Let us now look at these ingredients.

In order to write the most general Lagrangian one may use the matrix field U introduced above with the following symmetry

$$U = exp \quad \left(i \frac{\lambda^A \phi^A}{F} \right) \tag{25}$$

$$U \rightarrow LUR^+$$

where in $SU(2)$, $\lambda^A \rightarrow \tau^k$, $A = 1, 2, 3$, while in $SU(3)$, λ^A are the 8 Gell-Mann matrices. The field ϕ^A is either simply the pions in $SU(2)$ or the full pseudoscalar octet in chiral $SU(3)$. One wants all possible terms invariant under this symmetry. There is no term with zero derivatives since

$$Tr(UU^+) = 2 = \text{constant}$$

while with two or more derivatives, we have terms like

$$\begin{aligned}
&Tr(\partial_\mu U \partial^\mu U^+) \\
&Tr(\partial_\mu U \partial^\mu U^+ \partial_\nu U \partial^\nu U^+) \\
&Tr \partial_\mu U \partial_\nu U^+ \partial^\mu U \partial^\nu U^+)
\end{aligned} \tag{26}$$

The most general Lagrangian will contain all of these terms

$$\mathcal{L} = A tr(\partial_\mu U \partial^\mu U^+) + \alpha_1 \; Tr(\partial_\mu U \partial^\mu U^+ \partial_\nu U \partial^\nu U^+) + \alpha_2 \; Tr(\partial_\mu U \partial_\nu U^+ \partial^\mu U \partial^\nu U^+) + ... \tag{27}$$

where we are temporarily neglecting quark masses. There is an infinite number of terms.

The most important ingredient is the idea of the energy expansions. When we take matrix elements, the derivatives turn into factors of the pions momentum. Then, schematically we have a matrix element

$$\mathcal{M} = A q^2 + \alpha_1 q^4 + \alpha_2 q^4 + ... \tag{28}$$

However, at low enough energies $\alpha_i q^4 \ll A q^2$. Similary, yet higher order terms at order (q^6) are yet smaller. Out of the infinite number of terms we need only one if we work at low enough energy, i.e.

$$\begin{aligned}
\mathcal{L} &= \frac{F^2}{4} \; Tr(\partial_\mu U \partial^\mu U^+) + ... \\
&= 1/2(\partial_\mu \vec{\pi} \partial^\mu \vec{\pi}) + ...
\end{aligned} \tag{29}$$

where the normalization has been chosen to correctly normalize the kinetic energy term in the second line. Consideration of the axial current allows the identification $F = F_\pi = 93$ MeV. This term (plus the effect of quark masses, once properly included) reproduces the soft pion results of the 1960's. At somewhat higher energies there will be corrections to these results, parametrized by the low energy constants α_1, α_2, etc. At moderate energies, these corrections are under control in that each subsequent order is smaller. The perturbative parameter is q^2/Λ^2 where Λ is some mass scale. [In practice $\Lambda \sim m_\rho$ or 1 GeV.] At high energy, all terms in the energy expansion become equal and there is no possibility of prediction. Working to first non-leading order, i.e. order (q^4), is the present state of the art.

One must calculate all diagrams for full generality. This includes loop diagrams. One might worry about loop diagrams as the Lagrangian is formally non-renormalizable. However, the solution is simple: renormalization can be carried out order by order in the energy expansion. One-loop diagrams with an order (q^2) Lagrangian produce divergences at order (q^4). These can be absorbed into renormalized coefficients at order (q^4). Higher numbers of loops would renormalize higher order coefficients. While to work to all orders in the energy and loop expansions would require an infinite number of renormalizations (this is the sense in which it is non-renormalizable), at low energies one need only consider a finite number of these because of the finite number of relevant terms.

The summary of the method to order q^4 is then a subset of the rules given above:

1) Write out the most general Lagrangian at order q^2 and q^4, $\mathcal{L}_2$ and $\mathcal{L}_4$. This has been done by Gasser and Leutwyler [6] and will be displayed below.

2) Calculate tree-diagrams with $\mathcal{L}_2$ and $\mathcal{L}_4$ plus one-loop diagrams with $\mathcal{L}_2$.

3) Renormalize the parameters, using experiment.

4) Apply at low and moderate energy.

The Lagrangian at order q^2 including quark masses is known from the 60's:

$$\mathcal{L}_2 = \frac{F^2}{4} \quad Tr(D_\mu U D^\mu U^+) + \frac{F^2}{2} B_0 \quad Tr(mU + U^+ m) \tag{30}$$

where m is the quark mass matrix and B_0 is a constant. In $SU(2)$ the general form at order (q^4) is [6]

$$
\begin{aligned}
\mathcal{L}_4 = {} & \bar{L}_1 \quad Tr(D_\mu U D_\nu U^+ D^\mu U D^\nu U^+ - D_\mu U D^\mu U^+ D_\nu U D^\nu U^+) \\
& + \bar{L}_2 \quad [Tr(D_\mu U D^\mu U^+)]^2 \\
& + 2B_0 L_4 \quad Tr(D_\mu U D^\mu U^+) \quad Tr(mU + U^+ m) \\
& + 4B_0^2 L_2 \quad [Tr(mU + U^+ m)]^2 \\
& + 4B_0^2 L_1 \quad [Tr(mU - U^+ m)]^2 \\
& - iL_1 \quad Tr(L_{\mu\nu} D^\mu U D^\nu U^+ + R_{\mu\nu} D^\mu U^+ D^\nu U) \\
& + L_{10} \quad Tr(L_{\mu\nu} U R^{\mu\nu} U^+)
\end{aligned}
\tag{31}
$$

Here D_μ is a covariant derivative containing photon and/or W fields, and $L_{\mu\nu}$ and $R_{\mu\nu}$ are the appropriate left/right field strength tensors

$$
\begin{aligned}
D_\mu U &= \partial_\mu - i\ell_\mu + iU r_\mu \\
L_{\mu\nu} &= \partial_\mu \ell_\nu - \partial_\nu \ell_\mu + [\ell_\mu, \ell_\nu]
\end{aligned}
\tag{32}
$$

The L_i are the low-energy constants. In $SU(3)$, there are three more terms (hence the numbering scheme). The precise form of this is not important to anyone except those doing the calculations. The important thing to take away from this discussion is that one parametrizes the possible behaviour of all processes to order q^4 in terms of a few low-energy constants. Given these constants, a well-defined method exists for making predictions.

7. CHIRAL SYMMETRY IN ACTION

In this section, I want to show you a couple of examples of chiral perturbation theory applied phenomenologically. This is important both to get a feel for the nature of the method and to learn about the limits of validity of the energy expansion.

The simplest process that I can think of to illustrate the method is the pion electromagnetic form factor

$$
\langle \pi^+(p') | J_\mu^{e.m.} | \pi(p) \rangle = f(q^2)(p + p')_\mu
$$
$$
q_\mu = (p' - p)_\mu \tag{33}
$$
$$
f(q^2) = 1 + \frac{1}{6} \langle r^2 \rangle_\pi q^2 + \dots
$$

Using the Lagrangian $\mathcal{L}_2$, one obtains the result $f(q^2) = 1$, not a surprising answer. At order q^4 there is more content. One includes the effect of $\mathcal{L}_4$ plus the loop diagram of Fig. 2. The result after defining a renormalized parameter L_9^r

$$
f(q^2) = 1 + q^2 \left[\frac{2L_9^{ren}}{F_\pi^2} - \frac{1}{96\pi^2 F_\pi^2} (\ell n \; \frac{m_\pi^2}{\mu^2} - \frac{1}{3}) \right] + \frac{1}{96\pi^2 F_\pi^2}(q^2 - 4m_\pi^2)H(\frac{q^2}{m_\pi^2}) \tag{34}
$$

with

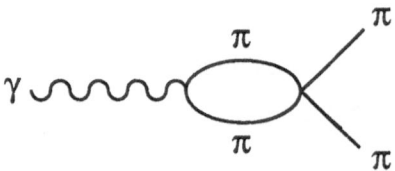

$$Fig.\ 2.\ \text{Loop diagram for } \gamma \to \pi\pi.$$

$$H(\frac{q^2}{m_\pi^2}) \equiv \left\{ 2 + \sigma \ell n \left| \frac{\sigma - 1}{\sigma + 1} \right| + i\pi\theta(q^2 - 4m_\pi^2) \right\}$$

$$\sigma \equiv \sqrt{1 - \frac{4m_\pi^2}{q^2}}$$

(35)

This latter function contains the imaginary parts required by unitarity (to this order in the energy expansion), and the kinematic effects of low energy rescattering. This process by itself is not predictive. One is required to simply fit the low energy constant L_q^r. Before going on, however, let us use this example to learn more about the energy expansion. We know from the success of vector dominance ideas, and from a direct look at the data, that the form factor behaves roughly as

$$f(q^2) = \frac{1}{1 - \frac{q^2}{m_\rho^2}}$$

$$= 1 + \frac{q^2}{m_\rho^2} + \frac{q^4}{m_\rho^4} + \dots$$

(36)

(Let us for clarity drop the imaginary parts and loop effects.) The first term in the expansion requires $2L_9^{ren} \approx F_\pi^2/m_\rho^2$. However, it also shows the nature of the expansion, which in this case is in q^2/m_ρ^2. As $q^2 \to m_\rho^2$, keeping only the first terms becomes a bad approximation, but for $q^2/m_\rho^2 \ll 1$ the expansion is well behaved.

This example becomes more predictive if we go to other processes [6,7]. There exists a radiative complex of pionic transitions ($\gamma \to \pi^+\pi^-; \pi^+ \to e^+\nu\gamma; \pi^+ \to e^+\nu e^+ e^-; \gamma\pi \to \gamma\pi$), that describes six form-factors in terms of two low-energy constants (L_9, L_{10}). Besides the charge radius with pion form factor described above, one has three form factors in weak radiative decays

$$M_{\mu\nu}(q,p) = \text{pion pole} + h_v \varepsilon_{\mu\nu\alpha\beta} p p^\alpha q^\beta$$
$$+ h_A (p-q)_\mu q_\nu - g_{\mu\nu}(p-q) \cdot q)$$
$$+ r_A A(g_{\mu\nu}q^2 - q_\mu q_\nu)$$

(37)

and two in $\gamma\pi$ reactions

$$\mathcal{M}(\gamma\pi \to \gamma\pi) = \vec{\varepsilon_1} \cdot \vec{\varepsilon_2} \left[-\frac{\alpha}{m_\pi}(1 - \frac{1}{6}\langle r_m^2 \rangle(q_1^2 + q_2^2)) + \omega_1 \omega_2 \alpha_E + \vec{\varepsilon_1} \times \vec{q_1} \cdot \vec{\varepsilon_2} \times \vec{q_2} \quad \beta_M \right]$$

(38)

The latter two are the electron and magnetic polarizabilities, describing the polarization of the pion in E and B fields

$$\mathcal{L} \sim \text{Born} + (\alpha_E E^2 + \beta_M B^2)\vec{\pi} \cdot \vec{\pi}$$

(39)

In a parameter-free manner one predicts (here I do not display the small effects from loops):

$$\alpha_E + \beta_\mu = 0$$

$$h_v = \frac{m_\pi}{4\sqrt{2}\pi^2 f_\pi^2} = \frac{0.026}{m_\pi} \tag{40}$$

in agreement with measurements

$$(\alpha_E + \beta_\mu)_{expt} = (1.4 \pm 3.1) \times 10^{-4} \; fm^2$$

$$h_v = \left(0.029^{+0.019}_{-0.014}\right) \tag{41}$$

The determination of L_9^r from the pion form factor above allows the predictions

$$\frac{r_A}{h_v} = 32\pi^2 L_9 = 2.6 \tag{42}$$

compared to the experimental value

$$\left(\frac{r_A}{h_v}\right)_{expt} = 2.3 \pm 0.6 \tag{43}$$

A new low-energy constant L_{10} enters in h_A

$$\frac{h_A}{h_v} = 32\pi^2(L_9 + L_{10}) \tag{44}$$

and the experimental value

$$\left(\frac{h_A}{h_v}\right)_{expt} = 0.46 \pm 0.02 \tag{45}$$

serves to determine this. Finally, this then serves to predict the electric polarizability

$$\alpha_E = \frac{4\pi}{m_\pi F_\pi^2}(L_9 + L_{10}) = 2.8 \cdot 10^{-4} \; fm^2 \tag{46}$$

Interestingly, this result does not seem to work, with the only experimental measurement being $\alpha_E = (6.8 \pm 1.4) \times 10^{-4} \; fm^2$. The experiment is a difficult one, done by a Soviet group [8] scattering pions off heavy Z atoms $\pi Z \to Z\pi\gamma$ where one uses Coulomb exchange to provide the initial photon. It involves a delicate procedure of subtracting off the Born terms and studying the remainder. Before drawing any conclusion from the apparent disagreement, the parameter should be remeasured.

One can also look at the applications of chiral Lagrangians to $\pi\pi$ scattering [6,9]. Here the amplitudes are expressed in a partial wave expansion as scattering amplitudes T_J^I with I = isospin and J = angular momentum. At low energies the five possible amplitudes $(T_0^0, T_1^0, T_1^1, T_0^2, T_2^2)$ can be expressed with 10 parameters (scattering lengths, slopes, curvatures), which in principle are independent. Chiral symmetry relates them with two low-energy constants ($i.e. L_1, L_2$). The analysis works remarkably well with sample results shown in Fig. 3.

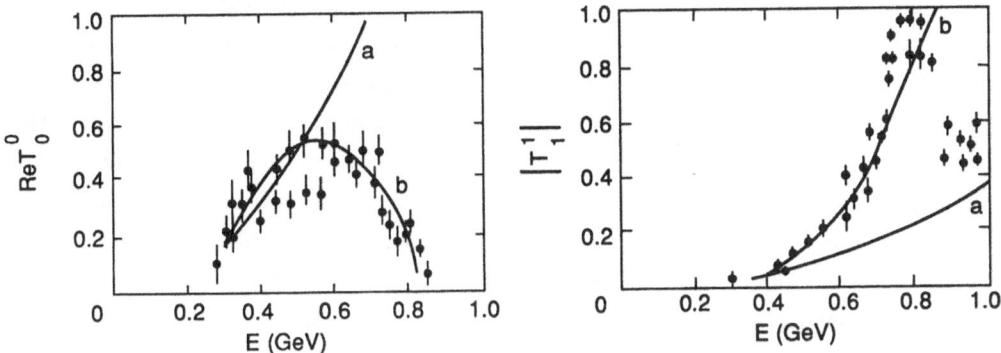

Fig. 3. (left) Tree-level $\pi\pi$ amplitudes. a) Lowest order, b) order q^4 ; (right) Full theory to one loop, order q^4.

Here curve a is the lowest order Weinberg prediction [10], *i.e.* from $\mathcal{L}_2$. It works at low energy but has clear problems as the energy increases, even violating unitarity below 700 MeV. The full theory to order q^4, curve b, does significantly better at correlating the various channels. Again one can see the nature of the energy expansion, for example by looking at T_1^1. The bump here is the ρ resonance, and keeping only two terms in the energy expansion can never reproduce the full ρ Breit-Wigner shape. It does, however, match on to the low energy tail of the ρ.

Let me just mention, without formulae, an analysis [11] which is coming out soon. The process $K^+ \to \pi^+\pi^-e^+\nu$ ($K_{\ell 4}$ is determined by the same low-energy constants which are relevant for $\pi\pi$ scattering, with our addition. However, the large N_c limit of QCD predicts that one combination of the three constants must vanish. (This is similar to a Zweig rule argument.) By the phenomenological study of $K_{\ell 4}$, one can then test this large N_c selection rule. The result seems to work at about the 1σ level, although higher experimental precision could tighter this considerably. The chiral constraints between $\pi\pi$ scattering and $K_{\ell 4}$ are especially stringent, and the low-energy constants can become well known.

There are many other applications of those ideas in other systems. Space and energy (*i.e.* my energy) limitations prevent me from continuing this survey. Hopefully, these examples have been sufficient for you to see both the beauty and the limitations of the theory.

8. THEORETICAL DIRECTIONS

There is a wide range of motivations of theorists using chiral symmetry. Even among those who apply chiral perturbation theory to phenomenology, there are many underlying interests. Most of us do not have as our primary interest phenomenology for its own sake. Rather the phenomenology serves a larger purpose. For many, the main motivation is to "tame" low energy QCD. Here we have a tool that is well defined, and the phenomenology lets us use it to uncover the ingredients important to QCD in this region. Others want to use chiral symmetry to solve other interesting problems. For example, the accurate prediction of CP violating amplitudes is an important goal for the testing of this unproved aspect of the standard model, or of theories beyond the standard model. For kaons, chiral symmetry is a main tool and we need to master it in order to test CP theories. In order to understand what phenomenology is to be desired, it helps to know the nature of questions which theorists are presently asking. The following is probably an incomplete list, but hopefully will give an indication of the range of questions.

i) What is the origin of chiral Lagrangians in QCD? We have seen above at that order q^2, the form of the chiral Lagrangian is universal, given F_π and m_π. One cannot tell QCD from the linear σ model at this level. However at order q^4 the situation is different. The low-energy constants L_i differ from one theory to the next, and by determining them from experiment we are obtaining the low energy imprint of QCD. There are half a dozen or so different suggestions in the literature as to what the important physics is for the effective Lagrangians. These amount to attempts to obtain the L_i from QCD. When compared to the data, there is a clear winner [12]. This is a generalized form of vector dominance. The low-lying spectrum determines the low energy constants, to a remarkable accuracy. Further work is needed to make the connection to QCD more precise.

ii) How does the energy expansion break down? There is often much information contained in the pattern of breakdown of the procedure. The symmetry ideas work best for pions at low energy, and have larger corrections for kaons or for pions at $q^2 = m_K^2$. The η brings in yet other corrections. If we regard the order q^2 results as fixed (which we should) it is the corrections at order q^4 and higher which are the interesting feature.

iii) What is the connections with dispersion relations and other techniques? Chiral perturbation theory satisfies all of the general properties of field theory, such as analyticity, crossing and unitarity, order by order in the energy expansion. It can therefore match on nicely to other techniques, such as dispersion relations, which make use of these properties [13]. The combination is more powerful than either one by itself. One hopes that these ideas will allow the results of chiral perturbation theory to be extended to higher energies. Correspondingly, dispersion theory by itself is filled with problems, such as the values of subtraction constants at low energy, which chiral symmetry can resolve. However, this marriage of ideas needs a long and careful courtship in order to determine the suitability of the union. One cannot just do anything, as for example, there are many ways to build exact unitarity into the theory, each yielding somewhat different results. One is risking giving up the rigour of the procedure. However, it seems likely that chiral perturbation theory will gradually absorb ideas from dispersion theory, to its ultimate improvement.

iv) What is the physics of the chiral anomaly? The axial anomaly is a fascinating subject which merges nicely with chiral perturbation theory [14]. The phenomenology of this has not been explored much beyond $\pi^0 \to \gamma\gamma, \eta \to \gamma\gamma$, but this is starting to change.

v) What is going on with the non-leptonic weak interactions? The weak non-leptonic decays are theoretically obscure. While their symmetry structure is known and well verified, the dynamics are not understood. Several applications come to mind. The higher order weak Lagrangians are now known [15]. If we can predict their relative coefficients, as in i) above, we can have widespread application. On the other hand, if we determine them from experiment, these coefficients contain information on the physics which drives the $\Delta I = 1/2$ enhancement. Finally, there is some hope of using chiral ideas, plus other physics, as an attempt to calculate the non-leptonic amplitudes.

vi) Can we apply what we have learned in new systems? I have already mentioned CP violation as an important application of chiral Lagrangians. Another one, in a far different area, is the study of $W_L W_L$ scattering [16]. There is an equivalence theorem which says that under certain conditions (a range of energies and a strongly interacting symmetry breaking sector), the scattering of longitudinal gauge bosons mirrors the Goldstone bosons of the symmetry breaking. These couplings are described by symmetry, and one obtains universal predictions at order q^2 and non-universal behaviour at order q^4, with the latter in principle being able to distinguish between the different possible underlying origins of symmetry breaking.

9. OPEN EXPERIMENTAL QUESTIONS

My survey here will be even less complete than the previous theory survey. If I try to list the reactions for which chiral symmetry has something to say, I come up with Table I. Each of these reactions has its own notation, jargon, phenomenology and literature. The underlying feature is that they describe the interactions of pions, kaons and etas with themselves or with other fields, in the limits where the π, K, η energies are small. For some of these reactions, especially those with heavy fields, this involves only a portion of the available phase space. However, given the right conditions chiral ideas can be applied. Here, I describe a subset of possible reactions for which I know that interest exists.

Table 1

Chiral Reactions

$\pi \rightarrow e\nu$	$K \rightarrow \mu\nu$	$\eta \rightarrow 3\pi$
$\pi \rightarrow e\nu\gamma$	$K \rightarrow \pi\pi$	$\eta \rightarrow \gamma\gamma$
$\pi \rightarrow \pi^0 e\nu$	$K \rightarrow 3\pi$	$\eta \rightarrow 2\pi\gamma$
$\pi \rightarrow e\nu e^+ e^-$	$K \rightarrow \pi e\nu$	$\eta \rightarrow \pi^0 \mu^+ \mu^-$
$\pi \rightarrow \gamma\gamma$	$K \rightarrow \pi e\nu$	$\eta \rightarrow \pi\gamma\gamma$
$\pi \rightarrow \gamma e^+ e^-$	$K \rightarrow \pi\gamma\gamma$	$\eta \rightarrow \gamma\mu^+\mu^-$
$\gamma\gamma \rightarrow \pi^+\pi^-$	$K \rightarrow e\nu e^+ e^-$	$\gamma\gamma^* \rightarrow \eta$
$\gamma\gamma \rightarrow \pi^0\pi^0$	$K \rightarrow e\nu\gamma$	$\eta \rightarrow \pi\pi\gamma$
$\gamma \rightarrow \pi^+\pi^-\pi^0$	$K \rightarrow \pi\pi\gamma$	$B_u \rightarrow \tau\nu$
		$B \rightarrow \pi e\nu$
$\gamma\gamma^* \rightarrow \pi^0$	$K \rightarrow \pi e^+ e^-$	$B \rightarrow \pi\pi e\nu$
$\tau \rightarrow 3\pi\nu_\tau$	$\gamma K \rightarrow \gamma K$	$D \rightarrow K e\nu$
$\tau \rightarrow 2\pi\nu_\tau$	$K \rightarrow 3\pi\gamma$	$D \rightarrow K\pi e\nu$
$\tau \rightarrow 4\pi\nu_\tau$	$K_s \rightarrow \gamma\gamma$	$D_s \rightarrow \mu\nu$
$\pi\pi \rightarrow \pi\pi$	$K_L \rightarrow \gamma\gamma$	$(N\bar{N}) \rightarrow M + \pi's$
		$(N\bar{N}) \rightarrow (N\bar{N}) + \pi's$
$\gamma p \rightarrow \pi p$	$\Sigma \rightarrow p\pi$	$\psi' \rightarrow \psi + \pi\pi$
$\pi p \rightarrow \pi p$	$\Xi \rightarrow \Lambda\pi$	$\Upsilon \rightarrow \Upsilon + \pi\pi$
$\pi p \rightarrow \pi\pi p$		$\Upsilon' \rightarrow \Upsilon + \pi^0\eta^0$

i) <u>Pions</u> - Here a remeasurement of the $\gamma\pi \to \gamma\pi$ amplitude is needed in order to clear up the disagreement mentioned in Section 6. The two photon reactions $\gamma\gamma \to \pi^+\pi^-, \pi^0\pi^0$ are related, but one is interested in the threshold region. Some effort tying together these reactions would be worthwhile. Data on $\gamma\gamma \to 3\pi$ would prove interesting in connection with the axial anomaly. These are related to $\gamma\gamma \to \pi$, but because of the three-body final state, contain more kinematic information.

ii) <u>Kaons</u> - The reaction $K \to \pi\pi e\nu$ contains within its form factors a test of the large N_c selection rule, as described briefly above. The last experiment was performed 15 years ago [17], but kaon physics has developed immensely since then. The last experiment was analyzed in a way which does not match well with the chiral prediction and the sensitivity hints at a small violation of the large N_c prediction, but is only at 1σ accuracy. This reaction can also yield the most model independent results on $\pi\pi$ scattering. A higher precision experiment would help these issues.

There exists a kaon radiative complex of decays ($K \to e\nu\gamma, K \to e\nu e^+e^-, K \to \pi e\nu, K \to \pi e\nu\gamma, \gamma K \to \gamma K$) analogous to the pionic reactions described above. These are all described without any new parameters, to order q^4. Predictions then exist for all reactions, most of which have not been tested. In $K \to 3\pi$, one can uncover the effects of higher order weak chiral Lagrangians by studying the quadratic dependence of the Dalitz plot distribution [15,18]. The full theory has recently been worked out. There are connections of this to other interesting issues in the weak interactions, such as the B parameter, the origin of the $\Delta I = 1/2$ rule and the ability of lattice methods to predict it, and rare K decays.

The reaction $K_L \to \pi^0\gamma\gamma$ is at the centre of a controversy right now [19,20]. There is a calculation of the amplitude and energy distribution using chiral loops. There are no tree level contributions at order q^4, so the result is purely loop dominated. However, from another study [20],it was suggested that there is a q^6 contribution from the ρ which is comparable. This is important to sort out because it influences strongly the search for CP violation in $K_L \to \pi^0 e^+e^-$ (see below). It is also interesting in its own right as to the relative roles of resonances and loops in the structure of chiral Lagrangians. The reaction $K_L \to \pi^0 e^+e^-$ is one of the prime modes for the observation of direct CP violation. The best signal involves an interference of the CP-conserving channel with a two-photon exchange (hence the relevance of $K_L \to \pi^0\gamma\gamma$) and the CP-violating one-photon mode. Although considerable theoretical work has been done, I feel that more can be accomplished. The observation of $K_s \to \pi^0 e^+e^-$ would be a major step as it would provide an absolute normalization of the effect of the ε parameter, allowing one to clearly see a non-ε effect.

iii) <u>Eta</u> - Eta physics has not yet been as thoroughly explored as that of π, K. More theoretical work will be forthcoming if the experiments are to be done. Here the rare modes $\eta \to \pi\gamma\gamma$ and $\eta \to \pi e^+ e^-$ share many of the same issues as $K \to \pi\gamma\gamma$ in relation to kaon CP violation. However, the analysis is in some ways clearer in η decay, as the weak interactions are not present. The study of both $K \to \pi\gamma\gamma$ and $\eta \to \pi\gamma\gamma$ would allow one to separate the weak and strong dynamics.

Decays such as $\eta \to \gamma e^+ e^-$ or $\gamma\gamma* \to \eta$ test the q^2 dependence of the anomaly analysis. The formalism has been worked out but needs better data to be applied [17]. The ultimate issue is again the understanding of the physics underlying these processes.

iv) <u>Nucleons</u> - There are several possible applications here, but let me just mention the one with the most outside interest at present [21]. The sigma term in $\pi N \to \pi N$ scattering measures the presence of explicit chiral symmetry breaking, i.e. quark masses. Combined with the Gell-Mann-Oakes-Renner analysis of baryon masses, it appears to indicate a sizeable strange quark content of the proton. There are a host of experimental and theoretical issues which have to be resolved before this question is settled definitively. However, it is important enough to justify further study.

v) <u>Direct $p\bar{p}$</u> - Since this talk is being given at a $p\bar{p}$ school, I thought it worthwhile to display how one can generate chiral predictions even in systems like this. One requires the pions to have low energy, which most often is not the case in $p\bar{p}$ annihilation. However, if one studies $p\bar{p} \to M + \pi$'s, with M being a heavy resonance, the mass of M can use up most of the $p\bar{p}$ energy, leaving little for the πs. In this case chiral relations enter. The $N\bar{N}$ can be in $I = 0, 1$ state and $J = 0, 1, ...$ Let me for simplicity, treat just one case here. The general formalism will be given elsewhere. I will look at $(\bar{N}N)(I = 0, J = 0) \to M(I = 0, J = 0) + \pi$'s such as $p\bar{p} \to \eta(1440) + \pi\pi$. The $\bar{N}N$ is treated as a heavy source, Φ. The M mass is also heavy. One expands an effective Lagrangian in terms of the energy of the pions. There is no term with zero derivatives, and two possible terms with two derivatives on the pions;

$$\mathcal{L} = a_1 \Phi M \quad Tr(\partial_\mu U \partial^\mu U^+) = a_2 \partial_\mu \Phi \partial_\nu M \quad Tr(\partial^\mu U \partial^\nu U^+) \qquad (47)$$

The extra derivatives on the heavy fields in the second term do not count in the energy expansion, as they yield energies which are not small. A possibility such as

$$\mathcal{L}' = b_1 \partial_\lambda \Phi \partial^\lambda M \quad Tr(\partial_\mu U \partial^\mu U^+) \tag{48}$$

yields the same results as the a_1 form given above, to leading order in the pion momenta since $p_\Phi \cdot p_M = p_\Phi \cdot (p_\Phi - p_{\pi\pi}) = \mathcal{M}_\Phi^2 - \theta(p_{\pi\pi})$. I have not included a quark mass term, as this is generally small for pions. The type of predictions that follows is that

$$\mathcal{M}(\Phi \rightarrow M\pi^+\pi^-) = \bar{a}_1 p_+ \cdot p_- + \bar{a}_2 E_+ E_- + ... \tag{49}$$

which says that the Dalitz plot distribution has a special kinematic suppression as E_π gets small. This is an "Adler zero" and forces the amplitude to grow with $\pi\pi$ energy in a way clearly distinguished from the phase space distribution. One also predicts $p\bar{p} \rightarrow M + 4\pi$ in terms of $p\bar{p} \rightarrow M + 2\pi$. At the moment, I do not see detailed uses for these types of predictions. Right now they are more of a curiosity as an application of chiral symmetry. However, perhaps with some study they can be put to more use, perhaps as flavour tags for the properties of the meson states. They also may be useful as a probe of how well chiral predictions do with heavy sources.

10. FINAL COMMENTS

The low energy experimental frontier is devoted to rare processes or high statistics. The field is active, with several low energy hadron facilities under way or being planned. While there are many motivations for such facilities, one component of the program should be chiral symmetry studies. These issues reflect QCD, there are open questions, and a community of theorists exists. While there are extremely compelling reasons for doing experiments to find processes which are forbidden by the standard model, any given experiment may not find such events. When one gets tired of measuring zero to higher precision, the same experiment may be sensitive to some of the rare process relevant for chiral studies. As a by-product of rare meson experiments, one can then contribute to the phenomenology described above. The mix of high states, high risk searches with safer and more conventional reactions provides a richer program for low energies.

I have tried to impress on you that the chiral framework is well defined and solid within its known limits. It needs to be more widely familiar as it is one of the foundations of phenomenology of the standard model. Theorists have returned to it in the hopes of taming QCD. I hope that at least some of the many experimenters who want to do something to study QCD will learn more about the field and keep it in mind when planning/doing experiments. Perhaps the 90's will see a renewal of experimental input to the field. If so, we will get closer to the goal of understand low energy QCD.

REFERENCES

[1] Many books describe this phenomena. See in particular S. Coleman, "Aspects of Symmetry", (Cambridge Univ. Press, 1985).

[2] Early developments are summarized in S. Adler and R. Dashen, Current Algebras and Applications to Particle Physics, (Benjamin, 1968).

[3] S. Weinberg, *Physica* **96A** (1979) 327.

[4] D. Geffen and S. Gasiorowicz, *Rev. Mod. Phys.* **41** (1969) 531.

[5] For book style treatment, try H. Georgi, "Modern Weak Interaction Theory" (Benjamin, 1984).

[6] J. Gasser and H. Leutwyler, *Nucl. Phys.* **B250**, (1985) 465; *Ann. Phys.* (NY) **158** (1984) 142.

[7] J.F. Donoghue and B.R. Holstein, *Phys. Rev.* **D40** (1989) 2378.

[8] Yu.M. Antipov et al., *Z. Phys.* **C26** (1985) 495.

[9] J.F. Donoghue, C. Ramirez and G. Valencia, *Phys. Rev.* **D38** (1988) 2195.

[10] S. Weinberg, *Phys. Rev. Lett.* **17** (1966) 616; *Phys. Rev. Lett.* **18** (1986) 188, 507.

[11] C. Riggenbach, J. Gasser, J. Donoghue and B.R. Holstein, to appear.

[12] J.F. Donoghue, C. Ramirez and G. Valencia, *Phys. Rev.* **D39** (1989) 1947;
G. Ecker, J. Gasser, A. Pich and E. de Rafael, *Nucl. Phys.* **B321** (1989) 311;
G. Ecker, J. Gasser, H. Leutwyler, A. Pich and E. de Rafael *Phys. Lett.* **B223** (1989) 425;
M. Praszalowicz and G. Valencia, Brookhaven preprint BNL 43148 (1989).

[13] Much of this is being explored by Gasser and Leutwyler (private communication).

[14] J.F. Donoghue and D. Wyler, *Nucl. Phys.* **B316** (1989) 289;
D. Issler, SLAC preprint (1989), to appear;
J. Bijnen, A. Bramon and F. Corner, *Phys. Rev. Lett.* **61** (1988) 1453;
J. Bijnen and F. Cornet, *Nucl. Phys.* **B296** (1988) 1557;
J.F. Donoghue, B.R. Holstein and Y.C. Lin, *Phys. Rev. Lett.* **55** (1985) 2766.

[15] J. Kambor, J. Missimer and D. Wyler, Zurich preprint ETH-TH/89-42 (1989).

[16] M.S. Chanowitz and M.K. Gaillard, *Nucl. Phys.* **B261** (1985) 379;
M.S. Chanowitz, M. Golden and H. Georgi, *Phys. Rev.* **D36** (1987) 1490;
M.S. Chanowitz, *Ann. Rev. Nucl. Part. Sci.* **38** (1988) 323;
A. Dobado and M. Herrero, *Phys. Lett.* **B228** (1989) 495; **B235** (1990) 129;
J.F. Donoghue and C. Ramirez, *Phys. Lett.* **B234** (1990) 361.

[17] L. Rosselet et al., *Phys. Rev.* **D15** (1977) 574.

[18] J.F. Donoghue, E. Golowich and B.R. Holstein, *Phys. Rev.* **D30** (1984) 587.

[19] G. Ecker, A. Pich and E. de Rafael, *Nucl. Phys.* **B303** (1988) 665 and Vienna preprint UWTh Ph-1989-65 (1989).

[20] L. Sehgal, *Phys. Rev.* **D38** (1988) 808;
J. Flynn and L. Randall, *Phys. Lett.* **B216** (1989) 221.

[21] J. Gasser, H. Leutwyler, M. Locher and M. Sainio, *Phys. Lett.* **B213** (1988) 85;
R. Jaffe and C. Korpa, *Comm. Nucl. Part. Phys.* **17** (1981) 163;
J.F. Donoghue and C. Nappi, *Phys. Lett.* **168B** (1986) 105;
T.P. Cheng, *Phys. Rev.* **D13** (1976) 2161;
T.P. Cheng and R. Dashen, *Phys. Rev. Lett.* **26** (1971) 594.

UNITARIZED CHIRAL PERTURBATION THEORY
AND RARE DECAY OF MESONS

Tran N. Truong

Enrico Fermi Institute and
Department of Physics
University of Chicago
Chicago, IL 60637

and

Department of Physics
University of California
Santa Barbara, CA 93106

and

Centre de Physique Theorique de
l'Ecole Polytechnique
91128 Palaiseau
FRANCE

INTRODUCTION

In the following three lectures I would like to discuss the following topics:

1 What is the Chiral Perturbation Theory?

2 Is it a reliable theory?

3 What is its role in the calculation of the rare decays of $\pi, \eta, \eta' \to \gamma e^+ e^-$; $\eta, \eta' \to \gamma\gamma\pi, \pi e^+ e^-$ and also in $K_L \to \gamma\gamma\pi$, $K_L \to \pi e^+ e^-$?

4 What are experiments to be done in η, η' and K decays, and what type of physics that we can learn from them?

LECTURE 1

Unitarized Chiral Perturbation Theory vs. Chiral Perturbation Theory

It is not possible to talk about rare decay of mesons without discussing the Effective Chiral Lagrangian and the current attempt to incorporate the unitarity constraint by the Chiral Perturbation Theory (C.P.Th.). Everybody agrees that the Effective Lagrangian method is important in describing the low energy phenomena involving

Medium-Energy Antiprotons and the Quark–Gluon Structure of Hadrons
Edited by R. Landua *et al.*, Plenum Press, New York, 1991

65

pions. The question is the usefulness of the loop correction to the tree Lagrangian. The effective Lagrangian method is very well covered by Professor Donoghue's lectures. I am addressing in my lectures the C.P.Th.

C.P.Th. is an application of the perturbation theory to the non-linear σ-model chiral Lagrangian taking into account the chiral symmetry breaking and the unitarity relation in the *perturbative manner*. It was invented a long time ago by Li and Pagels[1] but has recently been made more respectable by Weinberg[2] who laid down a fairly precise program. This program is nowadays very popular with world-wide followers. It is so popular that if you did not know what C.P.Th. was, you could be considered uneducated. This is why I started to learn about this field a few years ago, although many years before C.P.Th. became popular, dispersion relation and unitarity were used to find solutions to many problems where there is disagreement between the experimental data and the theoretical predictions made by *Current Algebra Low Energy Theorems* such as $K_{e_4}, \eta \to 3\pi$ problems[3]. Many current algebra results were beautifully demonstrated by Adler, Weissberger and Weinberg[4] and others in 1966, but unfortunately, they are only valid in the world where pions have a zero mass where the chiral symmetry is exact. This prevents us from confronting straightforwardly current algebra low energy theorems with experimental data. Corrections must therefore be made to these theorems. A typical example is the well-known Adler-Weissberger relation where assumptions were made about the smoothness of the extrapolations from the chiral world to the physical worlds. It was found that the corrections for the deviations from the chiral world for this relation are not important. This is also true for the Callan-Treiman relation involving the K_{L3} decay and other processes[4]. It was pointed out in 1981[3] that there are, however, other processes such as the $K \to \pi\pi e\nu, \eta \to 3\pi, \pi\pi \to \pi\pi$ low energy scattering where the corrections due to the pion pion S-wave unitarity (final state interaction) are important; the corrections to the low energy current algebra theorems for these processes could attain 30% to 50% in the amplitude which are indeed observed by experimental data.

The technique which was used was the application of the dispersion relation in combination with the exact elastic unitarity (to all order of the strong interaction) to correct the current algebra theorems. One then arrived at some singular integral equations of the Muskhelishvili-Omnès type[5] whose solution is straightforward. I wish to show in the first two lectures that if you carried out the program of C.P.Th. correctly with the *elastic unitarity to be satisfied to all orders, not perturbatively as suggested by the C.P.Th. program* then one would arrive at old fashioned dispersion relation results. Because the dispersion language is not now familiar, I shall give an approximate method to derive these results using field theoretic language. This last method has even more predictive power than the dispersion method.

The strong point of the C.P.Th. is its systematic study of the chiral symmetry breaking effect such as how the pseudoscalars acquire a finite mass and how the decay constants of π differs from that of K. In this scheme, one can do the calculation without using the assumption of the smoothness to extrapolate chiral theorems to the physical world. C.P.Th. confirms the assumption of the smoothness which is frequently used with dispersion approach.

The weak point of C.P.Th. is its perturbative approach to the unitarity correction. If one takes some time to think over the C.P.Th. program, one can see that it has little chance of being successful. This is so because the C.P.Th. is a power series expansion in the number of loops. For each higher loop, one increases the power of the pion momenta as one deals with the unrenormalizable non-linear σ-model for the pions. The C.P.Th. is therefore a power series expansion in pion momenta, apart from some unimportant logarithmic corrections. Because the low energy two-pion phenomenology is dominated by resonances, the vector meson ρ, or by a strong S-wave $I = 0$ pion pion interaction, it is difficult to see how a power series expansion in momenta can approximate these low energy phenomena. Because the ρ resonance is an unstable particle, it must have a Breit-Wigner representation:

$$\frac{m^2}{m^2 - s - im\Gamma} \neq \sum_n a_n s^n$$

It is clear that this Breit-Wigner amplitude cannot be approximated by a power series in momenta for s near m^2. One frequent excuse made by chiral perturbators is that C.P.Th. is not expected to work at the ρ mass. This is true, but then to what energy is the theory expected to be good? How many loops has one to calculate in order to approximately describe the main feature of our low energy physics? From the fundamental viewpoint, we only expect the power series to converge *below* the unitarity cut. From a more practical viewpoint because of the failure of the power series in describing the Breit-Wigner shape, it is not useful to do a higher loop calculation which involves a large amount of work and undetermined parameters to generate the ρ resonance. Furthermore, because the C.P.Th. series do not provide damping at high energy (in fact, the higher order terms have much worse behavior than the lower order terms), for such a problem as the calculation of the form factor of the virtual photon in $\pi \to \gamma e^+ e^-$, one must introduce an arbitrary counterterm which cannot be determined from independent experiments. The necessity of the C.P.Th. of explaining one experimental fact by one undetermined parameter is undesirable and points out the weakness of the theory. This is more so when the calculable terms by C.P.Th. represent a tiny correction to the undetermined parameters.

In the following, I would like to show explicitly the weakness of the C.P.Th. and then point out a method to remedy the problem. I shall first discuss the final state phase theorem and then use it to show the shortcomings of the standard C.P.Th.

calculation of the pion form factor, and show how to modify it and into a predictive theory[6].

1. Unitarity of the S-matrix and the Final State Phase Theorem

The following theorem is commonly attributed to Watson[7] but was also known to many people including Fermi. It relates the phase of the "weak" amplitude $a + b \rightarrow c + d$ or simply $b \rightarrow c + d$ to the phase of the strong elastic $c + d \rightarrow c + d$ where $c + d$ are in the eigenstates of the strong interaction S-matrix. This result is of fundamental importance because it is a consequence of the elastic unitarity of the S-matrix and time reversal invariance. No approximation scheme is involved in the demonstration of this theorem. It will be used to show the inadequacy of the C.P.Th. series and to show how this program can be modified such that the elastic unitarity is satisfied to all order of the perturbation in an approximate way.

Let us consider a two-channel problem, for example, $\gamma \pi^0 \rightarrow \pi^+ \pi^-$. Let $\mid \alpha >= \pi^+ \pi^-$ state in $I = 1, J = 1$ and $\mid \beta >= \gamma \pi^0$ state with the same I and J. Let us denote the S-matrix $S_{\alpha\beta} =< \alpha \mid S \mid \beta >$.

The unitarity of the S-matrix gives $SS^+ = S^+S = 1$. Time reversal invariance yields $S_{\alpha\beta} = S_{\beta\alpha}$. Let us define $S_{\alpha\alpha} = \eta e^{2i\delta}$ and $S_{\alpha\beta} = iAe^{i\phi}$ where η and A are real. From the unitarity of the S-matrix for strong interaction, δ is the usual P-wave phase shift. Define the T-matrix as $S_{\beta\alpha} = \delta_{\beta\alpha} + iT_{\beta\alpha}$, then using the unitarity of S-matrix (diagonal and off diagonal elements) we have

$$T_{\beta\alpha} =\mid T_{\beta\alpha} \mid e^{i\delta} \qquad \text{(modulo of } \pi\text{).} \qquad (1)$$

in other words below the inelastic threshold $16\, m_\pi^2$ the phase of $\gamma \pi^0 \rightarrow \pi^+ \pi^-$ is the P-wave $\pi\pi$ phase shift. Because the inelastic effect only becomes important beyond 1 GeV, we can safely use this phase theorem below 1 GeV.

Eq (1) can also be generalised to include initial and final state interaction, $T_{\beta\alpha} = (T_{\beta\alpha})e^{i(\delta_i + \delta_f)}$ where δ_i, δ_f are respectively initial and final state eigenphases.

2. Calculation of the Pion Form Factor

To show explicitly the shortcomings of C.P.Th., I would like to demonstrate that it gives results for the pion form factor which disagree strongly with the experimental facts. To see this let us write down the dispersion relation for the pion form factor $F(s)$:

$$F(s) = 1 + \frac{s}{\pi} \int_{4m_\pi^2}^\infty \frac{ImF(s')ds'}{s'(s' - s - i\epsilon)} \qquad (2)$$

To calculate $ImF(s)$, we use the elastic unitarity condition:

$$ImF(s) = F(s)e^{-i\delta} \sin\delta \qquad (3)$$

where δ is the P-wave pion pion phase shift. Putting the expression for $Im\ F(s)$ into the dispersion relation we get the integral equation of the Muskhelishvili-Omnès Integral Equation:

$$F(s) = 1 + \frac{s}{\pi} \int_{4m_\pi^2}^{\infty} \frac{F(s')e^{-i\delta} \sin \delta(s')ds'}{s'(s' - s - i\epsilon)} \qquad (4)$$

The solution of this integral Eq in terms of δ is well-known:

$$F(s) = P_n(s)D^{-1}(s)$$

with

$$D^{-1}(s) = exp\frac{s}{\pi} \int_{4m_\pi^2}^{\infty} \frac{\delta(s')ds'}{s'(s' - s - i\epsilon)} \qquad (5)$$

with $P_n(0) = 1$. It is usually assumed that $P_n(s) = 1$. This is a non-perturbative result. Because δ passes through 90^0 at the ρ mass, the pion form factor has the Breit-Wigner form as observed by experimental data. A more precise description is given by the so-called Gounaris-Sakurai formula.

How does the C.P.Th. for $F(s)$ look? The C.P.Th. result can be obtained by calculating the one loop Feynman graphs where the strong P-wave amplitude is given by the minimal term in the chiral lagrangian for pion-pion scattering or the linear term in the Weinberg expansion for the pion-pion scattering[8]:

$$f(s)^w = e^{i\delta} \sin \delta / \rho(s) = (96\,\pi f_\pi^2)^{-1}(s - 4m_\pi^2) \qquad (6)$$

where $\rho(s)$ is the phase space factor $\sqrt{\frac{s - 4m_\pi^2}{s}}$. It is clear, because of the high energy behavior of the scattering amplitude Eq (6), two subtractions must be made:

$$F(s) = 1 + sF'(0) + \frac{s^2}{\pi} \int_{4m_\pi^2}^{\infty} \frac{e^{-i\delta} \sin \delta(s')F(s')ds'}{s'^2(s' - s - i\epsilon)}.$$

Instead of solving the integral equation, one iterates it once by setting $F(s') = 1$ and $e^{-i\delta} \sin \delta = \rho(s)f^w(s)$ in the integrand. We have then the C.P.Th. result:

$$F(s) = 1 + sF'(0) - (96\,\pi f_\pi^2)^{-1}[(s - 4m_\pi^2)(h(s) - h(0)) + 4m_\pi^2 h'(0)s] \qquad (7)$$

where

$$h(s) = \frac{2}{\pi}\sqrt{\frac{s - 4m_\pi^2}{s}} \ln \frac{\sqrt{s} + \sqrt{s - 4m_\pi^2}}{2m_\pi} - i\rho(s).$$

$f_\pi = 93$ MeV, and $F'(0)$ is proportional to the pion r.m.s. radius. This result can be compared with the experimental data as shown in Fig. 1.

As can be seen, there is a strong disagreement between C.P.Th. results and experimental data on the phase and the modulus of $F(s)$. The reason for this dis-

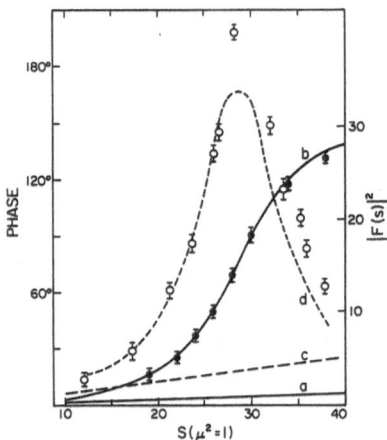

Figure 1. Solid curves: Phases of Pion Form Factor $F(s)$ calculated by (a) C.P.Th. Eq (7) and (b) U.C.P.Th. Eq (10). Solid circles are some experimental P-wave pion phase shifts. Dashed curves: $|F(s)|^2$ calculated by (c) C.P.Th. Eq (7) and (d) U.C.P.Th. Eq (10). Open circles are some experimental values of $|F(s)|^2$.

crepancy is that the *unitarity*, in the form of the *final state theorem*, is *violated* by the *perturbative* treatment of the M.O. integral equation.

Can one find another integral equation which can be handled in a perturbative manner? The answer is yes[6]. Let us write down the dispersion relation for the inverse of the form factor $F(s)$:

$$F^{-1}(s) = 1 - F'(0)s - \frac{s^2}{\pi} \int_{4m_\pi^2}^{\infty} \frac{e^{-i\delta} \sin \delta \ F^{*-1}(s')ds'}{s'^2(s'-s-i\varepsilon)}$$

Now

$$e^{i\delta} \sin \delta / \rho(s) = f^w(s)D(s)^{-1}N(s) \tag{8}$$

where $f(s)^w$ is the Weinberg amplitude and $N(s)$ is the so-called left-hand cut contribution[9] to the P-wave amplitude with $N(0) = 1$. It is a slowly varying function of s and can be approximated by a constant. Then

$$F(s) = F_1(s)/[1 - F_2(s)/F_1(s)] \tag{9}$$

with $F_1 = 1$ and $F_2(s)$ is the one-loop perturbative amplitude, i.e. the remaining terms on the R.H.S. of Eq (7). It satisfies the elastic unitarity relation to all orders of the strong interaction. Eq (9) is the diagonal [1, 1] *Padé approximant*[10]. It is a way to resum the perturbation series into an infinite geometric series so that unitarity is respected. This procedure is exactly the same method of summing up the infinite geometric series of the self energy operator for the propagator in order to satisfy unitarity. This leads to the inverse propagator method. We are using the same procedure here for the form factor and later for the scattering amplitude.

The comparison with the experimental data shows a dramatic improvement both in the *modulus* and in the *phase*. We make an extra prediction on the ρ width which is 142 MeV. This is the KSRF relation[11].

From the tree and the one-loop amplitudes of the C.P.Th. calculation which satisfy the unitarity relation perturbatively, one reconstructs a new amplitude by the Padé approximant method which satisfies approximately the unitarity relation to all orders. The final result for the form factor is:

$$F(s) = 1/\left\{1 - sF'(0) + (96\,\pi f_\pi^2)^{-1}[(s - 4m_\pi^2)(h(s) - h(0)) + 4m_\pi^2 h'(0)s]\right\}$$
$$= (s_R - 8m_\pi^2\gamma/\pi)/\left\{s_R - s + \gamma(s - 4m_\pi^2)h(s)\right\} \tag{10}$$

The unknown parameter s_R can be expressed either in terms of the pion r.m.s. radius or can be *determined* by requiring $F(s)$ satisfy the final state phase theorem. This second method is very useful, because it offers a unique and economical way of determining the counterterms in the C.P.Th. Using this method we can evaluate s_R by requiring the phase of $F(s)$ to be 90^0, the phase of the strong interaction, at the ρ mass (s_R is approximately s_ρ). In the following we refer to the combination of C.P.Th. with the Padé method as the *Unitarized Chiral Perturbation Theory* (U.C.P.Th.). The pion form factor in Eq (10) is proportional to the unstable ρ propagator and is the generalization of the Gell-Mann - Sharp - Wagner Vector Meson Dominance model (V.M.D.)[12] with the self energy correction due to the pion loop contribution taken into account to make the vector meson ρ unstable. In many low energy problems, far from the resonance, we can neglect the loop contribution. Unlike the Gell-Mann, Sharp, Wagner model where the ρ width has to be put in by hand, U.C.P.Th. automatically includes the ρ width. It should be emphasized that in Eq (5), the phase shift must be known over a wide range of s in order to calculate $F(s)$, in contrast with Eq (10) where it is sufficient to know δ at one energy.

The difference between Eqs (7) and (10) points out an important fact. Whereas Eq (7) satisfies unitarity perturbatively to $1st$ order, Eq (10) satisfies unitarity (approximately) to all orders. Although they have identical power series expansion around $s = 0$ up to and including terms which are linear in s, they differ widely in the physical region. This leads to the conclusion that perturbative unitarity is inadequate for the strong interaction problem.

The C.P.Th. at one-loop level without using the Padé or another unitarized scheme is inadequate. The momentum power counting method which is frequently used in the C.P.Th. is often unreliable and may lead to large errors.

We can illustrate the problem with the momentum power counting by studying the electromagnetic form factor of the kaon. We would think that the 3π state which corresponds to the 2-pion loop should be negligible according to the C.P.Th. power

71

counting method. This is not the case at all, because we know the 3π in $I = 0$, $J = 1$ interact strongly with each other and form the ω resonance which, together with the ϕ resonance ($K\bar{K}$, $I = 0, J = 1$), dominate the isoscalar kaon form factor. In other words, there is no basis for supporting the perturbative loop expansion of C.P.Th., treating the pions as if there were no strong interaction among themselves.

As was shown above and will be shown later, for physical problems where the pseudoscalar loops are in the $J = 1^-$ state, it is much more precise to use the V.M.D. model. One gains in addition, physical insight into the problem. (for example when we calculate the kaon loop contribution to the isoscalar electromagnetic kaon form factor, we cannot distinguish the relative contribution of ω from that of ϕ vector meson).

Let me now illustrate how to use the final state phase theorem to determine of the counterterms in the calculation of the scalar pion form factor. Because the existence of a low energy isoscalar-scalar meson is debatable, we cannot generalize the V.M.D. physics to this situation. Here the U.C.P.Th. can be useful. After applying the Padé technique to the C.P.Th. calculation of the scalar form factor $S(s)$, one arrives at:

$$S(s) = \frac{S(0)}{1 - S'(0)s + \frac{1}{16\pi f_\pi^2} \left[(s - \mu^2/2)(h(s) - h(0)) + (\mu^2/2)s\, h'(0)\right]} \tag{11}$$

Because Eq (11) satisfies the elastic unitarity we can use the final state phase theorem to fix the magnitude of $S'(0)$. Setting $s = m_K^2$, the experimental data on the S-wave pion-pion phase shift gives $\delta_0 \simeq 40^0$ at this energy, hence $S'(0) = 1/25\,\mu^2$. The magnitude and phase of $S(s)$ is shown in Figs. 2 and 3. Notice $S(s)$ has a complicated energy dependence around $s = 4\mu^2$ due to the square root singularity (it is usually misunderstood as the logarithmic singularity) and that above the physical threshold $s > 4\mu^2$, $|\,S(s)\,| \sim 1.35$ or larger; this enhancement factor was used to explain the discrepancy between current algebra prediction and experimental results on K_{e4} and $\eta \rightarrow 3\pi$. It can be shown that the scattering lengths calculated from Eq (10) and (11) by taking the limit of their phase are larger than those calculated by the usual C.P.Th. by, respectively 25% and 45% which reflect the C.P.Th. difficulty even at the threshold.

Similarly to the application of the C.P.Th. to the calculation of the vector and scalar pion form factor, one can calculate the vector and scalar $K - \pi$ form factors in $K \rightarrow \pi\mu\nu$ decays using U.C.P.Th. Using the final state phase theorem for the S and P wave $K - \pi$ system, we can similarly determine the corresponding counterterms in the C.P.Th. approach. This enables us to predict not only the r.m.s. radii of the vector and scalar form factors in $K_{\mu 3}$ decay but also their energy dependence over a much larger energy scale such as in $\tau \rightarrow \pi K\nu$ decays. (For the Pwave form factor, we would get the SU(3) relation between $\tau \rightarrow \rho\nu$ and $\tau \rightarrow K^*\nu$.)

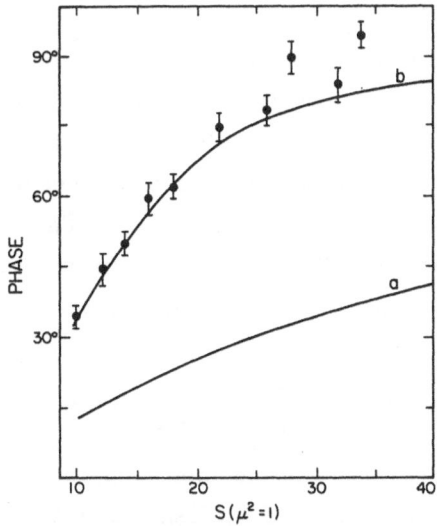

Figure 2. Solid curves: Phases of the scalar form factors calculated by (a), C.P.Th. and (b) U.C.P.Th. Eq (11)

The $K \to \pi \pi e \nu$ decays are much more complicated to handle, because we have to deal here with not only the pion loop but also the $K\pi$ loop. The use of the unitarity of the S matrix together with the Padé method enables us to determine the counterterms (subtraction constants). This is more or less done in Ref. 3 using the dispersion language.

The $\eta \to 3\pi$ problem can be similarly done in the U.C.P.Th. approach. Without $\eta\eta'$ mixing and switching off the pion-pion interaction, the $\eta \to 3\pi$ amplitude can be written as

$$\mathcal{M}(\eta_8 \to \pi^+ \pi^- \pi^0) = \frac{1}{f_\pi^2} \frac{\delta m^2}{\sqrt{3}} (1 - \frac{2E_0}{m_\eta}) \tag{13}$$

where $f_\pi = 93$ MeV and $\delta m^2 = m^2(K^+) - m^2(K^0) - m^2(\pi^+) + m^2(\pi^2)$.

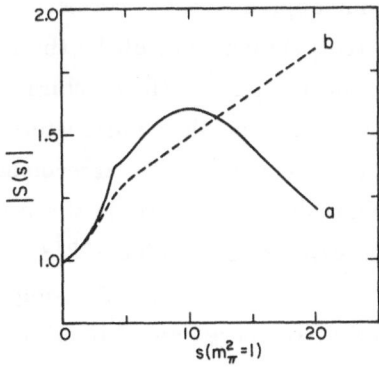

Figure 3. Modulus of scalar pion form factor. (a) U.C.P.Th. calculation Eq (11) and (b) C.P.Th. calculation

This matrix element yields $\Gamma(\eta_8 \to \pi^+\pi^-\pi^0) = 65$ eV which is much smaller than the observed width of 310 eV. The $\eta\eta'$ mixing can change the theoretical prediction to at most $80 \sim 90$ eV; we must therefore explain why the theoretical rate is too low by a factor of 3.5. We have shown a long time ago that the unitarity condition can account for this discrepancy[13]. Although the solution is complex due to the 3-body unitarity, the main idea can be understood, by applying the U.C.P.Th. to the limit $q_{\pi^0} \to 0$ for the matrix element $\eta \to \pi^+\pi^-\pi^0$. We can show by current algebra that it is related to $\frac{\delta m^2}{\sqrt{3}} \frac{1}{f_\pi} < \pi^+\pi^- \mid v \mid \eta >$, where v is a pseudoscalar quantity $v = \frac{\bar{u}\gamma_5 u + \bar{d}\gamma_5 d}{\sqrt{2}}$:

$$\lim_{q_{\pi^0}} \mathcal{M}(\eta_8 \to \pi^+\pi^-\pi^0) = \frac{\delta m^2}{\sqrt{3}} g_8(s = m_\eta^2)$$

where we have used Fig. (11) for the one-pion loop correction in the U.C.P.Th. scheme to calculate $g_8(s)$, $g_8(s) = S(s)/S(0)$. Using Eq (11), we have

$$g_8(s = m_\eta^2) = 1.50 e^{i45^0} \tag{14}$$

The magnitude of $g_\eta(s)$ can be obtained from the "odd" (i.e. π^0) pion spectra and extrapolate the $\eta \to \pi^+\pi^-\pi^0$ matrix element to the point $E_{\pi^0} \to 0$.

$$\mid g_\eta(s = m_\eta^2) \mid \simeq 2.0 \tag{15}$$

Taking into account the $\eta\eta'$ mixing there is an adequate agreement between theory and experiment.

The necessity for the unitarization of the perturbative calculation applies not only to the form factors involving hadrons but also to the strong interaction itself. A typical example is the pion-pion scattering problem. It is unfortunate that Weinberg[2] and others did not appreciate the importance of the unitarization of the $\pi\pi$ scattering problem. This important point was noticed much earlier by Lehmann who used the effective range method to unitarize the C.P.Th.[14]. Lehmann's article was often quoted but was apparently not appreciated by workers in chiral perturbation. We have independently rediscovered Lehmann's result by the Padé method[15]. Using the one-loop result of Leutwyler and Gasser[16] with the chiral symmetry breaking effect taken into account, based on two independent parameters, all the low energy pion-pion data up to 900 MeV can be calculated. The agreement between the theory and the experimental data is remarkable. For example the ρ width is found to be 155 $\pm$ 5 MeV compared with the experimental value 153 $\pm$ 2 MeV, the $I = 0$ S-wave scattering length is predicted to be $0.27 \pm .02\ m_\pi^{-1}$ compared to the experimental value $0.26 \pm .05\ m_\pi^{-1}$ and to the Weinberg prediction of $0.16\ m_\pi^{-1}$.

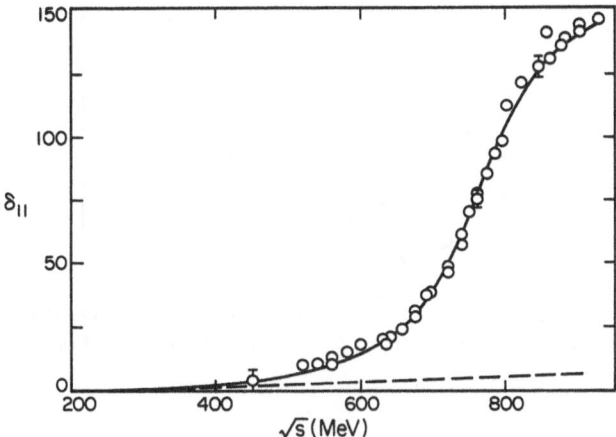

Figure 4. Two Parameter calculation of $\pi\pi$ scattering amplitudes. Solid curve: U.C.P.Th. P-wave phase shift. Dashed curve: C.P.Th. P-wave phase.

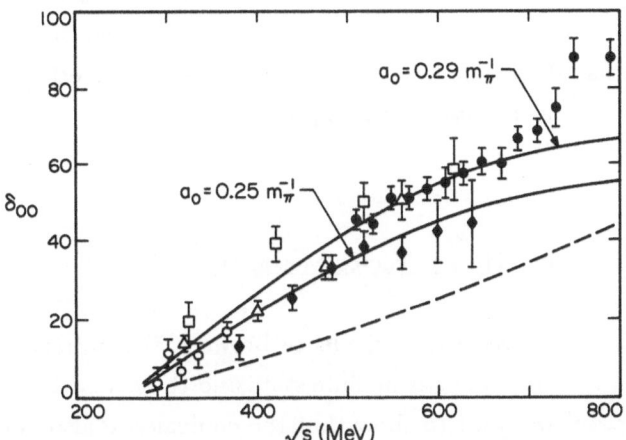

Figure 5. Two Parameter calculation of $\pi\pi$ scattering amplitudes. Solid curves: U.C.P.Th. S-wave, $I = 0$ phase shift. Upper curve corresponds to the scattering length, $a_o = 0.29 m_\pi^{-1}$, lower curve, $a_o = 0.25 m_\pi^{-1}$. Best fit $a_o = 0.27 \pm .03 m_\pi^{-1}$. Dashed curve: C.P.Th. phase.

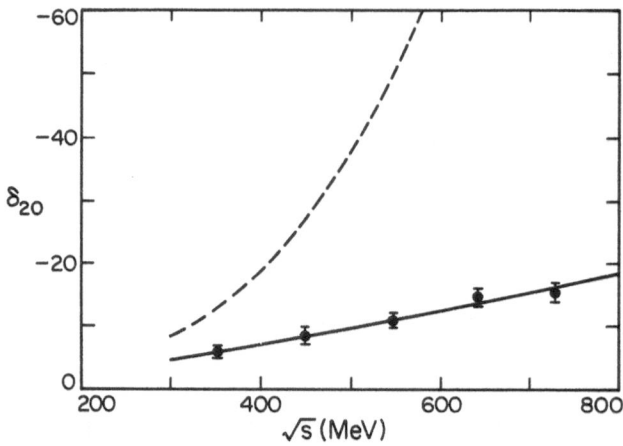

Figure 6. Two parameter calculation of $\pi\pi$ scattering amplitudes. Solid curve: U.C.P.Th. S-wave $I = 2$ $\pi\pi$ phase shift. Dashed curve: C.P.Th. phase.

The same technique can be applied to the strongly interacting Higgs sector of the standard model to get the solution for this problem which was not solved for many years. It was shown that it is not possible to use the perturbation theory and perturbative unitarity as it is usually done in the literature, to get an upper bound on the Higgs mass. Similarly to the low energy pion-pion scattering using chiral invariance arguments, there must be two independent parameters in the problem which are essentially related to the Higgs mass and the heavy ρ (technirho). Once these parameters are fixed, their widths can be predicted[15].

LECTURE II

Unitarized Chiral Perturbation Theory, Vector Meson Dominance and Chiral Anomalies

In the last lecture, we showed the equivalence between the unitarized C.P.Th. calculation for the P-wave two-pion intermediate state and the ρ V.M.D. for the pion form factor. In this lecture we want to show that the equivalence also holds for the virtual processes such as $P \rightarrow \gamma\gamma$, $P \rightarrow \gamma\gamma^*$ and $P \rightarrow \pi^+\pi^-\gamma$ where P stands for π^0, η and η'.

Similarly to the well-known problem of the discrepancy between the pure V.M.D. à la Gell-Mann, Sharp, Wagner and the chiral anomaly low energy theorem for $P \rightarrow \pi^+\pi^-\gamma$[17], this problem also exists in the dispersion or the U.C.P.Th. method. We present here the solution to this problem by requiring self-consistency in the $P \rightarrow \gamma\gamma$ calculation[18]. Our result is similar to the prescription of Fujiwara et al[19]. The main difference is that our $P \rightarrow \pi^+\pi^-\gamma$ amplitudes satisfy well the unitarity relation, while those given by Fujiwara et al do not; the difference is however not too important.

Let us define

$$\mathcal{M}\left(P(p) \rightarrow \gamma(\epsilon_1, k_1) + \gamma(\epsilon_2, k_2)\right) = -\epsilon_{\mu\nu\sigma\tau}\epsilon_1^\mu k_1^\nu \epsilon_2^\sigma k_2^\tau F_{P\gamma\gamma}(p^2) \qquad (16)$$

where P stands for π^0, η_8, η_0. The axial chiral anomaly predicts[20]:

$$F_{\pi\gamma\gamma}(0) = \frac{e^2}{4\pi^2 f_\pi} \qquad (17a)$$

$$F_{\eta_8\gamma\gamma}(0) = \frac{1}{\sqrt{3}} F_{\pi\gamma\gamma}(0) \qquad (17b)$$

$$F_{\eta_0\gamma\gamma}(0) = \frac{2\sqrt{2}}{\sqrt{3}} F_{\pi\gamma\gamma}(0) \qquad (17c)$$

where $f_\pi = 93$ MeV. Eq (17-a) gives $\Gamma(\pi^0 \rightarrow 2\gamma) = 7.64$ eV compared with the experimental value $\Gamma(\pi^0 \rightarrow 2\gamma) = 7.57 \pm 0.32$ eV. To calculate $\Gamma(\eta \rightarrow 2\gamma)$ and $\Gamma(\eta' \rightarrow 2\gamma)$ we must include the $SU(3)$ breaking effect.

The $P \rightarrow \pi^+\pi^-\gamma$ are also given by the chiral anomalies. Define

$$\mathcal{M}(P(p) \rightarrow \pi^+(q_+)\pi^-(q_-)\gamma(k)) =$$

$$-i\epsilon_{\mu\nu\sigma\tau}\epsilon^\mu k^\nu q_+^\sigma q_-^\tau F(s,t,u)$$

where $s = (q_+ + q_-)^2$ etc. The chiral anomaly relations are[21]:

$$F_{3\pi}(0,0,0) = \lambda \qquad (18-a)$$

$$F_{\eta_8\pi\pi}(0,0,0) = \frac{\lambda}{\sqrt{3}} \qquad (18-b)$$

$$F_{\eta_0\pi\pi}(0,0,0) = \sqrt{\frac{2}{3}}\lambda \qquad (18-c)$$

with $\lambda = F_{\pi\gamma\gamma}(0)/ef_\pi^2 = e/4\pi^2 f_\pi^3 = 9.54 \times 10^{-9}$ MeV^{-3}. The amplitudes $F_{P\pi\pi}$ can also be calculated by the V.M.D. model à la Gell-Mann, Sharp, Wagner which will be denoted as $F_{P\pi\pi}^{VMD}(s,t,u)$. There is a well-known discrepancy between the chiral anomalies as given by Eqs (18) and the low energy limit of the V.M.D. model[17]. The problem can be summarized by the following relations:

$$\frac{F_{3\pi}(0,0,0)}{F_{3\pi}^{VMD}(0,0,0)} = \frac{F_{\eta_8\pi\pi}(0,0,0)}{F_{\eta_8\pi\pi}^{VMD}(0,0,0)} = \frac{F_{\eta_0\pi\pi}(0,0,0)}{F_{\eta_8\pi\pi}^{VMD}(0,0,0)} = \frac{2}{3} \qquad (19)$$

This discrepancy arises because it is required that the V.M.D. model gives the correct $\rho \rightarrow P\gamma$ amplitudes in addition to the $\rho \rightarrow \pi\pi$ amplitude.

We want now to show that this same problem occurs in the dispersion or the U.C.P.Th. method and how to solve it. To do this, let us consider the process $\eta_8 \rightarrow \gamma + \gamma_V^*$ whose amplitude is denoted by $F_{\eta_8\gamma_V\gamma_V}$. We have

$$F_{\eta_8\gamma_V\gamma_V} = \frac{3}{2} F_{\eta_8\gamma\gamma}(0). \qquad (20)$$

Taking one of the two photons off its mass shell $k_1^2 \equiv s$ and using the unitarity condition, keeping only the 2π intermediate state, we have (omitting the p^2 dependence):

$$Im\, F_{\eta_8 \gamma_V \gamma_V}(s) = \frac{e}{96\pi} \sqrt{\frac{(s - 4m_\pi^2)^3}{s}} F_{\eta_8 \pi\pi}(s) F^*(s) \tag{21}$$

where $F(s)$ is the pion form factor calculated by the U.C.P.Th. Eq (10). Similarly to the calculation of the pion form factor, we can calculate first $F_{\eta_8 \pi\pi}$ at the one-loop level using the C.P.Th. technique. A counterterm is required which introduces a new subtraction constant. We then unitarize it by the Padé technique to get

$$F_{\eta_8 \pi\pi}(s) = \frac{\lambda}{\sqrt{3}} F(s, \tilde{s}_R) \tag{22}$$

where $F(s, \tilde{s}_R)$ is given by Eq (10) with $\tilde{s}_R$ replacing s_R. We now use the final state phase theorem requiring $F_{\eta_8 \pi\pi}(s)$ to have the same phase of the strong P-wave $\pi\pi$ phase shift. This condition is satisfied if $\tilde{s}_R = s_R$, and hence

$$Im\, F_{\eta_8 \gamma_V \gamma_V^*}(s) = \frac{e}{96\pi} \frac{\lambda}{\sqrt{3}} \sqrt{\frac{(s - 4\mu^2)^3}{s}} \mid F(s) \mid^2 \tag{23}$$

Using the unsubtracted dispersion relation for $F_{\eta_8 \gamma_V \gamma_V}$ and using Eq (23) we have:

$$F_{\eta_8 \gamma_V \gamma_V^*}(s) = \frac{2}{3} F_{\eta_8 \gamma_V \gamma_V}(0) F(s) \tag{24}$$

Because $F(0) = 1$, Eq (24) lead to an inconsistency. The resolution of this problem consists in multiplying the R.H.S. of Eq (22) by $1 + \alpha s$ where α is real, which satisfies the unitarity and the low energy theorem provided $\tilde{s}_R = s_R$. Using this factor in Eq (23), the consistency requirement is satisfied if $\alpha = (1/2)s_\rho$ and hence

$$F_{\eta_8 \pi\pi}(s) = \frac{\lambda}{\sqrt{3}} F(s, s_R)(1 + \frac{s}{2s_\rho}) \tag{25}$$

and we have approximately:

$$F_{\eta_8 \gamma_V \gamma_V}(s) = F_{\eta_8 \gamma_V \gamma_V}(0) F(s, s_R) \tag{26}$$

which is the V.M.D. relation. Similarly we have

$$F_{\eta_0 \pi\pi}(s) = \sqrt{\frac{2}{3}} \lambda F(s, s_R)(1 + \frac{s}{2s_\rho}) \tag{27}$$

and

$$F_{3\pi}(s, t, u) = \frac{\lambda}{3} [F(s, s_R)(1 + \frac{s}{2s_\rho}) + (s \leftrightarrow t) + (s \leftrightarrow u)] \tag{28}$$

For clarity we can approximate

$$F(s, s_R) \simeq \frac{s_\rho}{s_\rho - s - i\Gamma(s)\sqrt{s_\rho}} \tag{29}$$

Eqs (25), (27) and (28) are the usual V.M.D. result but modified by a calculable factor $(1 + \frac{s}{2s_\rho})$ as required by the consistency condition. Eq (26) is the standard V.M.D. result for $\eta_8 \rightarrow \gamma_V \gamma_V^*$. It is seen that the U.C.P.Th. calculation for $F_{P\pi\pi}$ given by Eqs (25), (27) and (28) satisfy the chiral anomaly theorems and the unitarity. Putting $s = s_\rho$ in Eq (28) and neglecting the t and u contributions which are small, we can calculate $\Gamma(\rho \rightarrow \pi\gamma) = 80$ keV, which agrees well with the experimental data.

Let us now compare our results with those given by Fujiwara et al[19] whose treatment was based on the study of the general solutions to the Wess-Zumino anomaly equation treating vector mesons as dynamical gauge bosons in the hidden local symmetry combining with the nonlinear chiral lagrangian. Their treatment is unnecessarily complicated for this problem; we give here a simple proof. By examining Eq (19), which shows the discrepancy between the low energy limit of the V.M.D. models and the chiral anomalies, we can blame the fault on the assumption of a complete V.M.D. picture in calculating the $P \rightarrow \pi^+\pi^-\gamma$ amplitude. This is so because the low energy chiral anomaly theorems should be considered as more fundamental than the dynamical approximation scheme of the V.M.D. It is simple to modify $F_{P\pi\pi}^{VMD}$ by subtracting from its amplitude a contact term whose strength is such that the RHS of Eq (19) is unity. Taking this condition into account the $P \rightarrow \pi^+\pi^-\gamma$ amplitude can now be written as

$$F_{3\pi}(s,t,u) = \lambda \left[1 + \frac{1}{2} \left(\frac{s}{s_\rho - s} + \frac{t}{s_\rho - t} + \frac{u}{s_\rho - u} \right) \right] \qquad (30a)$$

$$F_{\eta_8\pi\pi}(s,t,u) = \frac{\lambda}{\sqrt{3}} \left(1 + \frac{3}{2} \frac{s}{s_\rho - s} \right) \qquad (30b)$$

$$F_{\eta_0\pi\pi}(s,t,u) = \sqrt{\frac{2}{3}} \lambda \left(1 + \frac{3}{2} \frac{s}{s_\rho - s} \right) \qquad (30c)$$

Eq (30a) was obtained by Fujiwara et al. Eqs (30b) and (30c) are obtained by a straightforward generalization. Taking into account of Eq (29) in Eqs (25), (27) and (28), the low energy limit of these equations are the same as those of Eqs (30). The main difference between them is that the amplitudes given by Eqs (25), (27) and (28) satisfy the unitarity while those of Eqs (30) do not. To see how large the violation is, let us calculate the phase of $F_{\eta_0\pi\pi}$ at $s = s_\rho$ where the strong P-wave $\pi\pi$ phase shift is 90^0. The phase of $F_{\eta_0\pi\pi}$ calculated from Eq (27) is indeed 90^0, while that given by Eq (30c) is only 77^0, using the usual prescription of replacing s_ρ by $s_\rho - i\sqrt{s_\rho}\Gamma_\rho$ in the denominator of Eq (30c). The difference of the magnitude squared calculated from Eqs (27) and (30c) is however negligible.

The dipion spectrum calculated by either Eq (23) or (30c) is

$$|\mathcal{M}|^2 \sim 1 + 3s/s_\rho \qquad (31a)$$

which is slightly larger than the experimental data[22] which indicates

$$| \mathcal{M} |^2 \sim 1 + 2s/s_\rho \qquad (31b)$$

There must by systematic errors in this experiment because the above formula fits better at the low value of s but not at the higher end of the dipion spectrum. Let us now suppose future experiments show that the theoretical prediction (Eq (31a)) was incorrect, could be fix it by some method? This could be done by noticing that the polynomial ambiguity in Eq (22) or the contact term in Eqs (30) represent some high energy contribution such as the inelastic effect or the higher excited ρ' which we cannot calculate. We can phenomenologically replace the factor $1 + s/2s_\rho$ by $(1 + s/2(s_{\rho'} - s))$ and can take $s_{\rho'} \simeq 2s_\rho$ (plus an imaginary part of $\sqrt{s} \geq 1$ GeV). In this case, at low energy

$$| \mathcal{M} |^2 \sim 1 + 2.5\, s/s_\rho \qquad (31c)$$

This is about the lowest slope that one can expect from our model. It is very important to remeasure the dipion spectrum from $\eta \to \pi^+\pi^-\gamma$ and also the low energy end of this spectrum in $\eta' \to \pi^+\pi^-\gamma$ decay to test this rather fundamental problem.

Using the experimental data $\Gamma(\eta \to \pi^+\pi^-\gamma) = (64 \pm 6)eV$, $\Gamma(\eta' \to \pi^+\pi^-\gamma) = 72 \pm 7$ keV and Eqs (25) and (27), we determine using the dipion slope given by Eq (31c)[23]

$$
\begin{aligned}
| F_{\eta\pi\pi}(0) | &= (6.8 \pm .4) \times 10^{-9}\text{MeV}^{-3} \\
| F_{\eta'\pi\pi}(0) | &= (5.3 \pm .3) \times 10^{-9}\text{MeV}^{-3}
\end{aligned}
\qquad (32)
$$

and also $\Gamma(\pi^0 \to 2\gamma) = (7.57 \pm 0.32)$ eV, $\Gamma(\eta \to 2\gamma) = (0.51 \pm 0.04)$ keV, $\Gamma(\eta' \to 2\gamma) = 4.7 \pm .7)$ keV, we have

$$
\begin{aligned}
| F_{\pi\gamma\gamma}(0) | &= (2.50 \pm 0.05) \times 10^{-5}\text{MeV}^{-1} \\
| F_{\eta\gamma\gamma}(0) | &= (2.54 \pm 0.12) \times 10^{-5}\text{MeV}^{-1} \\
| F_{\eta'\gamma\gamma}(0) | &= (3.27 \pm .20) \times 10^{-5}\text{MeV}^{-1}.
\end{aligned}
\qquad (33)
$$

These numbers fit very well in the nonet scheme:

$$
\begin{aligned}
F_{\eta 2\gamma} &= \frac{F_{\pi\gamma\gamma}(0)}{\sqrt{3}}\left(\frac{f_\pi}{f_8}\cos\theta - 2\sqrt{2}\frac{f_\pi}{f_0}\sin\theta\right) \\
F_{\eta' 2\gamma} &= 2\sqrt{\frac{2}{3}}F_{\pi\gamma\gamma}(0)\left(\frac{1}{2\sqrt{2}}\frac{f_\pi}{f_8}\sin\theta + \frac{f_\pi}{f_0}\cos\theta\right) \\
F_{\eta\pi\pi}(0) &= \frac{\lambda}{\sqrt{3}}\left(\frac{f_\pi}{f_8}\cos\theta - \sqrt{2}\frac{f_\pi}{f_0}\sin\theta\right) \\
F_{\eta'\pi\pi}(0) &= \sqrt{\frac{2}{3}}\lambda\left(\frac{1}{\sqrt{2}}\frac{f_\pi}{f_8}\sin\theta + \frac{f_\pi}{f_0}\cos\theta\right)
\end{aligned}
\qquad (33)
$$

with

$$\frac{f_8}{f_\pi} = 1.25, \ \frac{f_0}{f_\pi} = 1.04, \ \theta = -21^0. \tag{34}$$

Because we have 4 pieces of the experimental data, Eqs (32) and (33), and 3 unknowns, Eq (34) are calculated values.

Let us now discuss $\eta \rightarrow \gamma\gamma^*$ decay. Eq (26) also gives information on the e^+e^- spectrum in $\eta \rightarrow \gamma e^+ e^-$ decay. It shows that the e^+e^- spectrum is proportional to the ρ propogator. The pion loop contributes very little to the physics at this low energy because the imaginary part of the ρ propogator is small compared to the real part. It plays an important role however in $\eta' \rightarrow \gamma e^+ e^-$. Eq (26) shows that this decay can be considered as a two-step process $\eta' \rightarrow \gamma + \rho$ and $\rho \rightarrow e^+ e^-$. Because Eq (26) gives only the isovector photon contribution, we must add to it the isoscalar contribution of ω and ϕ. Using the pseudoscalar mixing angle $\theta_P = -21^0$ and $\frac{f_8}{f_\pi} = 1.25$, $\frac{f_0}{f_\pi} = 1.04$, the $\eta \rightarrow \gamma\gamma^*$ amplitude is dominated by the ρ contribution. This result is also valid for $\eta' \rightarrow \gamma\gamma^*$. Recent space-like form factor measurements in $\eta \rightarrow \gamma\gamma^*$ and $\eta' \rightarrow \gamma\gamma^*$ show that the form factor agrees with the ρ V.M.D. or equivalently our result of the U.C.P.Th. There have been attempts to calculate this dependence within the framework of the C.P.Th. These attempts fail because in this approach $F(s)$ and $F_{\eta_8\pi\pi}(s)$ are treated as point-like (i.e., no mutual strong interaction between the pion pair) which is in conflict with the experimental data. We show here that strong interaction between the P-wave pion pair does provide damping at high energy for $F(s)$ and $F_{P\pi\pi}(s)$ to arrive at the V.M.D. result of Eq (26).

Let us end this lecture by making two remarks concerning the use of the Padé method in resumming the perturbation series. The first one deals with some warnings and the second one with the ghost problem. Let us suppose that the dipion spectrum in $\eta \rightarrow \pi^+\pi^-\gamma$ was measured and found to obey, for example, Eq (31a). Could one use this information to determine the subtraction term in Eq (10), assuming that one can write $F_{\eta\pi\pi}(s) = F_{\eta\pi\pi}(0)F(s)$ as prescribed naively by the Padé method. One would have in the case of $F'(0) \sim \frac{3}{2s_\rho}$ and the phase of $F_{\eta\pi\pi}(s)$ would be 90^0 at $s = \frac{2}{3}s_\rho$ which is clearly in the violation of the final state theorem. The correct choice is given by Eq (25) which has the correct low energy slope and satisfies the unitarity relation. Eq (5) can also be generalized to include also the initial state interaction by taking the product of Eq (5) with the corresponding expression obtained with the initial state eigenphase δ_i. The corresponding Padé version is obvious.

Eqs (10) and (11) develop some unwanted poles at some negative value of s. They can be gotten rid of by multiplying their RHS by $(s_g + s)/s_g$. For Eq (10), $s_g \simeq 10^6 m_\pi^2$, but for Eq (11) $s_g \simeq 150 m_\pi^2$. At low energy, the ghost killing factor is unimportant.

LECTURE III

We now discuss the calculation of $\eta \to \pi^0\gamma\gamma$ and $\eta \to \pi^0 e^+ e^-$ which were first done by Cheng and others in 1967 by the V.M.D. model[24,25]. It is instructive to see how one would approach this problem within the framework of C.P.Th.

From the power counting method one would try to calculate the $\eta \to \pi^0\gamma\gamma$ as the analytic continuation of the amplitude $\eta\pi \to K^+K^- \to 2\gamma$, i.e. the charged S-wave kaon loop contribution. This amplitude is of the $0(p^4)$. It should be much more important than the $\eta\gamma \to \pi^+\pi^- \to \pi^0\gamma$, the charge P-wave pion loop contribution which is of the order $0(P^6)$. Using the U.C.P.Th. as discussed previously the S-wave kaon loop contribution is effectively the process $\eta\pi \to a_1(980) \to \gamma\gamma$ where $a_1(980)$ is the isospin $I = 1$ scalar meson. Using the experimental data $\Gamma(a_0 \to 2\gamma) = 0.2$ keV, this contribution is negligible. In the U.C.P.Th. calculation for the P-wave pion loop (order $0(P^6)$ term), this amplitude is simply $\gamma\eta \to \rho \to \pi^0\gamma$ which can be straightforwardly calculated by the V.M.D. model. We must add to this amplitude the ω contribution. Define

$$\mathcal{M}\left(\eta(k) \to \pi^0 + \gamma(\epsilon_1, q_1) + \gamma(\epsilon_2, q_2)\right) = \epsilon_1^\mu \epsilon_2^\nu \mathcal{M}_{\mu\nu}$$

and decomposing $\mathcal{M}_{\mu\nu}$ into the invariant amplitudes.

$$\mathcal{M}_{\mu\nu} = A(s,t,u)(q_{2\mu}q_{1\nu} - q_1 \cdot q_2 g_{\mu\nu})$$
$$+B(s,t,u)(\frac{k \cdot q_1 k \cdot q_2}{q_1 \cdot q_2} + k_\mu k_\nu - \frac{k \cdot q_1}{q_1 \cdot q_2}q_{2\mu}k_\nu$$
$$-\frac{k \cdot q_2}{q_1 \cdot q_2}q_{1\nu}k_\mu)$$

the ρ, ω exchange diagram gives[25]

$$A(s,t,u) = \frac{2}{9}g_{\omega\pi\gamma}^2 \left(\frac{t + M^2}{m_\rho^2 - t} + \frac{u + M^2}{m_\rho^2 - u}\right) \times$$
$$\left(\frac{f_\pi}{f_8}\sqrt{3}\cos\theta - \frac{f_\pi}{f_0}\sqrt{6}\sin\theta\right)$$

(35a)

$$B(s,t,u) = \frac{2}{9}g_{\omega\pi\gamma}^2 (\frac{s}{2})\left[\frac{1}{m_\rho^2 - t} + \frac{1}{m_\rho^2 - u}\right] \times$$
$$\left(\frac{f_\pi}{f_8}\sqrt{3}\cos\theta - \frac{f_\pi}{f_0}\sqrt{6}\sin\theta\right)$$

(35b)

where $M = m_\eta$ and for simplicity we have set $m_\omega = m_\rho$. Using $\frac{f_8}{f_\pi} = 1.25, \frac{f_0}{f_\pi} = 1.04, \theta = -21^0$ the $\eta \to \pi^0\gamma\gamma$ width is

$$\Gamma(\eta \to \pi^0\gamma\gamma) = 1.2 \text{ eV}$$

(36)

which compares favorably with the experimental value

$$\Gamma(\eta \to \pi^0\gamma\gamma) = 0.89 \pm 0.28 \text{ eV}.$$

The CP conserving decay $\eta \to \pi^0 e^+ e^-$ via 2γ exchange was estimated by Cheng and was found to be $B(\eta \to \pi^0 e^+ e^-) \leq 4 \times 10^{-8}$ which is way below the current limit $B(\eta \to \pi^0 e^+ e^-) < 5 \times 10^{-5}$.

Before going on to discuss the rare weak decay of K mesons, let us summarize important experiments to be done on π^0, η and η' system. There is an interest in performing experiments on rare π^0, η and η' decays with a branching ratio of 10^{-10} to 10^{-11} to study exotic decays. These are, however, extremely difficult experiments due to the huge background problem. In my opinion, experimentalists should not be encouraged to think along these lines. There are other useful experiments to be done which are easier, from which we can learn a great deal of good physics. I should like to point out a few obvious experiments.

i) π^0 decays.

This is an old subject, but there is a lack of adequate experimental information on

a) $\pi^0 \to \gamma e^+ e^-$: there are conflicting experimental results on the $e^+ e^-$ spectrum as a test of V.M.D. and U.C.P.Th.

b) $\pi^0 \to e^+ e^-$. What exactly is the B.R. for this decay? The answer will probably be given in a year or two by the rare K meson decay study groups as a byproduct of their experiment.

ii) η, η' decays:

a) It is extremely important to measure accurately the dipion spectra in $\eta \to \pi^+ \pi^- \gamma$ as an indirect test of the chiral anomalies as discussed in Lecture II. This experiment should be supplemented by an independent experiment on the low energy dipion spectrum in $\eta' \to \pi^+ \pi^- \gamma$ to test again the low energy chiral anomalies.

b) Present experimental data on the space-like γ^* of $\eta, \eta' \to \gamma\gamma^*$ using the 2-photon physics, yields results which are consistent with the V.M.D. or U.C.P.Th.. Experimental data on $\eta \to \gamma e^+ e^-$ are also quite consistent with V.M.D. or U.C.P.Th. It is very desirable to perform more precise experiments to test these theories. Because the $\eta' \to \gamma e^+ e^-$ process is dominated by the vector meson ρ in the intermediate state $\eta' \to \gamma + \rho \to \gamma e^+ e^-$, it will be extremely difficult to measure the low energy $e^+ e^-$ spectrum.

c) More precise experimental data on $\eta \to \pi^0 \gamma\gamma$ is needed. This is needed to check if the S-wave kaon pair contribution is negligible. The experimental data on $\eta' \to \pi^0 \gamma\gamma$ is also needed to test V.M.D. or U.C.P.Th. which gives this decay as a two-step process $\eta' \to V\gamma \to \pi\gamma\gamma$ where $V = \rho, \omega$.

iii) A very important experiment on the production of pion in the Coulomb field of a heavy nucleus but with a much higher incident pion momentum is needed to test not only the low energy chiral anomaly but also at intermediate energy and at the ρ resonance. Present experimental data on[26]

$$\pi^+ + Z \rightarrow \pi^+ + \pi^0$$

gives a value of $F_{3\pi}(0) \simeq (1.29 \pm .09 \pm .05) \times 10^{-9}$ MeV^{-3} which is higher than the chiral anomaly prediction

$$F_{3\pi}(0) = (0.95 \pm .02) \times 10^{-5} \text{ MeV}^{-2}.$$

Experiments on

$$\pi^\pm + Z \rightarrow \pi^\pm + \eta(\eta')$$

are more difficult to perform due to their relative lower cross sections, but can provide valuable information on the low-energy chiral anomalies.

Rare Decay of K Mesons

Let us now discuss the rare decay of K mesons. The physics is very rich here. The main interests are

[i] CP violation effect in $K_L \rightarrow \pi\pi$ decays

[ii] Search for CP violation other than 2π modes.

[iii] Search for strangeness changing neutral current in $K^+ \rightarrow \pi^+ \nu \bar{\nu}$

[iv] Search for exotic physics, e.g., lepton non-conservation, etc.

These subjects are very well covered in the literature and lectures. We are concerned here with the application of the U.C.P.Th. in the study of $K_S \rightarrow 2\gamma, K_L \rightarrow \pi^0 \gamma\gamma$ and $K_L \rightarrow \pi^0 e^+ e^-$.

Let us discuss first the $K_S \rightarrow 2\gamma$ amplitude. In the C.P.Th. approach one must calculate first the $K_S \rightarrow 2\pi$ amplitude which, under the assumption that H_w behaves as an octet ($\Delta I = \frac{1}{2}$ rule), is given by[27]

$$\mathcal{M}(K_S(p) \rightarrow \pi^+(k_1) + \pi^-(k_2)) = \frac{1}{2}iC(2p^2 - k_1^2 - k_2^2) \qquad (37)$$

C is assumed to be real and independent of $s = p^2$. Unitarity requires C to have the phase δ of the $I = 0$, S-wave pion-pion interaction. Disregarding this problem, C can be calculated from $K_S \rightarrow \pi^+\pi^-$, $C = 1.26 \times 10^{-11}$ MeV^{-2}. Following the same analysis that leads to Eq (11), the U.C.P.Th. enables us to write

$$\mathcal{M}(K_S \rightarrow \pi^+(k_1) + \pi^-(k_2)) = iC'(s - m_\pi^2)S(s) \qquad (38)$$

where we have put $k_1^2 = k_2^2 = m_\pi^2$. Using the experimental data on $K_S \rightarrow \pi^+\pi^-$ and $| S(m_K^2) |$ calculated from Eq (11), we have $C' = 0.90 \times 10^{-11}$ MeV^{-2}. It is seen that

Eq (38) has a very much different behavior than Eq (37) due to the energy dependence of $S(s)$. We have in fact $S(m_K^2) = 1.45 \, e^{i40^0}$ which cannot be approximated by unity as in Eq (37).

To calculate the $K_S \rightarrow 2\gamma$ amplitude, we must also know the $\gamma\gamma \rightarrow 2\pi$, with 2π in $I = 0$, S-wave. This amplitude can be easily calculated in the C.P.Th. under the assumption that the S-wave pion-pion interaction can be neglected. It is much more complicated to calculate in the U.C.P.Th. because of the requirement of the unitarity. Fortunately a general method to handle this problem was given by Goble and Goble et al[28]. From their work it was possible to show that the ratio $R = \Gamma(K_S \rightarrow 2\gamma)/\Gamma(K_S \rightarrow \pi^+\pi^-) = 3.0 \times 10^{-6}$ independently of the strong $I = 0$ S-wave $\pi\pi$ interaction. This result depends on the assumption of the absence of the polynomial ambiguity in studying the $\gamma\gamma \rightarrow \pi\pi$ amplitude. This assumption can be removed by using the experiment data on the axial form factor of $\pi \rightarrow e\nu\gamma$ which is related to the charged pion polarizability. This effect is presumably small. The experimental number for R is $R = (3.5 \pm 1.8) \times 10^{-6}$.

The C.P.Th. calculation, where it is assumed that there is no strong S-wave pion-pion interaction, gives $R = 3.5 \times 10^{-6}$, which is very much in agreement with the result of Goble and Goble et al. The C.P.Th. result is therefore a special case when the S-wave pion pion interaction is switched off[29].

Let us see how the related amplitude $K_L \rightarrow \pi^0\gamma\gamma$ can be calculated. This amplitude is related to $K_L \rightarrow \pi^0\pi^+\pi^-$ with a rescattering of $\pi^+\pi^-$ into 2γ. In this calculation, the π^0 acts as a spectator. Taking into account the $\pi^+\pi^-$ $I = 0$ S-wave interaction in both $K_L \rightarrow (\pi^0)\pi^+\pi^-$ and $\pi^+\pi^- \rightarrow 2\gamma$ and following the method of Goble and Goble et al.[28], the ratio of $\frac{d\Gamma}{ds}(K_L \rightarrow \pi^0\gamma\gamma)/\frac{d\Gamma}{ds}(K_L \rightarrow \pi^0\pi^+\pi^-)$, where $s = (p_+ + p_-)^2$ should also be independent of the strong interaction. This is the result of Ko and Rosner[30]. One should get approximately the same result in the C.P.Th. approach and the rescattering model. Detailed calculations show that there is a difference between C.P.Th. result and that of Ko and Rosner. This is due to the fact that Ko and Rosner use the experimental data on $K_L \rightarrow \pi^+\pi^-\pi^0$ rate and the "odd" pion spectra, while Ecker, Pich, de Rafael[31] deduce these quantities from the $\Delta I = 1/2$ effective chiral lagrangian. It is well known that the chiral lagrangian yields results which are in disagreement with the experimental data on $K_L \rightarrow \pi^+\pi^-\pi^0$ rate and spectra. The $\Delta I = 1/2$, $K_L \rightarrow \pi^+\pi^-\pi^0$ amplitude deduced from the chiral lagrangian Eq (37), using as input the $K_S \rightarrow \pi^+\pi^-$ rate is

$$\mathcal{M}_{CL}(K_L \rightarrow \pi^+\pi^-\pi^0) = 7.43 \times 10^{-7}(1 + 0.233\frac{s - s_0}{m_\pi^2}) \qquad (39a)$$

while the $\Delta I = 1/2$ experimental value is:

$$| \mathcal{M}_{exp}(K_L \rightarrow \pi^+\pi^-\pi^0) | = 9.10 \times 10^{-7}(1 + 0.264\frac{s - s_0}{m_\pi^2}) \qquad (39b)$$

which shows a large discrepancy in rate and "odd" pion slope. It was pointed out that the discrepancy was because of the neglect of the unitarity condition due to the final state $I = 0$ S-wave $\pi\pi$ interaction[32]. Taking into account of the unitarity condition, it was shown that the $K_L \to \pi^+\pi^-\pi^0$ rate is now given by

$$| \mathcal{M}(K_L \to \pi^+\pi^-\pi^0) | = 8.90 \times 10^{-7}(1 + 0.250\frac{s - s_0}{m_\pi^2}) \qquad (39c)$$

which is in good agreement with the experimental data Eq (39b). This shows again the importance of the unitarity constraint.

Returning now to the $K_L \to \pi^0\gamma\gamma$ calculation, we note that in the C.P.Th. approach, one uses Eq (39a) which is smaller than the observed $K_L \to \pi^+\pi^-\pi^0$ amplitude, hence from the previous discussion of the ratio R, the result of Ko and Rosner should be more reliable than that given by the C.P.Th., pending further investigation of the pion polarizability condition on their study of the $\gamma\gamma \to \pi\pi$ amplitude.

Similarly to the calculation of $\eta \to \pi^0\gamma\gamma$, we must add to the S-wave pion loop contribution, the P-wave pion loop contribution from the $K_L \to \gamma(\pi^+\pi^-) \to \gamma(\gamma\pi^0)$. It can be shown by dispersion theory or the U.C.P.Th., the unitarity requires that the $\pi^+\pi^-$ pair interact strongly with each other to form the ρ resonance. As emphasized above, the V.M.D. model is more suitable for this purpose[33]. The experimental data on $K_L \to \gamma\pi^+\pi^-$ can be used to fix the $K_L\rho\gamma$ coupling. To calculate the complete V.M.D. contributions to $K_L \to \pi\gamma\gamma$ we must also take into account of the ω contribution. There is, however, no experimental knowledge on the $K_L\omega\gamma$ coupling. We must therefore make use of the pseudoscalar meson π^0, η, η' poles assumption, i.e., $K_L \to P \to \gamma(V) \to \gamma(\pi^0\gamma)$ where V stands for ρ and ω. This leads us to pursue the question of how to study the amplitudes $a(K_L P) \equiv < P \mid H_w \mid K_L >$ where P stands for π^0, η and η'.

Although the assumption of the pseudoscalar meson pole dominance is widely used in the literature[34], there is no firm basis for this hypothesis. For example in the $K_L \to 3\pi$ study, one has the $K_L \to (\pi^+\rho^- + \pi^-\rho^+)$ via the pion pole but also corresponding contact terms where ρ's are emitted from the weak vertex. The possibility of having a contact term in $K_L \to \gamma V$, in addition to those from the pseudoscalar pole, has recently been raised by Ecker et al[35]. This is a legitimate objection to the widely used pseudoscalar meson pole dominance model[33,34]. It is unlikely that one can have a clear-cut answer to this problem because the nature of the non-leptonic weak hamiltonian is not well-understood.

We therefore want to make a detailed study of the $K_L \to \gamma\gamma$ and $K_L \to \pi^+\pi^-\gamma$ decays under the assumption of the pseudoscalar pole dominance, and use the parameters deduced from this analysis to calculate $K_L \to \pi^0\gamma\gamma$ and $K_L \to \pi^0 e^+ e^-$

under the same hypothesis. This study is done by Ko and Truong[36]. The parameters to be determined are $a_{K_L P} \equiv\, < P \mid H_w \mid K_L >$ where $P = \pi^0, \eta, \eta'$. $a_{K_L \pi^0}$ is determined from current algebra with the $I = 0$ S-wave pion-pion interaction taken into account, Eq (38). $A_{K_L \eta}$ and $a_{K_L \eta'}$ are to be determined using $\Gamma(K_L \rightarrow 2\gamma)$ and $\Gamma(K_L \rightarrow \pi^+ \pi^- \gamma)$.

Similarly to the notation and the analysis of $P \rightarrow 2\gamma$ and $P \rightarrow \pi^+ \pi^- \gamma$ given previously, Eqs (16), (17), 25)-(28), the $K_L \rightarrow 2\gamma$ is now given by

$$F_{K_L \gamma\gamma} = \frac{a_{K_L \pi^0}}{m_k^2} \left[\frac{F_{\pi\gamma\gamma}}{1 - r_\pi} + \alpha \frac{F_{\eta\gamma\gamma}}{1 - r_\eta} + \beta \frac{F_{\eta'\gamma\gamma}}{1 - r_{\eta'}} \right] \tag{40}$$

and

$$\begin{aligned} F_{K_L \pi\pi}(s,t,u) = &\frac{a_{K_L \pi^0}}{m_K^2} \left[\frac{F_{3\pi}(0)}{1 - r_\pi} \left(1 + \frac{m_K^2 + 2m_\pi^2}{2s_\rho} \right) \right. \\ &\left. + \left(\alpha \frac{F_{\eta\pi\pi}(0)}{1 - r_\eta} + \beta \frac{F_{\eta'\pi\pi}(0)}{1 - r_{\eta'}} \right) \left(\frac{s_\rho}{s_\rho - s} \right) \left(1 + \frac{s}{2s_\rho} \right) \right] \end{aligned} \tag{41}$$

Using $A_{K_L \pi^0} = 2.9 \times 10^{-2}$ MeV^2 as given by Ref. 32, and using the experimental data $B(K_L \rightarrow 2\gamma) = (4.41 \pm 3.2) \times 10^{-5}$ and the direct emission branching ratio $B(K_L \rightarrow \pi^+ \pi^- \gamma) = (2.89 \pm .28) \times 10^{-5}$, we have four possible solutions for α and β.

(i) $\alpha = 0.068,\ \beta = -0.088, \Delta m = 0.14$

(ii) $\alpha = 1.53,\ \beta = -15.7, \Delta m = -100$

(iii) $\alpha = 2.47,\ \beta = -20.6, \Delta m = -182$

(iv) $\alpha = 0.94,\ \beta = -4.72, \Delta m = -11$

where Δm is the pseudoscalar pole contribution to the mass shift $K_L - K_S$ in the unit of the experimental mass shift 3.5×10^{-6} eV, calculated by the π^0, η, η' pole contributions:

$$\Delta m = 1 - 4.3\alpha^2 - 0.37\beta^2.$$

Solution (i) and (iv) give rise to $\pi^+ \pi^-$ distribution $\mid 1 - 0.04 \frac{s}{s_\rho - s} \mid^2$ which is acceptable, while (ii) and (iii) have a dipion spectrum proportional to $\mid 1 + 6.5 \frac{s}{s_\rho - s} \mid^2$ which is excluded by experimental data. Solution (iv) yields $\Delta m = -11$ which is too large. In sum the solution (i) is the only acceptable one. It coresponds to the dominance π^0 pole contribution. The η and η' contributions cancel each other out.

It is useful to redo our calculation with the "standard" value, $a_{K_L \pi^0} = 4.0 \times 10^{-2}$ MeV^2 which is frequently used in the literature (neglecting the S-wave pion-pion interaction.) We have in this case

$i)$ $\alpha = 0.36,\ \beta = -3.18, \Delta m = -3.3$

$$ii) \ \alpha = 1.45, \ \beta = -14.5, \Delta m = -86$$

$$iii) \ \alpha = 2.12, \ \beta = -17.8, \Delta m = -136$$

$$iv) \ \alpha = 1.025, \ \beta = -6.4 \ \Delta m = -18.9$$

(i) and (iv) give rise to a $\pi^+\pi^-$ energy distribution $\mid 1 - 0.4\frac{s}{s_\rho - s} \mid^2$ and (ii) and (iii) a $\pi^+\pi^-$ energy distribution $\mid 1 + 13\frac{s}{s_\rho - s} \mid^2$ which are not consistent with the present experimental data[37]. Furthermore the solution (i) now gives a too large value of Δm which should also be excluded.

In conclusion, there is only one possible solution corresponding to the corrected value of $a_{K_L\pi^0} = (2.9 \pm .6) \times 10^{-2}$ MeV^2 and $\alpha \cong 0.068$ and $\beta \cong -0.88$ with large errors. These solutions correspond to a small η and η' contribution to $K_L \to \pi^+\pi^-\gamma$ amplitude. If we take for example, $\alpha = 0.12$ and $\beta = -1.7$, then there is a large cancellation between η and η' pole. It corresponds roughly to the nonet scheme of $K_L\pi^0, K_L\eta$ and $K_L\eta'$ mixing. Using the standard notation in the analysis of $K_L \to \pi^0\gamma\gamma$ we have now $G_\rho m_K^2 = 0.12 \times 10^{-8}$ and $G_w m_K^2 = 1.5 \times 10^{-8}$ where they refer, respectively, to the ρ and ω contributions. It should be emphasized the ω contribution is about a factor of 10 larger than those of ρ and other contributions. Because the $\omega\eta_8\gamma$ and $\omega\eta'_0\gamma$ couplings are a factor of 3 smaller than that of $\omega\pi\gamma$ and that the $K_L \to \pi^0\gamma\gamma$ amplitude is proportional to the square of these couplings, this calculation is fairly reliable.

The relative sign between the S-wave pion loop amplitude and those of the V.M.D. can be fixed by using the chiral lagrangian method. It is found that these two amplitudes give destructive interference, which leads to a branching ratio of $B(K_L \to \pi^0\gamma\gamma) \simeq 0.7 \times 10^{-6}$. Although this branching ratio is comparable to the chiral perturbation theory and the rescattering model (S-wave pion loop), the two-photon spectra are quite different. In the present calculation the low $m_{\gamma\gamma}^2$ region is enhanced by the vector meson exchange, but the region between $0.3 m_K^2 \leq m_{\gamma\gamma}^2 \leq 0.4 m_K^2$ is suppressed, due to the destructive interference, so that the total decay rate does not change very much Fig. 7. (If the interference is constructive, in the situation where there is a large direct vector meson emission amplitude with an opposite sign from the π^0 pole term, the branching ratio would be $B(K_L \to \pi^0\gamma\gamma) \simeq 1.4 \times 10^{-6}$). The absorptive conribution to $B(K_L \to \pi^0 e^+ e^-)$ is 3×10^{-12} independent of the sign of the $K_L \to \pi^0\gamma\gamma$ amplitudes.

It is very important to measure accurately the $K_L \to \pi^0\gamma\gamma$ rate and spectra to test the predictions of the pole model. Independently of any model, the low mass $m_{\gamma\gamma}$ measurement will help us to calculate in a model independent way the CP conserving $K_L \to \pi^0 e^+ e^-$ amplitude. If the pion pole model discussed above was correct, one would expect a larger signal from the CP violation one photon exchange diagram which is estimated to be of the order of 10^{-11}.

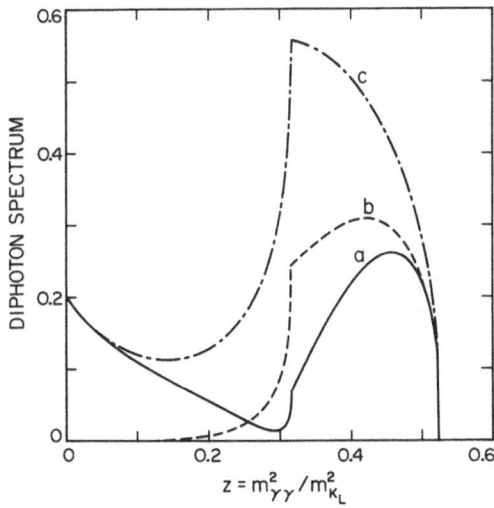

Figure 7. Predicted Di photon spectrum in $K_L \rightarrow \pi^0 \gamma \gamma$: (a) Solid curve: Prediction of calculation discussed in text (destructive interference) (b) Dashed curve: C.P.Th. prediction based on the S-wave pion loop contribution (c) Dashed dot curve for constructive interference (shown for illustration purposes).

We have shown in these lectures, the usefulness of the unitarity relation. C.P.Th. at one or many loop level without being supplemented by some unitarization scheme is misleading. We advocate here that C.P.Th. at the one-loop level should be used in combination with the Padé method to satisfy the elastic unitarity relation. This is the Unitarized Chiral Perturbation Theory. U.C.P.Th. is a field theoretical approach to the use of the dispersion theory corection to the low energy current algebra theorems advocated for many years by this author. It could be considered as the generalisation of the inverse propogator method to the form factor and scattering problems. For P-wave pion and kaon loop U.C.P.Th. leads to the usual V.M.D. with its virtue and defect. The resolution of the disagreement between the V.M.D. model and the low energy chiral anomaly can also be done with the U.C.P.Th. or dispersion theory approach yielding a $P \rightarrow \pi^+\pi^-\gamma$ amplitude satisfying the unitarity relation. U.C.P.Th. has its strength also in the treatment of the S-wave pion-pion interaction. It leads to the resolution of the $\pi\pi \rightarrow \pi\pi$, $\eta \rightarrow 3\pi$, $K \rightarrow \pi\pi e\nu$ on the same footing. The very fashionable problem of the strongy interacting Higgs sector of the standard model is also recently solved by the U.C.P.Th. method.

Because U.C.P.Th. is a modification of C.P.Th. it is as complex a method as the C.P.Th. approach. In many physical applications it is more desirable to make the approximation of the smoothness which was usually done in the old-fashioned current algebra appraoch, e.g. the Adler-Weissberger relation, and to treat the problem of chiral symmetry breaking by hand. This approach appears not be be systematic, but one gains in transparancy. Throughout these three lectures we have more or less followed this approach to emphasize the importance of the non-perturabtive treatment of the unitarity relation.

Soft strong interaction physics is a difficult subject. It must be guided by some general principles such as the unitarity of the S-matrix to check our approximation scheme. This is the main point of our approach. Current development of the Chiral Perturbation Theory puts strong emphasis on the mathematical aspects which are relevant to some fine details, but ignore the main issue of implementing the unitarity relation. There have recently been some efforts to apply the one-loop C.P.Th. to the non-leptonic weak decay of mesons. The number of counterterms involved is staggering, more than forty! It is suggested that the method used in these lectures, namely the Padé approximant method and the unitairy of the S-matrix can reduce consideraby the number of counterterms. The most economical method is still to implement the unitarity relation by hand as discussed in these lectures.

This remark points out the possibility of reformulating the Effective Chiral Lagrangian as an effective power series expansion in s, t, u, etc., of the physical amplitude wherever it is possible to do so, for example, below the unitarity cut. The coefficients of the different terms in the power series are then related to the renormalised physical quantities but evaluated at unphysical points (e.g. below the unitarity cut). In this approach, we do not have to worry about loop calculation and renormalisation, because unitarity can be implemented by construction in such a way that the constructed amplitudes satisfy not only unitarity but also the boundary condition given by the power series expansion. Many methods developed thirty years ago for strong interaction physics can be useful for this purpose.

ACKNOWLEDGEMENTS

It is a pleasure to thank the International School of Physics organizers for their invitation to lecture and the Enrico Fermi Institute and the Physics Departments of the University of Chicago and of the University of California at Santa Barbara for Visiting Professorships which enabled the author to complete these lectures. He would like to thank J. Rosner and T.N. Pham for useful dicussions and to thank Pyungwon Ko for collaborating on the investigation of $K_L \rightarrow \pi^0 \gamma\gamma$ and $K_L \rightarrow \pi e^+ e^-$ decays which were discussed in this lecture. Last but not least, he would like to apologize to many of his friends with whom he expressed disagreement on their Chiral Perturbation Theory program for the sake of clarity in physics. He hopes that, with time, they will recognize that he has helped their program by being outspoken. This work is supported in part by the U. S. Department of Energy, DOE DE-AC02 80ER-10587 at the University of Chicago and NSF PHY86-14185 at the University of California at Santa Barbara.

REFERENCES

1. L.F. Li and H. Pagels, Phys. Rev. Lett. 27:1089 (1971).

2. S. Weinberg, Physica 96A:327 (1979).

3. T.N. Truong, Phys. Lett. 99B:228 (1981). For a review, see T.N. Truong in *Wandering in the Fields*, edited by K. Kawarabayashi and A. Ukawa (World Scientific Singapore, 1987).

4. For a review, see: S.L. Adler and R.F. Dashen, "Current Algebra," W.A. Benjamin, New York (1968).

5. N.I. Muskhelishvili, "Singular Integral Equations," P. Noordhoff, Gromingen (1953). R. Omnès, Nuovo Cimento 8:316 (1958).

6. T.N. Truong, Phys. Rev. Lett. 61:2526 (1988).

7. K.M. Watson, Phys. Rev. 95:228 (1955).

8. S. Weinberg, Phys. Rev. Lett. 17:616 (1966).

9. See, for example, K. Nishijima, "Fields and Particles," W.A. Benjamin, New York (1968).

10. For a review, see, for example, J. Basdevant, Fortsehr. Phys. 20:283 (1972).

11. K. Kawarabayashi and M Suzuki, Phys. Rev. Lett. 16:225 (1966). Riazuddin and Fayazuddin, Phys. Rev. 147:1071 (1966).

12. M. Gell-Mann, D. Sharp, and W.G. Wagner, Phys. Rev. Lett. 8:261 (1962).

13. C. Roiesnel and T.N. Truong, Nucl. Phys. B187:293 (1981). C. Roiesnel, Thèse d'Etat, Université de Paris Sud (Orsay), (1982), (unpublished).

14. H. Lehmann, Phys. Lett. 41B:529 (1972).

15. A. Dobado, M.J. Herrero, and T.N. Truong, Phys. Lett. 235:134 (1990). Ibid. 235:129 (1990).

16. H. Leutwyler and J. Gasser, Ann. Phys. (N.Y.) 158:142 (1984).

17. P.G.O. Freund and A. Zee, Phys. Lett. 132B:419 (1983). Ibid. 144B:455E (1984).

18. T.N. Truong (in preparation).

19. T. Fujiwara et al, Prog. Theor. Phys. 73:926 (1985).

20. S.L. Adler, Phys. Rev. 162:1734 (1967).

21. R. Aviv, N.D. Hari Dass, and R.F. Sawyer, Phys. Rev. Lett. 26:591 (1971); S.L. Adler, B.W. Lee, S.B. Treiman, and A. Zee, Phys. Rev. D4:3497 (1971); M. Terentev, J.E.T.P. Lett. 14:94 (1971); J.Wess and B. Zumino, Phys. Lett. 37B:95 (1971); W.A. Bardeen, Phys. Rev. 184:1848 (1969); Nucl. Phys. B223:422 (1983); R.Aviv and A. Zee, Phys. Rev. D5:2372 (1971).

22. J.G. Layter et al, <u>Phys. Rev.</u> D7:2565 (1973).

23. P. Ko and T.N. Truong, EFI preprint (1990).

24. T.P. Cheng, <u>Phys. Rev.</u> 162:1734 (1967).

25. G. Oppo and S. Oneda, <u>Phys. Rev.</u> 160:1397 (1967).

26. Y.N. Antipov et al, <u>Phys. Rev.</u> D36:21 (1987).

27. J.A. Cronin, <u>Phys. Rev.</u> 161:1483 (1967).

28. R.L. Goble, <u>Phys. Rev.</u> D7:931 (1973); R.L. Goble, R. Rosenfeld, and J.L. Rosner, <u>Phys. Rev.</u> D39:3264 (1989).

29. G. D'Ambrosio and D. Espriu, <u>Phys. Lett.</u> B175:237 (1986); J.L. Goity, <u>Z. Phys.</u> C34:341 (1987).

30. P. Ko and J.L. Rosner, <u>Phys. Rev.</u> D40:3775 (1989).

31. G. Ecker, A. Pick, and E. de Rafael, <u>Phys. Lett.</u> B189:363 (1987); <u>Nucl. Phys.</u> B303:665 (1988).

32. T. N. Truong, <u>Phys. Lett.</u> B207:495 (1988).

33. T.N. Truong, EFI 89-57 preprint (1989).

34. L.M. Sehgal, <u>Phys. Rev.</u> D38:808 (1988). T. Morozumi and H. Iwasaki, KEK preprint Th-206 (1988) unpublished. J. Flynn and L. Randall, <u>Phys. Lett.</u> B216:221 (1989).

35. Talk presented by A. Pieh at the Europhysics Conference on High Energy Physics, Madrid, Sept. 13-16, 1989.

37. S. Carroll et al, <u>Phys. Rev. Lett.</u> 44:529 (1980).

Proton Spin Puzzle for Cyclists

Marek Karliner

School of Physics and Astronomy
Tel Aviv University
Tel Aviv 69978, Israel[*]

ABSTRACT

A brief review is given of the current experimental and theoretical situation concerning the distribution of the proton spin between the constituents. A connection with recently suggested OZI violation in baryonic physics is also discussed. Special attention is given to the question of the possible gluon polarization in the proton, its contribution to the $g_1^p(x)$ structure function and its significance for future experiments.

Most properties of hadrons, and protons in particular, are well described by the non-relativistic quark model (NRQM) in which the proton contains nothing but three slowly-moving quarks, analogous to nucleons in nuclei, or electrons in atoms. Despite the successes of the NRQM we have known for quite a few years that the quarks in a fast moving proton carry only about 50% of proton's *linear* momentum, the rest being carried by gluons. Nevertheless, until very recently, nearly everyone was convinced that quarks carry most of proton's spin, or *angular* momentum. In 1987-88, experiments of the EMC collaboration at CERN[1,2,3] and subsequent[4,5,6] theoretical analysis have caused a major revision of this viewpoint. The CERN data leads to a rather surprising conclusion that *the valence- and sea-quarks together carry only a small fraction of proton's spin*, in sharp disagreement with expectations based on the otherwise enormously successful quark model.

Perhaps the single most important reason for the surprise is the popular misconception whereby the difference between *current* and *constituent* quarks is blurred. This difference, although in principle only a question of language, is a crucial one: while the proton contains three constituent quarks and nothing else, it contains an *infinite* number of current quarks and gluons. The constituent quarks are heavy objects, weighting about 300 MeV; current u and d quarks are very light, with masses $\sim$ few MeV. It is the current quarks which correspond to the quark fields in the QCD Lagrangian.

In the Non-Relativistic Quark Model (NRQM) the spin of the proton is due to the

---

[*] *e-mail: MAREK@TAUNIVM.BITNET*

Medium-Energy Antiprotons and the Quark–Gluon Structure of Hadrons
Edited by R. Landua *et al.*, Plenum Press, New York, 1991

93

coupling of the spins of the three constituent quarks. How is this to be understood in the language of QCD? Loosely speaking, each constituent quark can be thought of as a (current) valence quark, surrounded by a sea of gluons and of virtual quark-antiquark pairs. Both the gluons and the sea-quarks carry spin and can contribute to the spin of the constituent quark and through it, to the spin of the proton. The question how much of the spin is due to the quarks is therefore an *experimental question*.

In the following I will give a brief and necessarily heuristic description of how one can answer this question with the help of the EMC data. I will then discuss the attempts to derive the experimental results from theory and explore some of the more interesting implications of the emerging picture.

What do we know about the possible contributions to the proton spin? To get an intuition, it is useful to think about the proton as a relativistic bound state of a large number of partons – quarks and gluons. Both quarks and gluons carry spin and therefore contribute to the to the total spin through their polarization. In addition, if we go to a Lorentz frame in which the proton is moving very fast ("infinite momentum frame") and quantize the spin along the direction of motion there will be a contribution from the average *orbital* angular momentum of the partons. The total spin must of course be $\frac{1}{2}$.[†] In the infinite momentum frame, one usually introduces parton distributions $q(x)$, $0 \leq x \leq 1$, giving the probability for finding a quark (or anti-quark) of flavor q carrying a fraction x of proton's linear momentum. Thus $\int_0^1 dx\, u(x) = 2$, $\int_0^1 dx\, d(x) = 1$, etc. One can also introduce *helicity dependent* parton distributions $q_{\uparrow(\downarrow)}(x)$ which describe the probability of finding a quark of flavor q and helicity $+\frac{1}{2}(-\frac{1}{2})$. Then $q(x) = q_{\uparrow}(x) + q_{\downarrow}(x)$, while the net helicity distribution is given by $\Delta q(x) = q_{\uparrow}(x) - q_{\downarrow}(x)$. The total helicity carried by a given flavor q is denoted by Δq and is given by the first moment of $\Delta q(x)$:

$$\Delta q = \int_0^1 dx\, \Delta q(x) = \int_0^1 dx\, \left[q_{\uparrow}(x) - q_{\downarrow}(x) \right] \tag{1}$$

Similarly, the net helicity carried by the gluons, ΔG, is given by

$$\Delta G = \int_0^1 dx\, \Delta G(x) = \int_0^1 dx\, \left[G_{\uparrow}(x) - G_{\downarrow}(x) \right] \tag{2}$$

The contribution of the proton constituents to its spin can now be concisely encoded in the sum rule:

$$\frac{1}{2} \sum_q \Delta q + \Delta G + \langle L_z \rangle = \frac{1}{2}. \tag{3}$$

where $\langle L_z \rangle$ is the expectation value of the orbital angular momentum of all partons along the direction of motion. Of the three terms in (3) the only one on which there's firm experimental information is the first one, quark contribution. To understand how this information is obtained, recall that in the non-relativistic limit the space-like

---

† Strictly speaking, in the infinite momentum frame one ought to be speaking in terms of the proton *helicity* rather than spin. The whole discussion can be carried through in terms of helicity, but we shall proceed as if the two concepts are interchangeable.

components of the axial current $\bar{q}\gamma_5\gamma_\mu q$ yield the quark spin operator. In relativistic kinematics, the net quark helicities Δq are related to the matrix element of the appropriate axial current between proton states at zero momentum transfer,

$$\langle p| \bar{q}\gamma_5\gamma_\mu q |p\rangle = \Delta q \cdot \Sigma_\mu(p) \tag{4}$$

where $\Sigma_\mu(p)$ is the proton spin.

Axial currents of quarks appear in weak decays of hadrons. Indeed, β-decay of the neutron $n \to p\,e^-\,\bar{\nu}_e$ involves a d quark in the neutron flipping its spin and decaying into a u quark in the proton. Using the isospin $SU(2)$ symmetry, one can obtain from the neutron lifetime the combination

$$g_A = \Delta u - \Delta d = 1.25 \tag{5}$$

Similarly, using the $SU(3)_f$ symmetry[‡] one can extract from the weak decays of hyperons the combination[8]

$$(\Delta u + \Delta d - 2\Delta s)/\sqrt{3} = 0.39 \tag{6}$$

The two matrix elements in (5) and (6) provide us with two linear equations for the three unknowns, Δu, Δd and Δs. In $SU(3)_f$ there are only two independent reduced matrix elements, so in order to obtain a third equation, we need to look elsewhere. We need experimental information about matrix elements of axial currents. On the first sight it would seem that QED-mediated processes are useless for that purpose, since the photon has only vector couplings to fermions. It turns out, however, that QED can provide us with an axial current matrix element in the proton, albeit indirectly. The trick is to look at a process where both the photon and the proton are polarized – deep inelastic scattering of polarized muons (or electrons) on a polarized proton target. There are two possible relative orientations of the photon and the proton spin, and correspondingly, two cross-sections: parallel ($\equiv \sigma_{3/2}$) and anti-parallel ($\equiv \sigma_{1/2}$). In the scaling limit the normalized difference A_1 between the two cross-sections can be expressed as ratio of the helicity-dependent structure functions:

$$A_1 \equiv \frac{\sigma_{1/2} - \sigma_{3/2}}{\sigma_{1/2} + \sigma_{3/2}} \xrightarrow[\substack{\text{Bjorken} \\ \text{limit}}]{} \frac{\sum_q e_q^2\,[q_\uparrow(x) - q_\downarrow(x)]}{\sum_q e_q^2\,[q_\uparrow(x) + q_\downarrow(x)]} \tag{7}$$

where e_q is the electric charge of flavor q. Using the measured values of the unpolarized structure function

$$F_2 = \sum_q e_q^2\, x\,[q_\uparrow(x) + q_\downarrow(x)] \tag{8}$$

one can extract from A_1 the proton spin structure function $g_1^p(x)$, which is the charge-

---

[‡] It has been pointed he that the $SU(3)_f$ symmetry is not particularly good for axial charges. Nevertheless, the conclusions of the following discussion remain unchanged, even in the presence of large systematic errors, due to $SU(3)_f$ breaking[7].

squared-weighted helicity distribution inside the proton:

$$g_1^p(x) = \frac{1}{2} \sum_q e_q^2 \left[q_\uparrow(x) - q_\downarrow(x) \right] = \frac{1}{2} \sum_q e_q^2 \Delta q(x). \tag{9}$$

$g_1^p(x)$ was obtained in this way by the SLAC-Yale collaboration in the 1970's[9] and in 1987, for a wider range of x, $0.01 \le x \le 0.7$, by the EMC collaboration.[1] Recently, the EMC collaboration has taken into account a small amount of additional data and re-examined all sources of systematic errors, especially the uncertainty on $F_2^p(x)$. Where SLAC data are available $(0.1 \le x \le 0.7)$, they can be combined with those of EMC. The final result is[2,3]

$$\int_0^1 dx g_1^p(x) = \frac{1}{2} \left(\frac{4}{9} \Delta u + \frac{1}{9} \Delta d + \frac{1}{9} \Delta s \right) = 0.126 \pm 0.019 \,(\text{stat}) \pm 0.015 \,(\text{sys}) \tag{10}$$

Thus the first moment of $g_1^p(x)$ yields[§] a third linear combination of axial-current matrix elements, independent of (5) and (6). In 1974 an attempt was made, by Ellis and Jaffe[10] to predict the r.h.s. of (10), using only the then-known low-energy data. Since (10) is independent of (5) and (6), making such a prediction obviously requires an extra dynamical assumption. The assumption in Ref. 10 seemed a safe one at the time: no strange-quark sea in the proton, therefore $\Delta s = 0$. The corresponding prediction is usually referred to as the Ellis-Jaffe sum rule:

$$\int_0^1 dx g_1^p(x) = 0.19 \tag{11}$$

Clearly, the sum rule is not obeyed by the data. We can understand why by looking at the solution of the three linear equations (5), (6) and (10):

$$\left. \begin{aligned} \Delta u &= 0.75 \pm 0.06 \\ \Delta d &= -0.50 \pm 0.06 \\ \Delta s &= -0.22 \pm 0.06 \end{aligned} \right\} \quad \Delta u + \Delta d + \Delta s = 0.03 \pm 0.18 \tag{12}$$

Eq. (12) contains two interesting results:
a) $\Delta s \neq 0$ and is quite large, about 50% of Δd.
b) sum of quark helicities is very small, consistent with zero.

   a) The quark sea contains a substantial $\bar{s}s$ component and the strange quarks are polarized. There have been some previous indications of a large strange sea at low-Q^2, most notably the πN σ-term[11] but little has been known about its polarization. Additional corroboration of the estimate (12) of Δs can be

---

   § When discussing experimentally determined moments of structure functions, such as (10), one should keep in mind that they necessarily involve some extrapolations, since experimentally always $0 < x < 1$.

obtained[12,13] from weak neutral current, elastic $\nu p \to \nu p$ and $\bar{\nu} p \to \bar{\nu} p$ scattering,[14] the basic observation being that the Z^0 couples to the axial current $\bar{u}\gamma_\mu\gamma_5 u - \bar{d}\gamma_\mu\gamma_5 d - \bar{s}\gamma_\mu\gamma_5 s$. The large strange sea in the proton implies a substantial violation of the OZI[15] rule. This has many interesting consequences, among them a possibility for the enhancement of OZI-forbidden couplings to baryons, such as of $\bar{s}s$ mesons ϕ and f' to the nucleon.[16] The main idea is that due to the presence of a large strange sea, such mesons can couple via connected, rather than disconnected diagrams. There is experimental evidence for the presence of such couplings and it is also possible to make quantitative predictions for certain OZI-violating processes that should be observed in the near future by LEAR.

b) This is a very important result. It answers the question that we have posted in the beginning, "*how much of the proton spin is due to quarks*" and the answer is rather surprising. There have been many theoretical attempts to understand it. As a first step note that $\Delta u + \Delta d + \Delta s$ is proportional to the matrix element of the flavor-singlet axial current, corresponding to the anomalous $U(1)$ symmetry of QCD:

$$\Delta u + \Delta d + \Delta s \propto \langle p| \, \bar{u}\gamma_\mu\gamma_5 u + \bar{d}\gamma_\mu\gamma_5 d + \bar{s}\gamma_\mu\gamma_5 s \, |p\rangle \equiv \langle p| \, J_\mu^{05} \, |p\rangle \qquad (13)$$

A natural way of computing the matrix element of a current is through an effective Lagrangian. There is a class of effective Lagrangians which correctly reproduce many low-energy properties of QCD and in which the relevant matrix elements can be computed rather easily. These are the non-linear chiral Lagrangians containing only scalar fields, of the non-linear σ-model type. Nucleons appear as classical solutions with non-trivial topology – chiral solitons or "skyrmions". The classical solutions have a rather unusual kind of symmetry, one which mixes rotations in ordinary and isospin space. This symmetry is associated with a large degeneracy of the classical solutions. When the degeneracy is removed, by means of collective-coordinate quantization, eigenstates of the collective-coordinate Hamiltonian have the quantum numbers of baryons, including half-integer spin. Nucleon wave-function is obtained in terms of the collective coordinates.[17] This makes it possible to compute matrix elements of the various currents. It turns out[6] that in this entire class of Lagrangians

$$\langle p| \, J_\mu^{05} \, |p\rangle = 0 \qquad (14)$$

in a *model-independent way*, by virtue of the symmetry of the classical solution, providing a first step towards a theoretical justification for (12).

The applicability to QCD of (14) depends on certain assumptions: there are good reasons[18] to think that the Skyrme-type chiral Lagrangians are in the same universality class as the large-N QCD. In this sense the result (14) is obtained in the large-N limit. In addition, (14) is exact only in the chiral limit, i.e. with zero current quark masses. $1/N$ corrections and explicit breaking of chiral symmetry could in principle change the prediction. In Ref. 6 we have looked at first-order chiral perturbation theory corrections, and found them to be small, but could not preclude possible contribution from higher-order terms. It is much more difficult to estimate the leading-order $1/N$ corrections. The only step so far in this direction has been taken in Ref. 19, where some $1/N$ terms were evaluated directly in a chiral Lagrangian, and were found to be roughly equal in magnitude but opposite in sign to the chiral perturbation theory corrections.

In addition to the effective Lagrangian approach, there have been attempts to understand the experimental results directly in the context of QCD. A crucial observation was made[20,21,22] that in QCD the first moments of helicity distributions are *not the same* as in the naïve parton model. The difference is due to the same mechanism which is responsible for the non-conservation of the singlet axial current, the triangle anomaly. The bare parton helicities Δu, Δd and Δs undergo an additive renormalization, and are shifted by a term proportional to the net gluon helicity ΔG of eq. (2). What we actually observe are the linear combinations of the renormalized helicities, Δq:

$$\Delta u \rightarrow \widetilde{\Delta u} = \Delta u - (\alpha_s/2\pi)\Delta G$$

$$\Delta d \rightarrow \widetilde{\Delta d} = \Delta d - (\alpha_s/2\pi)\Delta G \qquad (15)$$

$$\Delta s \rightarrow \widetilde{\Delta s} = \Delta s - (\alpha_s/2\pi)\Delta G$$

All previous expressions involving first moment of quark helicities, e.g. eq. (12) should now be reinterpreted as referring to $\widetilde{\Delta q}$-s and not Δq-s. Since the ΔG term is multiplied by α_s, one might be tempted to assume that it can be neglected at high energies, when α_s becomes small. In fact, that is not the case.[21] The reason is that $\Delta G \sim \log Q^2$, while $\alpha_s \sim 1/\ln Q^2$, and the product $(\alpha_s/2\pi)\Delta G \equiv \Delta\Gamma$ is independent of Q^2 to leading order in α_s. Another important thing to notice about the $\Delta\Gamma$ term is that it is the same for all flavors. Therefore it only contributes to $SU(3)$ singlet matrix elements, cancelling out for the non-singlet ones, for example $\widetilde{\Delta u} - \widetilde{\Delta d} = \Delta u - \Delta d$. Now, it is precisely the singlet matrix element in eq. (12) that we found so surprising, when we attempted to interpret it in terms of the bare parton helicities. The realization that glue could contribute in a non-trivial way seemed at first like a beautiful resolution of the spin puzzle. The glue contribution in (15) comes with a negative sign and given sufficiently large glue polarization it could in principle rescue the NRQM intuition and account for less than expected first moment of $g_1^p(x)$ in (10) and the failure of the Ellis-Jaffe sum rule (11). Upon closer inspection, however, it turns out that there is a price to pay, and it may be too high.

What is the price? A potentially elegant way to rescue the NRQM would be to say that $\Delta s = 0$, while $\widetilde{\Delta s} \approx -0.2$, as given by (12). The difference would be accounted for by the $\Delta\Gamma$ term, cf. eq. (15):

$$-0.2 \approx \widetilde{\Delta s} = \Delta s - \Delta\Gamma; \quad \Delta s = 0 \qquad (16)$$

so that

$$\Delta\Gamma = (\alpha_s/2\pi)\Delta G \approx 0.2 \qquad (17)$$

As mentioned, to leading order $\Delta\Gamma$ is independent of Q^2, but ΔG is not. Given that $\alpha_s(Q^2=10 \text{ GeV}^2)\sim 0.2$, with $\Delta\Gamma \approx 0.2$ implies $\Delta G \sim 6 \div 7$. This very large value of ΔG can be ruled out by the data on the *unpolarized* glue distribution, as follows. From positivity of $G_{\downarrow(\uparrow)}(x)$ we have $\Delta G(x) \leq G(x)$, where $G(x)$ is the usual unpolarized glue momentum distribution, $G(x)=G_{\uparrow}(x)+G_{\downarrow}(x)$. From deep inelastic neutrino scattering we know $G(x)$ at moderate and high values of x. The integral of $G(x)$ over that region, say $\int_{0.1}^1 dx G(x)$, is rather small. At low values of x $G(x)$ is singular $\sim 1/x$, due to pomeron exchange. Pomeron parametrizes the helicity-conserving part of $G(x)$ and therefore we do not expect the $1/x$ singularity in $\Delta G(x)$ which is the helicity-flip part of $G(x)$. On these grounds we expect $\Delta G(x)$ and $g_1^p(x)$ to have the same functional dependence on x, i.e. $\sim x^\delta$, $\delta \sim 0$. These two constraints taken together make it extremely unlikely[12] that ΔG could reach values as large as 6 or 7. Another argument against a large value of ΔG is provided by the sum rule (3). The *l.h.s.* is of course independent of Q^2, while the *r.h.s.* includes three terms which depend on Q^2 in essentially different way. At large Q^2 the first term is almost a constant, since its Q^2 dependence is a two-loop effect[23]. The other two terms, ΔG and $\langle L_z \rangle$ grow logarithmically with Q^2. If at $Q^2 = 10$ GeV^2 we have (say) $\Delta G \sim 6$ it will grow with Q^2 without limit and require a delicate cancellation with equally large $\langle L_z \rangle$.

Recently it has been realized[24] that the explanation of the EMC data through glue polarization suffers from an even more serious problem. Eq. (15) shows that the first moment of $\Delta q(x)$ is shifted by an amount proportional to the first moment of $\Delta G(x)$. This does not mean, however, than on an x by x basis $\Delta q(x)$ is shifted by an amount proportional to $\Delta G(x)$. In fact, it turns out that the contribution of $\Delta G(x)$ to the first moment of $\widetilde{\Delta q}$ is concentrated at low values of x. The physical reason for this phenomenon is not difficult to understand. The gluon contribution to the matrix element of the flavor single axial current comes in through the box diagram with two gluons and two photons.[21,22] Consider a gluon carrying a fraction x' of proton momentum. It splits into a quark anti-quark pair. The quark which is sensed by the external current carries a fraction x of the proton momentum, where $x \leq x'$, and gives rise to the extra contribution to the singlet part of $g_1^p(x)$. The corresponding shift in helicity distribution at a given x, $\delta q(x) \equiv \widetilde{\Delta q}(x) - \Delta q(x)$, results from a convolution of $\Delta G(x')$ with a non-local kernel $K(x, x')$, given by the box diagram:

$$\delta q(x) = \int_x^1 dx'\, K(x, x')\, \Delta G(x') \qquad (18)$$

The explicit form of $K(x, x')$ for massless and for massive quarks is given in Ref. 24. By consistency with eq. (15), for massless or light ($m_q \lesssim \Lambda_{QCD}$) quarks, $\delta q(x) \equiv \delta q_0(x)$ must satisfy

$$\int_0^1 dx\, \delta q_0(x) = -(\alpha_s/2\pi)\Delta G(x') \qquad (19)$$

while for massive quarks ($m_q \gg \Lambda_{QCD}$)[22]

$$\int_0^1 dx\, \delta q_m(x) = 0 \qquad (20)$$

To get numerical values of $\delta q(x)$ and see where the glue contribution is felt the most, it is necessary to use an explicit form of $\Delta G(x')$. Experimentally nothing is known about $\Delta G(x')$, except for the constraints which were discussed earlier. However, in the literature there are several generic parametrizations, in particular those that have been proposed in order to solve the spin puzzle along the lines of eq. (15).[25,26] For massless quarks we found, regardless of which particular $\Delta G(x')$ was used in eq. (18), that roughly 50% of the integral in (19) comes form the very low x region, $x < 10^{-2}$. Thus a very substantial part of the possible polarized gluon contribution to $g_1^p(x)$ is in the region $x < 0.01$ not measured by the EMC. This means that the polarized gluon distributions proposed so far are too small to explain the EMC data via their effect on light quarks alone. The normalizations of these distributions would apparently need to be more than doubled in order to fit the data. However, this is not possible without violating the bound on $\Delta G(x)$ given by the unpolarized gluon distribution $G(x)$. An additional difficulty for large $\Delta G(x)$ is that it implies, through eq. (18), a rather singular behavior of $g_1^p(x)$ at small x. On the other hand, as discussed previously, on general grounds we expect the helicity flip distributions to be non-singular at small x.[12,27]

When effects of charmed quarks are taken into account, the conclusion changes in a somewhat surprising way. The analysis of Ref. 22 shows that the contribution of heavy quarks to the first moment of $g_1^p(x)$ is zero, cf. eq. (20). One might be tempted to think this occurs by having $\delta q_m(x)=0$ identically. What actually happens is much more interesting: $\delta q_m(x)$ is positive at small x and negative in the x region surveyed by EMC, cf. Fig. 2 of Ref. 24. The box-diagram charm contribution to $g_1^p(x)$ is proportional to $\delta q_m(x)$. Thus it gives a net negative contribution to the EMC result (10). It works in the same direction as the light quarks and could help explain the data. In addition, the troublesome singular behavior introduced into $g_1^p(x)$ at low x by $\delta q_0(x)$ is largely offset by $\delta q_m(x)$.

Whether or not the polarized glue contribution can explain the EMC data is an experimental question. It seems unlikely that $\Delta G(x)$ could be so large as to entirely avoid the strange sea, along the lines of eqs. (16) and (17). However, even a moderately positive $\Delta G(x)$ would work in the right direction, decreasing the first moment of $g_1^p(x)$. Clearly, it is very important to determine $\Delta G(x)$ experimentally, especially since a priori we do not even know whether it is positive. We need a process with a strong dependence on gluon helicity. A generic amplitude of this type is $\bar{q}q$ production via gluon-gluon or photon-gluon fusion. In order to distinguish this type of processes from other $\bar{q}q$ production mechanisms, it is best to look at $\bar{c}c$ events. The photon-gluon fusion is a "cleaner" process, since it involves lepton-proton, rather than proton-proton scattering. The basic idea is to use a polarized lepton beam, with a polarized proton target, just as was done by the EMC. The main difference is that one would need measure polarization asymmetry in events involving charm production, rather than in the total cross section. Some relevant processes have been discussed in Refs. 6, 28 and 29.

The EMC measurements of the polarized structure functions are the key motivation for at least two experiments which are now in the planning stage. One at HERA, with a gas target in the storage ring and one at CERN. One of their main goals, in addition to refining the measurements of $g_1^p(x)$ will be to test the Bjorken sum rule[30]

$$\int\limits_0^1 dx[g_1^p(x) - g_1^n(x)] = \frac{g_A}{6}\left(1 + \mathcal{O}(\frac{\alpha_s}{2\pi})\right) \qquad (21)$$

which is an important testing ground for our understanding of QCD. The EMC result (10), together with (21) implies that the first moment of the neutron structure function $g_1^n(x)$ is much larger than was previously expected on the basis of the (11) and should be accessible experimentally.

On the theoretical front, there is an important challenge of understanding (12) from first principles.

REFERENCES

1. EMC collaboration, J. Ashman et al., Phys. Lett. **206B**(1988), 364.

2. G. Baum, V.W. Hughes, K.P. Schüler, V. Papavassiliou and R. Piegaia, Phys. Lett. **B212**(1988), 511.

3. J. Ashman et al., EMC Collaboration, CERN preprint, CERN-EP/89-73.

4. J. Ellis, R. Flores and S. Ritz, Phys. Lett. **198B**(1987), 393; M. Glück and E. Reya, Dortmund preprint, DO-TH-87/14, Aug. 1987.

5. J. Ellis and R. Flores, Nucl. Phys. **B307**(1988), 883.

6. S. Brodsky, J. Ellis and M. Karliner, Phys. Lett. **206**(1988), 309.

7. H. J. Lipkin, Argonne preprint ANL-HEP-114 and references therein.

8. M. Bourquin et al., Z. Phys. **C21**(1983), 27; for a review see: J.-M. Gaillard and G. Sauvage, Ann. Rev. Nucl. Part. Sci. **34**(1984), 351.

9. M. J. Alguard et al., Phys. Rev. Lett. **37** (1976), 1261, ibid **41**(1978)70; G. Baum et al., Phys. Rev. Lett. **51**(1983), 1135.

10. J. Ellis and R. Jaffe, Phys. Rev. **D9**(1974), 1444.

11. T.P. Cheng, Phys. Rev. **D13**(1976), 2161; see also J. Donoghue and C. Nappi, Phys. Lett. **B168**(1986), 105; J. Donoghue, in Proc. of II-nd Int. Conf. on πN Physics; V.M. Khatymovsky, I.B. Khriplovich and A.R. Zhitnitsky, Z. Phys. **C36**(1987), 455; Riazudin and Fayyazuddin, Dharan Univ. preprint, 1988.

12. J. Ellis and M. Karliner, Phys. Lett. **B213**(1988), 73.

13. D.B. Kaplan and A. Manohar, Nucl. Phys. **B310**(1988), 527.

14. L. A. Ahrens et al., Phys. Rev. **D35**(1987), 785.

15. S. Okubo, Phys. Lett. **5**(1963), 165;
    G. Zweig, CERN Report No. 8419/TH412, 1964 (unpublished);
    I. Iuzuka, Prog. Theor. Phys. Suppl. **37-38**(1966), 21;
    see also G. Alexander, H.J. Lipkin and P. Scheck, Phys. Rev. Lett. **17**(1966), 412.

16. J. Ellis, E. Gabathuler and M. Karliner, Phys. Lett. **217B**(1988), 173.

17. E. Witten, Nucl. Phys. **B223**(1983), 422, ibid 433; G. Adkins, C. Nappi and E. Witten, Nucl. Phys. **B228**(1983), 433; for the 3 flavor extension of the model see: E. Guadagnini, Nucl. Phys. **236**(1984), 35; P. O. Mazur, M. A. Nowak and M. Praszałowicz, Phys. Lett. **147B**(1984)137.

18. E. Witten in Lewes Workshop Proc.; A. Chodos et al., Eds; Singapore, World Scientific, 1984.

19. Z. Ryżak, Phys. Lett. **B217**(1989), 325; erratum, ibid, **B224**(1989), 450.

20. A.V. Efremov and O.V. Teryaev, Dubna preprint, JIN-E2-88-287, submitted to Münich Conference, 1988.

21. G. Altarelli and G. Ross, *Phys. Lett.* **B212**(1988), 391.

22. R. D. Carlitz, J.D. Collins and A.H. Mueller, *Phys. Lett.* **B214**(1988), 229.

23. J. Kodaira, S. Matsuda, T. Muta, T. Uematsu and K. Sasaki, *Phys. Rev.* **D20**(1979), 627; J. Kodaira, S. Matsuda, K. Sasaki and T. Uematsu, *Nucl. Phys.* **B159**(1979), 99; J. Kodaira, *Nucl. Phys.* **B165**(1979), 129.

24. J. Ellis, M. Karliner and C. Sachrajda, *Phys. Lett.* **B231**(1989), 497.

25. G. Altarelli and W.J. Stirling, *Particle World* **1**(1989), 40.

26. Zoltán Kunszt, *Phys. Lett.* **B218**(1989), 243.

27. R.L. Heimann, *Nucl. Phys.* **B64**(1973), 429.

28. P. Kalyniak, M.K. Sundaresan and P.J.S. Watson, *Phys. Lett.* **216B**(1989), 397.

29. J.L. Cortes and B. Pire, *Phys. Rev.* **D38**(1988), 3586.

30. J. Bjorken, *Phys. Rev.* **148**(1966), 1467.

T-DIQUONIA AND BARYON-ANTIBARYON SYSTEMS*

W. Roberts[1], B. Silvestre-Brac[2] and C. Gignoux[2]

[1] Department of Physics, Harvard University
Cambridge, MA 02138, USA

[2] Institut des Sciences Nucléaires, 53, Avenue des Martyrs
38026 Grenoble CEDEX France

ABSTRACT

The flavor SU(3) multiplets of T-diquonia ($q^2\bar{q}^2$) are classified, their masses estimated, and their decay widths into baryon-antibaryon pairs are evaluated. Using the same formalism, but with broken SU(2), the contributions of such states to some exclusive $p\bar{p}$ scattering cross sections are estimated, in an attempt to understand the experimental sparsity of diquonia candidates. It is found that if the meson-meson decay widths of these states are of the order of 50 MeV, they are difficult to detect experimentally. It is noted that, in general, except for the elastic scattering channel, the assumption of scattering via diquonia intermediates gives a reasonable description of physical reality.

I: INTRODUCTION AND MOTIVATION

One aspect of low energy $N\bar{N}$ physics that has received some interest in the past is the possible application to discovery of low-mass exotic hadrons. However, while many such states are predicted to exist or, more accurately, are allowed to exist in the standard model, very few have been positively identified experimentally.

Among the many exotic hadrons, the four-quark states commonly called diquonia or baryonia are quite closely related to baryon-antibaryon systems (hence the name baryonia). These states may be produced in such systems by the mechanism of figure 1. It is therefore conceivable that signatures of diquonia may show up in baryon-antibaryon scattering processes. Indeed, a few candidates have been identified in such systems[1].

---

* Research supported in part by the Natural Science Foundation under grant # PHY-87-14654, and by the Natural Sciences and Engineering Research Council of Canada. Preprint # HUTP-90/A008. 02/90.
Invited talk presented at the Erice Winter School on Medium Energy Antiproton Physics and Non-Perturbative QCD, 25 - 31 January, 1990.

Medium-Energy Antiprotons and the Quark–Gluon Structure of Hadrons
Edited by R. Landua *et al.*, Plenum Press, New York, 1991

103

For this reason, it is crucial that the prospects for discovery of diquonia in $N\bar{N}$ collisions, for example, be analyzed. With the assumption of the mechanism of figure 1, such an analysis is reasonably straightforward, but still requires some knowledge of, or assumptions about the masses and total widths of diquonia, as well as their couplings to $B\bar{B}$ systems. Some prior work has been done here[2-12].

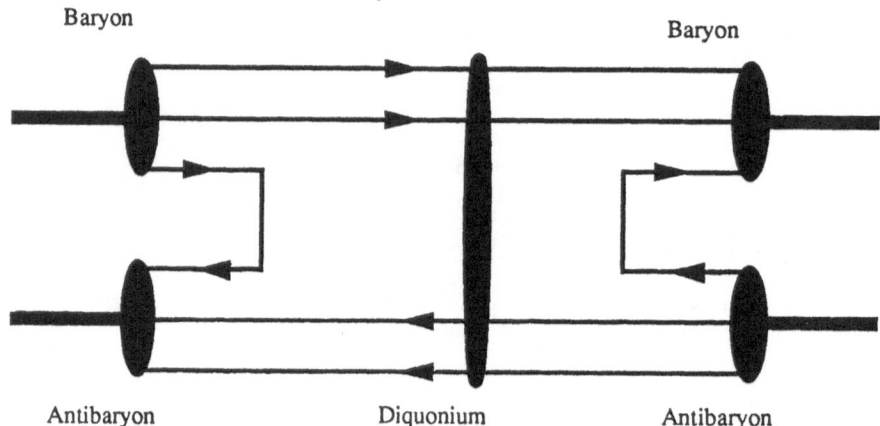

Figure 1. Mechanism for diquonia formation and decay in baryon-antibaryon systems.

In this presentation, we propose to provide some idea about the masses and widths of diquonia using simple models. We also obtain the couplings to $B\bar{B}$ systems in this way. We then use this information to determine whether or not any signatures of diquonia may be easily seen in $p\bar{p}$ scattering cross sections. Our approach is first to perform the analysis of masses and widths assuming that flavor SU(3) is valid. This will give an idea of whether there are any states that are reasonably narrow. We then perform a similar analysis using broken SU(2), concentrating on those states that can couple to $p\bar{p}$ systems and estimate their contributions to various $p\bar{p}$ scattering cross sections.

The rest of this paper is organized as follows. Section II lists the SU(3) diquonium multiplets in which we are interested, and outlines the procedure we use to estimate their masses. In section III, we discuss very briefly the decay model we use, as well as the results we obtain using that model. Section IV motivates the second part of this study, while section V lists the diquonia states of broken SU(2), and briefly discusses their masses and decay widths. In section VI we evaluate the contributions of these states to the cross sections of various proton-antiproton scattering processes, and discuss the results we obtain. In section VII we present some final comments.

II: SU(3) MULTIPLETS OF T-DIQUONIA AND MASS ESTIMATES

The diquonia states we discuss are assumed to consist of an S-wave diquark and an S-wave anti-diquark with some orbital angular momentum between them. The use of this basis is motivated by the need for simplicity in treating the decays into baryon-antibaryon pairs. The diquark may belong to a $\bar{3}$ or to a 6, while the anti-diquark may belong to a 3 or a $\bar{6}$. To form a color singlet object, a diquark in a $\bar{3}$ must combine with an anti-diquark in a **3**, or a **6** diquark must combine with a $\bar{6}$ anti-diquark.

The diquonia formed from the latter combination $(6,\bar{6})$ are the so-called "mock" or M-diquonia, and are of no interest to us here, as their decays into baryon-antibaryon pairs are forbidden without colour mixing. This is because the diquark must combine with a quark to create a color singlet baryon. This is only possible if the diquark belongs to the color $\bar{3}$.

The "true" or T-diquonia, $(\bar{3},3)$, can decay into $B\bar{B}$ pairs within the framework of the $^3P_0$ model, with no need for color mixing. The mechanism is illustrated in figure 2. We thus confine our discussion to states of T-diquonia. The color restriction is the only one that we place on the states that we study: we discuss all states compatible with the Pauli principle.

We classify the diquonia states by the flavor multiplet of the diquarks of which they are made, the flavor multiplet of the diquonia themselves, and the total spin and orbital angular momentum of the diquonia. To begin, we point out that the notation used in discussion of diquonia within flavor SU(2) finds a natural extension in SU(3)[5]. A diquark belonging to the $\bar{3}$ of SU(3) is denoted β, while one belonging to the **6** is denoted δ. Overall antisymmetry of the diquonium wave function under exchange of quarks or antiquarks constrains the spin of the β type diquark to be 0 and that of the δ type to be 1. The diquark states are therefore $\beta(\bar{3},0)$ and $\delta(6,1)$, where the numbers in parentheses are the flavor multiplet and the total spin, respectively.

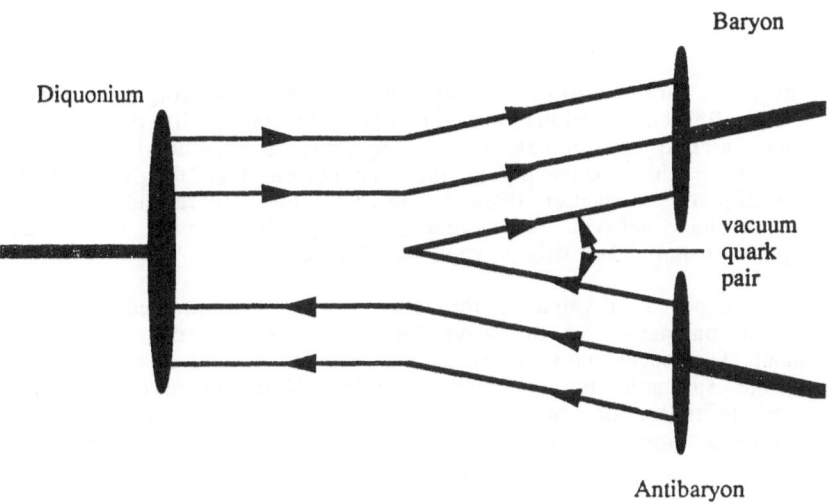

Figure 2. Diquonium decay into baryon-antibaryon pair via pair creation model.

The diquonia multiplets that can be formed from these diquarks and antidiquarks are shown in table 1. Note that these are not, in general, states of definite G-parity, but are, instead, states of definite flavor. The notation used in the table is the same as that used for SU(2): "A" denotes a diquonium with diquark content $\beta\bar{\beta}$, "B" corresponds to $\beta\bar{\delta}$ and $\delta\bar{\beta}$, while "C" denotes diquark content $\delta\bar{\delta}$.

To estimate the diquonia masses we use an additive potential similar to that used by Bhaduri[13], consisting of a linear confining term with Coulomb and short-range spin-spin terms. The values of some of the potential parameters we use are different from those of reference 13. These differences arise because we choose to treat the baryons and diquonia in the diquark approximation: the baryon consists of an S-wave diquark and a quark with relative orbital angular momentum L between the quark and the diquark, and the diquonium consists of an S-wave diquark and an S-wave anti-diquark with orbital angular momentum L. While the dynamical motivation for this approximation may be poor, especially in the case of baryons and diquonia with low orbital angular momentum, it is a convenient one that allows a simple treatment of both the $B\bar{B}$ decays of the diquonia, and calculation of their masses.

Table 1. SU(3) diquonia multiplets, spins, diquark content and masses.

State	Multiplets	Spin	Flavor-Spin Wave Function	M^2 (GeV^2)
A	1, 8	0	$\beta\bar{\beta}$	1.29 L + 1.55
B	8, 10	1	$\beta\bar{\delta}$	1.44 L + 2.56
B	8, 10	1	$\delta\bar{\beta}$	1.44 L + 2.56
C	1, 8, 27	0	$\delta\bar{\delta}$	1.65 L + 2.68
C	1, 8, 27	1	$\delta\bar{\delta}$	1.61 L + 2.97
C	1, 8, 27	2	$\delta\bar{\delta}$	1.43 L + 4.06

The diquark approximation is implemented in evaluating the diquonium mass by expanding the various position-dependent terms in the potential as series in the internal coordinates of the diquarks, and truncating at quadratic terms. Inherent in this expansion is the assumption that the internal dimension of the diquark is smaller than the inter-diquark separation in the diquonia, or the distance between the diquark and the third quark in the baryon. This assumption is not necessarily a good one for low L.

When the expectation value of the total energy is minimized as a function of the gaussian parameters of the wave functions, some of the masses obtained are very small. This occurs only for states with L = 0, 1, and is a consequence of the short-range spin-spin term of the potential. Note, however, that for the baryons, the minimization procedure gives masses that are in reasonable agreement with a more complete theoretical calculation[14] for baryons with L = 2.

To escape the problem of very light states, we modify the parameters of the spin-spin term of reference 13. The new values of these parameters are used in obtaining diquonia masses. Here, however, we are unable to perform a fit *per se*, so we must again minimize the total energy as a function of the gaussian widths. In doing this, we again run into very light states. We remedy this by taking the masses of the diquonia with large L ($2 \leq L \leq 8$), where the diquark approximation is expected to have a greater degree of validity, performing a Regge fit to these masses, and using this Regge fit to extract the masses of the states at lower L (L < 2). The results of these fits are given in column five of table 1.

Some comments on the masses obtained by this method are in order. First, it is worth pointing out that most of the masses obtained are similar to those given by Jaffe's corresponding equations for SU(2)[5], with an additional 300

McV to estimate the masses for the SU(3) multiplets. They are also similar to the masses estimated in reference 10, where strange quarks have been included. In the case of the A states, however, we find masses that are smaller. We expect that the masses of our A states are consistently too small by about 150 MeV. Note that the masses that we have obtained take into account the important spin-spin interaction, while those of references 5 and 10 do not.

Let us also point out that although the diquark approximation is questionable, and the minimization procedure leads to problems with very light states, the results for $L = 2$ in the baryon sector are reasonable. We therefore expect that the masses obtained for the diquonia with $L \geq 2$ are trustworthy and, subsequently, that the masses obtained for $L = 0, 1$, based on the Regge fit to states with higher L, are also dependable.

We conclude this section by pointing out that even though most of the diquonia with $L = 0, 1$ are below the lowest $B\bar{B}$ threshold, they are all well above MM thresholds (M denotes meson). None of these states are therefore bound, and will fall apart "easily", modulo angular momentum selection rules, into pairs of mesons.

III: DECAY MODEL AND TOTAL WIDTHS

The model used for description of the decays is the version of the $^3P_0$ model popularized by the Orsay group[16], illustrated in figure 2. This model has been discussed several times in the literature[17], so we will not include any details here. We will, however, comment briefly on two concepts related to this model.

First, the reader is reminded that there is a popular alternative description of the pair creation process in the $^3S_1$ model. This model is thought of as the first term in a "perturbative" series, where the quark pair is created from a single gluon. In keeping with this, one is tempted to think that the $^3P_0$ model represents the next term in this "perturbative" series. However, there is a problem with each of these ideas.

In the case of the $^3S_1$ model, while it does represent part of the process of pair creation from a single gluon, it is not complete. What is missing is a D-wave contribution that is not, *a priori*, negligible: a single gluon will also generate quark pairs with angular momenta corresponding to $^3D_1$.

With regard to $^3P_0$ as the next term in a "perturbative" expansion, we turn to perturbative QCD. There, the next set of diagrams contributing to pair creation are the box diagrams (vacuum polarization, vertex corrections and external leg renormalization diagrams do not change the allowed partial waves of the created quark pair). Here[18] it has been found that, near threshold, the dominant partial wave is $^1S_0$, while at high energies the leading partial waves are $^3P_1$, $^3P_2$ and $^3F_2$. $^3P_0$ is never *the* dominant partial wave. It is never even *a* dominant partial wave. It is therefore a fallacy to regard the $^3P_0$ model as the second term in a perturbation series: it is simply a successful phenomenological model, and is the one we have used in this work.

Figures 3 to 5 present the total $B\bar{B}$ decay widths for some of the diquonia states discussed in section II, for $L = 0$ to 4. In these figures, the widths are shown as functions of the mass of the decaying diquonium. The segment of each curve shown corresponds to the mass range M - 300 MeV to M + 300 MeV, where M is the mass obtained from the procedure outlined in the previous section, and is indicated by an arrow on each curve. The widths for the states not shown exhibit behavior similar to one of the graphs shown: the octet A states are similar to the singlet A's, the **10** and $\overline{\mathbf{10}}$ B's are similar to the **8** B's, and the **8** and **27** C's are similar to the singlet C's, as are the C's with S = 0 and 1.

It is clear from these figures that many of the total $B\overline{B}$ widths are very large and that the corresponding states would perhaps not be very interesting. Some of the states have narrow widths which remain narrow throughout the entire mass range shown. In contrast, some states have broad widths that become immense as the mass is increased. Such states are almost certainly without interest, unless their masses are significantly smaller than those calculated here.

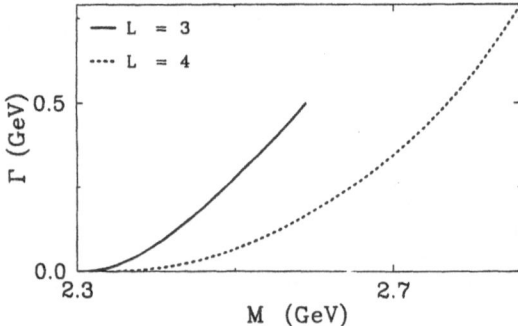

Figure 3. Total decay widths of singlet A states as functions of the mass of the decaying state.

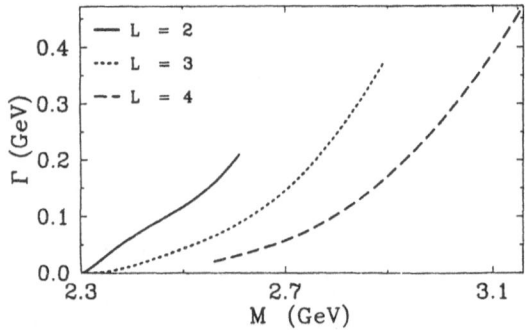

Figure 4. Total decay widths of the octet B states as functions of the mass of the decaying state.

The A states have small total widths corresponding to the masses calculated, but these widths increase rapidly when the mass of the state is increased. The reader is reminded that our A states are probably too light, so that the widths corresponding to more realistic masses are considerably larger. In contrast with the A states, the total widths of the B states increase less rapidly when the mass of the state is increased. The C states are perhaps the most intriguing in that their total widths remain quite small for a significant part of the mass range investigated, then become quite large as the mass is increased beyond about 2.8 GeV. This is the effect of a new threshold, and is seen most clearly in the curve for the C state with L = 2 in figures 5. This threshold is the $\Delta\bar{\Delta}$ threshold, where Δ signifies the lowest lying baryon decuplet. C states that lie below this threshold thus appear to be the best candidates for experimental discovery.

We conclude this section by pointing out that without resorting to any admixtures of M-diquonia, we obtain widths that are very narrow. We have also obtained widths that are quite broad, especially when channels involving baryons from the lowest lying decuplet are accessible. We expect this range of widths to persist for broken SU(3) and SU(2). There, however, the important question to be answered is whether the predicted states will exhibit the same narrow widths as the corresponding experimental candidates, without invoking significant admixtures of M-diquonia.

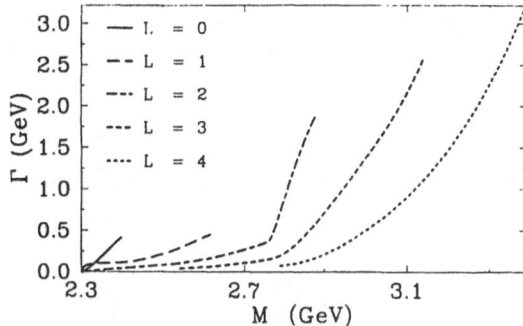

Figure 5. Total decay widths of the singlet C states, S = 2, as functions of the mass of the decaying state.

IV: MOTIVATION II: WHERE ARE ALL THESE STATES?

We have seen that the number of SU(3) multiplets of diquonia is quite large, which is consistent with the number of states investigated by other authors[2-12]. It is somewhat puzzling, therefore, that of the very large number of states predicted by many theorists, very few candidates have been seen experimentally[1].

We address this puzzle by estimating the lowest order contributions of diquonia to the cross sections of a few $B\bar{B}$ scattering processes. In doing this, we are assuming that the scattering process takes place via the mechanism illustrated in figure 1. We limit the discussion to $p\bar{p}$ scattering between 2 and 3 GeV, since this should be sufficient to illustrate why many more diquonia candidates have not been observed experimentally. Let us emphasize that we are not undertaking a complete calculation of $p\bar{p}$ scattering cross sections, as this would require the inclusion of many contributions that have no bearing on the point we are investigating. We comment further on this later.

V: DIQUONIA OF BROKEN SU(2) AND THEIR DECAY WIDTHS

Since we are interested in the process $p\bar{p} \rightarrow B\bar{B}$, we need consider only diquonia that can couple to $p\bar{p}$ pairs. These diquonia will therefore not possess any strange quarks, although decays into baryons with strangeness is allowed, through vacuum creation of a strange quark pair. We can therefore classify the states using the notation of SU(2). Note, however, that when we calculate the masses of the diquonia, we explicitly break this symmetry by choosing the d quark to be 6 MeV heavier than the u quark. This allows us to look at the effects due to nearly degenerate states. For this reason, we classify the states by their quark content. The 9 types of diquonia states that may decay into $p\bar{p}$ pairs are shown in table 2.

Let us note here that our states are flavor eigenstates, not isospin eigenstates. The effects of isospin mixing do not modify the results and conclusions that we present significantly. Such mixing appears only for the C states of table 2, and would change the magnitude of the cross sections, but would not modify the effects that we wish to demonstrate.

To evaluate the masses of these states, we use the procedure outlined earlier in section II. The masses obtained are shown as Regge trajectories in table 2. Note that the masses of our A states are again smaller than those reported elsewhere[5, 7, 8]. The consequences of this will be discussed in the next section. We discuss only the 41 states with masses greater than $p\bar{p}$ threshold, but less than 3.2 GeV.

Table 2. SU(2) diquonia states, quark content, spin and masses.

State	quark content	S	M^2 (GeV^2)
A	$(ud-du)(\bar{u}\bar{d}-\bar{d}\bar{u})/2$	0	$1.26L + 1.08$
B	$(ud+du)(\bar{u}\bar{d}-\bar{d}\bar{u})/2$	1	$1.46L + 2.23$
B	$(ud-du)(\bar{u}\bar{d}+\bar{d}\bar{u})/2$	1	$1.46L + 2.23$
C	$(ud+du)(\bar{u}\bar{d}+\bar{d}\bar{u})/2$	0	$1.742L + 2.270$
C	$uu\bar{u}\bar{u}$	0	$1.736L + 2.292$
C	$(ud+du)(\bar{u}\bar{d}+\bar{d}\bar{u})/2$	1	$1.692L + 2.627$
C	$uu\bar{u}\bar{u}$	1	$1.686L + 2.645$
C	$(ud+du)(\bar{u}\bar{d}+\bar{d}\bar{u})/2$	2	$1.470L + 3.957$
C	$uu\bar{u}\bar{u}$	2	$1.467L + 3.959$

To calculate the $B\bar{B}$ partial widths of these states, we again use the $^3P_0$ vacuum pair creation model, where we assume that u, d and s pairs may be created with equal probability. The results we obtain for the partial and total widths are similar to those obtained with unbroken SU(3), and are discussed in some detail in reference 19. For instance, the partial widths of the C states into $p\bar{p}$ are small, while the widths into $\Delta^{++}\bar{\Delta}^{++}$ are very large, and are by far the dominant contribution to the total widths of these states. In general, decays of C states into pairs of baryons from the decuplet dominate over decays into pairs from the octet. The A states have the largest partial widths into $p\bar{p}$ and $n\bar{n}$, while the B states are intermediate between the A and C states. In addition, we find that there are states of T-diquonia that have very small baryon-antibaryon total widths, with no admixture of M-diquonia.

VI: CONTRIBUTIONS TO SCATTERING CROSS SECTIONS

To evaluate the cross section for the process $p\bar{p} \to B\bar{B}$ via diquonia intermediates, we use the prescription of reference 8. The cross sections we obtain for $p\bar{p} \to p\bar{p}$, $n\bar{n}$, $\Sigma^+\bar{\Sigma}^+$, $\Lambda\bar{\Lambda}$, and $\Delta^{++}\bar{\Delta}^{++}$ are shown in figure 6 and figures 8 to 11, respectively. More channels are discussed in reference 19. Other channels show effects that are similar to those seen in one of the above channels. In these figures, the squares are the experimental data (where available) and the solid curves show the contribution to the cross section via diquonia intermediates with total widths exactly as calculated in the previous section. The long-dashed curve shows the contribution when the meson-meson decay widths of diquonia with L = 0 is assumed to be 100 MeV, and those with L = 1 is assumed to be 10 MeV. States with higher L may have negligible meson-meson decay widths, as the centrifugal barrier involved may be too large to be overcome. The dotted curves show the contribution when the meson-meson widths of all the diquonia are 50 MeV. Note that the solid curve and the long-dashed curve are indistinguishable for most of the cases shown.

Before discussing each of these figures individually, let us make a few general comments. First of all, we note that in all cases where experimental data[20] exist, the contribution to the cross section calculated herein is of the same order of magnitude as the data. In the case of $p\bar{p}$ final states, the theoretical contribution is always less than the experimental data, while for the other channels, it is much closer to the data and sometimes exceeds it. This may be understood in terms of the mechanism required for producing the specific final state. For $p\bar{p}$, the four quark intermediate state does not give the leading contribution, which may come from a six quark state. For other channels, on the other hand, the four quark state is necessarily present in some form, since one pair of quarks must be annihilated and another pair of different flavor created.

The fact that the diquonia contribution exceeds the data for $n\bar{n}$ and $\Lambda\bar{\Lambda}$ may be traced to two related factors. The first of these is that the decay amplitudes grow like $k^l \exp(-a^2 k^2)$, where a is some constant, and k is the 3-momentum magnitude of the baryon pair in the center-of-momentum frame. For large l, this form continues to grow for relatively large k, and k increases with energy since it is calculated off-shell. For $n\bar{n}$ and $\Lambda\bar{\Lambda}$, the main resonant contributions come from the A states with L = 3, 4, 5, 6 and 7, corresponding to l = 2 to 8, and hence the cross sections continue to grow in the energy range shown. Note that this effect is present to a lesser extent in the $p\bar{p}$ channel as well.

The second and related factor is that our A states are too light. This means that more realistic masses corresponding to a given L would be larger. The larger phase volume available for heavier states makes these states broader, so that resonant effects become more difficult to observe. These effects are illustrated in figure 10.

Perhaps the most striking feature of all these figures is the number of resonant features that are seen. For instance, in the elastic scattering channel, there are 27 diquonia states with masses between 2 and 3 GeV, but only 8 structures are present in the cross section. The main reason for this is that many of the states are very broad, so that resonant effects in the cross sections due to these states are not easily discernible. This is especially so in the case of the C states, and a little less so in the case of the B states. In keeping with this, note that most of the resonant features seen in the figures correspond to A states. Let us look at each channel in some more detail.

$p\bar{p} \rightarrow p\bar{p}$ (figure 6): As mentioned before, it is clear that diquonia intermediate states do not provide the major contribution to $p\bar{p}$ elastic scattering.
This, in fact, may be expected to come from processes in which one or more
pairs of quarks simply scatter off each other. Alternatively, this intermediate
may be described as "protonium" as shown in figure 7.

The noticeable features here are the remarkable disappearence or di-
minishing of resonant effects as the meson-meson widths of some or all of the
diquonia are made non-zero. The scenario that corresponds closest to physical
reality is that with all meson-meson widths set to 50 MeV. To see how this case
looks when compared with the data, we add an incoherent "background" of the
form $400/E^3$, where E is the total center-of-momentum energy of the baryon
pair. This is shown as the dash-dotted curve of figure 6. Note that this form
may correspond to no real physics.

Because the resonant features in the cross section are made to disappear
so easily, it is not surprising that not many diquonia candidates have been
identified in pp elastic scattering. Indeed, the above scenario suggests that
they may be impossible to find, unless techniques such as phase shift analyses
are employed. Even then, the number of nearly degenerate states will compli-
cate matters somewhat.

$p\bar{p} \rightarrow n\bar{n}$ (figure 8): Much of what has been said for the $p\bar{p}$ channel is also
applicable here, but with a few differences. The first difference is that the few
resonant effects that remain when all the meson-meson widths are set to 50
MeV are a little more clearly visible. This is largely because no background is
needed in this channel so that on the scale used, resonant effects are more
easily discernible. However, these can be made to disappear by increasing the
meson-meson widths a little further.

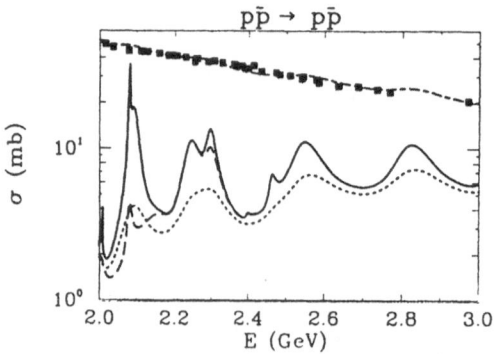

Figure 6. Contribution of diquonia states to $p\bar{p}$ elastic scattering.
For figure 6 and figures 8 to 11, the squares are the experimental
data (where available) and the solid curves show the contribution
to the cross section via diquonia intermediates with total widths
exactly as calculated in the previous section. The long-dashed
curve shows the contribution when the meson-meson decay
widths of diquonia with L = 0 is assumed to be 100 MeV, and those
with L = 1 is assumed to be 10 MeV. The dotted curves show the
contribution when the meson-meson widths of all the diquonia
are 50 MeV. Note that the solid curve and the long-dashed curve
are indistinguishable for most of the cases shown. In figure 6
only, the dash-dotted curve shows the contribution of diquonia,
plus an incoherent background as described in the text.

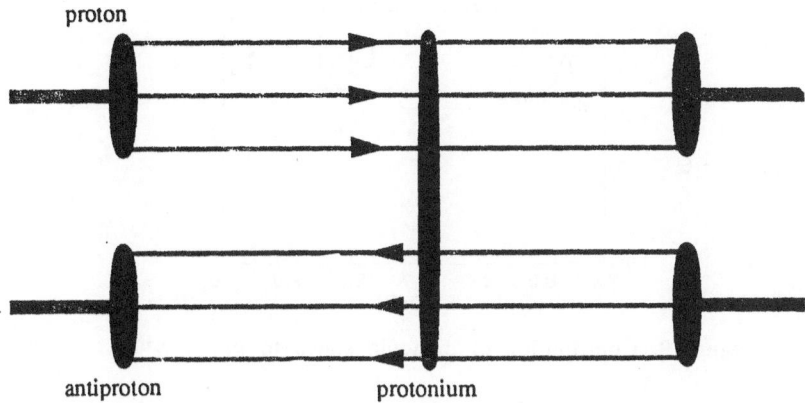

Figure 7. Possible dominant contribution to p$\bar{\text{p}}$ elastic scattering, via "protonium" intermediate.

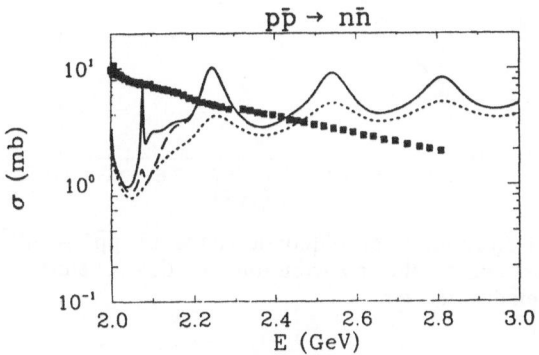

Figure 8. Contribution of diquonia states to p$\bar{\text{p}}$ → n$\bar{\text{n}}$.

The second difference between this channel and the $p\bar{p}$ channel is that the contribution from diquonia intermediates is comparable to the data without addition of any "background". This is because a four-quark intermediate state of some sort must play the leading role in $p\bar{p} \to n\bar{n}$. In fact, the theory exceeds the data above 2.5 GeV, but this is understood as a consequence of the light A states that we use. The rising cross section is also attributable to the light A states, as discussed earlier. Like the elastic scattering channel, this channel shows few resonant effects, and the effects that are seen can be made to vanish.

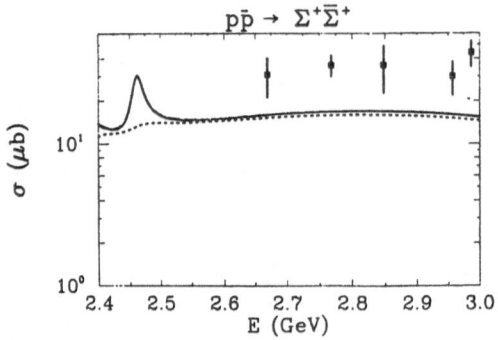

Figure 9. Contribution of diquonia states to $p\bar{p} \to \Sigma^+\bar{\Sigma}^+$.

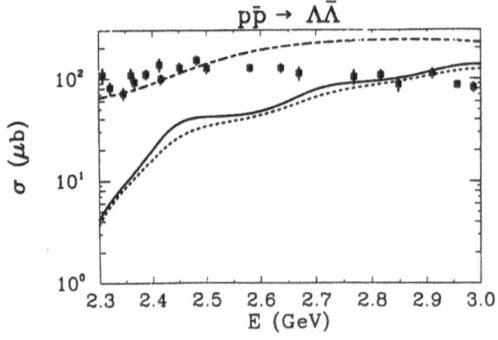

Figure 10. Contribution of diquonia states to $p\bar{p} \to \Lambda\bar{\Lambda}$. The dash-dotted curve shows the contribution to this channel when the A states are made heavier.

$p\bar{p} \to \Sigma^+\bar{\Sigma}^+$ (figure 9): In this channel, there is only one "observable" resonant feature in the energy range explored, which persists when the meson-meson widths of diquonia with L = 0, 1 are non-zero, but which disappears entirely when all meson-meson widths are non-zero. As in the previous channel, the theory is "consistent" with the data, but is consistently less than the data. This can be remedied by increasing the partial widths for decay into this channel by a factor of 2. This is not as drastic as it sounds: the largest partial width for this channel is less than 10 MeV, and most of the partial widths are less than 4 MeV.

$p\bar{p} \to \Lambda\bar{\Lambda}$ (figure 10): This channel shows very weak resonant effects even with all meson-meson widths set to zero, and these effects essentially vanish when these widths are all set to 50 MeV. Again, the theory has the right order of magnitude, but is wrong in details. This is another effect of our light Λ states, which are the only ones that contribute to this channel. If we make these states heavier and recalculate the total and partial widths, as well as the cross section, the result is the dash-dotted curve in figure 10. The form of this curve is more consistent with the trend of the data, although it exceeds the data beyond 2.5 GeV. Similar to the previous channel, this can be remedied by decreasing the partial widths into the $\Lambda\bar{\Lambda}$ channel by a factor slightly different from unity. Again, this is not a drastic condition, since the partial widths here are already small.

$p\bar{p} \to \Delta^{++}\bar{\Delta}^{++}$ (figure 11): This is one of the less interesting cases as no resonant effects are seen, even with meson-meson widths set to zero. The absence of observable resonance features here is easily understood, since all the states that contribute to this channel are extremely broad, with total widths covering most, if not all or more, of the energy range shown. The very large partial widths, and subsequently, the dominant branching ratios into this channel, lead to a large cross section for production of $\Delta^{++}\bar{\Delta}^{++}$ pairs: it is comparable to the cross section for production of nn pairs at the same energy. This appears a little surprising, but has not yet been tested experimentally.

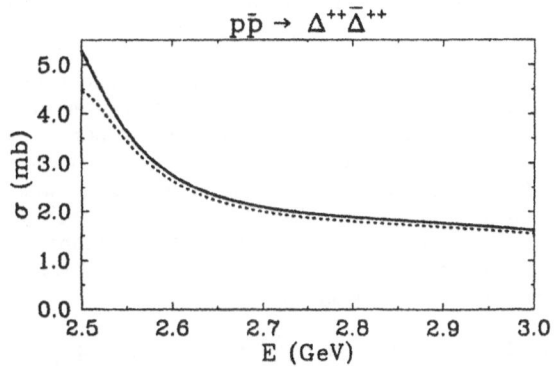

Figure 11. Contribution of diquonia states to $p\bar{p} \to \Delta^{++}\bar{\Delta}^{++}$.

VII: COMMENTS AND CONCLUSION

The figures and discussion of the previous section have illustrated that diquonia states can be difficult to observe experimentally, even though the theoretical spectrum is quite rich. Relatively "small" widths for decays of such states into meson-meson pairs, or baryon-antibaryon channels with excited baryons, can lead to the disappearance of detectable resonant effects in most baryon-antibaryon channels.

In addition, let us emphasize that we have considered only the lowest order effects due to diquonia. Since we are dealing with the strong force, higher order terms should be included for a complete treatment of the scattering process. Inclusion of such terms, some of which would be equivalent to "rescattering" terms, will smear any resonant effects that are still visible. This has the effect of making such resonant signals even more difficult to observe. The overall result is that *most diquonia may be extremely difficult to detect experimentally*

One very attractive by-product of this analysis is the reasonably good agreement between the cross sections obtained here, and the experimental data, where such data exist. We emphasize this point, especially since we have done absolutely no fitting of the theory to compare with the experimental data. Whether this agreement is merely coincidental is yet to be determined, but it is perhaps not too surprising, since a four-quark state of some sort must play a role in baryon-antibaryon production from proton-antiproton scattering. In addition, we note that duality arguments[21], as well as the P-matrix formalism of Jaffe and Low[22], suggest that an approach such as this is useful in shedding light on the dynamics of hadron scattering.

An important test of the mechanism described is the measurement of the cross sections for channels including Δ's such as $\Delta^{++}\bar{\Delta}^{++}$. Confirmation or contradiction of the predictions of the model for channels such as these will be useful in determining whether the order of magnitude agreement obtained so far is merely fortituous.

Other tests would include comparison of experiment with the model for other baryon-antibaryon scattering processes into baryon-antibaryon final states. Note, for instance, that in this model, creation of baryons with more than a single strange quark (Ω^- or Ξ, for example) from $p\bar{p}$ is suppressed, so that cross sections for such processes should be smaller. This is borne out by the little data available in channels containing such baryons. More stringent tests of the model would include the comparison of predicted angular distributions with experimental data.

If we take our results at face value, then the mechanism involving diquonia intermediates offers a viable alternative description of baryon-antibaryon scattering processes. This could be compared with other approaches where the scattering process has been characterized in terms of meson-exchange or potential scattering, for example[23], and there is certainly some overlap between the treatment described herein and that of meson exchange. Although a detailed discussion of such a comparison is beyond the scope of this presentation, we include a few comments on the similarity and differences between this mechanism and that of meson exchange.

A comparison between meson exchange and diquonia intermediates is valid because at the quark level, the diagrams for these two processes are very similar. The differences then appear simply to be one of language: in meson exchange, resonances are said to be exchanged in the t-channel, while in the mechanism proposed here the resonances are s-channel resonances.

In the case of the diquonia, however, there is some uncertainty involved in generating something like a 'potential' description, because we do not know the masses of the resonances. There is no such problem for the meson exchange approach: we know the meson masses very well. The uncertainties in the positions of the complex poles of the s-channel propagators will lead to uncertainties in the 'potential' parameters, and ultimately, to uncertainties in any predictions that the model may make. In contrast, in the case of the t-channel resonances, there would be much less uncertainty arising from uncertainties in the positions of the complex poles.

However, the spectrum of diquonia is very dense (inclusion of L-S coupling would lead to many more states than we have discussed, as would inclusion of radial excitations) and can perhaps be approximated by a continuum of four-quark states. This has been done by Furui and Faessler, reference 23. If no "convenient" assumptions are made about the structure of this continuum (dominant partial waves, quark content, etc.) some potentially attractive features of the discrete spectrum are lost.

For example, the momentum dependence observed near threshold in $\Lambda \bar{\Lambda}$ production is not what is expected[24]: the cross section grows like the cube of the momentum of the final states. With a discrete diquonium spectrum, this behavior can be accomodated in the model of scattering via diquonia intermediates by having a D-wave diquonium of type A sitting near $\Lambda \bar{\Lambda}$ threshold. Such a state would lead to P-wave (and of course, F-wave) $\Lambda \bar{\Lambda}$ pairs with the "correct" threshold behavior of the cross section. Such behavior is less easily accomodated with a four-quark continuum. Such a continuum may be more valid at higher energies.

Finally, we note that even though diquonia will be difficult to detect directly through resonant increases in cross sections, there is at least one other signature that can be sought. This was alluded to in the previous paragraph, namely exotic threshold behavior of scattering cross sections. This kind of signal will be clearest in scattering channels where only one type of diquonium state contributes: $\Lambda \bar{\Lambda}$ where only A states play a role, and $\Delta \bar{\Delta}$ where only C states are involved, are examples of such channels. Of course, for any such exotic behavior to be seen, the intermediate diquonium must not be P-wave, as such a state would lead to S-wave baryon pairs, with completely non-exotic threshold behavior in the cross section.

REFERENCES

1. L. Montanet, G. C. Rossi and G. Veneziano, Phys. Rep. **3C** (1980) 149.
2. Duality arguments were first used to predict four-quark states. H. Harari, Phys. Rev. Lett. **22** (1969) 562; J. L. Rosner, Phys. Rev. Lett. **21** (1968) 950; Phys. Rep. **11C** (1974) 89.
3. R. L. Jaffe, Phys. Rev. **D15** (1977) 267, 281.
4. S. Ono and S. Furui, Z. Phys. **C36** (1987) 651; S. Ono in Proceedings of IV European Antiproton Symposium, A. Fridman, ed., 1978.
5. R. L. Jaffe, Phys. Rev. **D17** (1978) 1444.
6. M. Fukugita and T. H. Hansson, Phys. Lett. **84B** (1979) 493.
7. J. P. Ader, B. Bonnier and S. Sood, Nuovo Cimento **68** (1982) 1; Z. Phys. **C5** (1980) 85; Phys. Lett. **84B** (1979) 488.
8. I. M. Barbour and J. P. Gilchrist, Z. Phys. **C7** (1981) 225; erratum, Z. Phys. **C8** (1981) 282.
9. C. Rosenszweig, Phys. Rev. Lett. **36** (1976) 697.
10. K. Igi and S. Yazaki, Phys. Lett. **74B** (1978) 257; Prog. Theor. Phys. **61** (1979) 487; C. Hong-Mo and H. Hogaasen, Phys. Lett. **72B** (1977) 121, 400; C. Hong-Mo in Proceedings of IV European Antiproton Symposium, A. Fridman, ed., 1978.
11. M. Imachi, S. Otsuki and F. Toyoda, Prog. Theor. Phys. **52** (1974) 346; **54** (1975) 280; **55** (1976) 551; **57** (1977) 517; M. Imachi and S. Otsuki, Prog. Theor. Phys. **58** (1977) 1657,1660; **59** (1978) 1290; M. Imachi, S. Ito and S. Otsuki, Prog. Theor. Phys. **61** (1979) 202.
12. M. B. Gavela, A. LeYaouanc, L. Olivier, O. Pène, J. C. Raynal and S. Sood, Phys. Lett. **79B** (1978) 459.
13. R. K. Bhaduri, "Models Of the Nucleon: From Quarks To Soliton", Addison-Wesley, Reading, MA (1988).
14. B. Silvestre-Brac and C. Gignoux, Phys. Rev. **D32** (1985) 743.
15. W. Roberts, B. Silvestre-Brac and C. Gignoux, Phys. Rev. **D41** (1990) 182.
16. For a full review see A. LeYaouanc, L. Olivier, O. Pène and J. C. Raynal, "Hadron Transitions In The Quark Model", Gordon and Breach, New York (1988). This work contains many references to other articles on the $^3P_0$ model.
17. M. Chaichian and R. Kögerler, Annals of Physics **124** (1980) 61; B. Silvestre-Brac, C. Gignoux and W. Roberts, unpublished report ISN 89.21, Institut des Sciences Nucléaires, Grenoble.

18. W. Roberts, Phys. Rev. **D39** (1989) 938.

19. W. Roberts, B. Silvestre-Brac and C. Gignoux, to appear in The Physical Review.

20. All data is taken from V. Flaminio, W. G. Moorehead, D. R. O. Morrison and N. Rivoire, "Compilation of Cross Sections III: p and $\bar{p}$ Induced Reactions", CERN-HERA 1984.

21. J. L. Rosner, Phys. Rev. Lett. **21** (1968) 950; Phys. Rep **11C** (1974) 89; H. Harari, Phys. Rev. Lett. **22** (1969) 562.

22. R. L. Jaffe and F. E. Low, Phys. Rev. **D19** (1979) 2105.

23. An exhaustive list of references on this topic would make this paper overly long. We give here a few representative references on work in this area. F. Tabakin and R. A. Eisenstein, Phys. Rev. **C31** (1985) 1857; M. Dillig and R. v. Fankenberg, Proceedings of the conference on Antinucleon and Nucleon-Nucleus Interactions, Plenum Press, 1985; S. Furui and A. Faessler, Nucl. Phys. **A468** (1987) 669; P. Kroll and W. Schweiger, Nucl. Phys. **A474** (1987) 608; M. Kohno and W. Weise, Phys. Lett. **179B** (1986) 15; P. Kroll and W. Schweiger, Proceedings of IV Lear Workshop, C. Amsler et al., eds., 1987; J. Niskanen, Proceedings of the Third Lake Louise Winter Institute, B. A. Campbell et al., eds., 1988; C. B. Dover and J. M. Richard, Phys. Rev. **D17** (1978) 1770.

24. P. D. Barnes et al., Phys. Lett. **189B** (1987) 249.

EXPERIMENTAL TRENDS IN THE STUDY OF THE ANTIPROTON-NUCLEON INTERACTION

Rolf Landua

PPE Division, CERN
1211 Geneva 23, Switzerland

FOREWORD

There is no lack of reviews about the various aspects of antiproton physics. These lectures are not intended to add another one, but rather as an introduction to the particle physics aspects of antiproton physics. They should help the interested reader in finding his way into (or through) the jungle of the literature. The basic theoretical concepts and the experimental techniques are illustrated by selecting three recent experiments at LEAR and by describing their motivations and some of their results. Since a comprehensive overview about antiproton physics, also covering experiments at laboratories like KEK, BNL, or Fermilab, and applications in atomic, nuclear, and solid state physics, was beyond the scope of these lectures, I did not attempt a systematical comparison of all the recent results in experiment and theory.

The lecture is organized as follows. Chapter 1 introduces to the basics of Quantum Chromodynamics (QCD) and models of the antiproton-nucleon interaction. Chapter 2 gives a short history of experiments with antiprotons to illustrate the technical and theoretical progress over the last 35 years. The following chapters are devoted to three recent experiments at LEAR representing the main lines of the present research with antiprotons. Chapter 3 deals with elastic scattering and inclusive cross-sections, chapter 4 with hyperon-pair production at threshold, and chapter 5 with $\bar{p}p$ annihilations and meson spectroscopy. Chapter 6 summarizes the current experimental programme at LEAR, and chapter 7 concludes with an outlook.

I INTRODUCTION

The proton and QCD

The proton is not an elementary particle. The anomalous magnetic moment, the finite size, the existence of excited states, and the appearance of three point-like scattering centres in deep inelastic scattering experiments leave no reasonable doubt. Within the non-relativistic SU(6) quark model, the proton is made of two up-quarks and one down-quark, and it fits nicely into an octet of baryons with spin 1/2 and positive parity. The other seven octet members are the neutron and the Λ, Σ and Ξ hyperons. For an introduction to the quark model, see ref. [1].

Medium-Energy Antiprotons and the Quark–Gluon Structure of Hadrons
Edited by R. Landua *et al.*, Plenum Press, New York, 1991

According to Quantum Chromodynamics (QCD), quarks have a strong ("colour") charge and interact by the exchange of gluons. The colour charge comes in three varieties (baptized "red", "green", and "blue" , in analogy to colour TV). The most convincing arguments in favour of the existence of three different colour charges are the existence of the Δ^{++} state, which would otherwise not obey the Fermi-Dirac statistics, the $\pi^{\circ} \to \gamma\gamma$ decay constant, and the ratio of cross-sections for hadron vs muon pair production in e^+e^- collisions.

The dogma of QCD is that only colour-neutral states are observable. These colour-neutral states are made by combining three quarks into a completely antisymmetric colour wave function (qqq = "baryon") or by a quark-antiquark pair of opposite colour charge (q $\bar{q}$= "meson"). All known particles fit nicely in either of these two categories. However, many other colour-neutral states could exist, e.g. bound states of two quarks and two antiquarks (qq $\bar{q}$ $\bar{q}$ = "4-quark states"), of four quarks and one antiquark (qqqq $\bar{q}$= "Pentaquarks"), of six quarks (qqqqqq = "di-baryons"), or of three quarks and three antiquarks (qqq $\bar{q}$ $\bar{q}$ $\bar{q}$ = "baryonium"). None of these "exotic" states has at present been identified with certainty.

Quarks interact by the exchange of gluons, which are - like the photon - massless field quanta with spin 1. Since QCD is based on the principle of gauge invariance, the existence of three colour charges ("generators of a gauge transformation") implies a non-Abelian structure of the theory. In other words, the gluons themselves carry a colour-charge, and two gluons can interact with each other. This gluon-gluon (gg) interaction is very likely the origin of "confinement", which is another way of saying that free quarks have never been observed. As an important consequence, there should be purely gluonic bound states (gg , ggg = "glueball"), or states where a gluon is bound to a q$\bar{q}$ pair (q$\bar{q}$g = "hybrid"). Unfortunately, these states can have the same quantum numbers as ordinary meson states. The observable meson states are therefore probably mixtures of mesons with glueballs and hybrids, and although the search for these new states has been going on for more than a decade now, no conclusive evidence has been obtained.

The strong coupling constant α_s (the equivalent of the electromagnetic coupling constant $\alpha_e = 1/ 137$) is large at low energies, but it decreases with increasing energy. For high momentum transfer the coupling constant is therefore small and the quarks behave almost like free particles ("asymptotic freedom"). In "perturbative" QCD, the basic Lagrangian is expanded in a power series of α_s, in analogy to QED. Perturbative QCD has been rather successful in describing total and differential cross-sections of many different "hard" processes quantitatively (Drell-Yan, deep inelastic scattering, two- and three-jet-events), where the production of "jets" reflects the fundamental quark-quark, quark-gluon and gluon-gluon scattering. The jets are a remainder of the original two-body process and they are formed by "hadronization", i.e. breaking the gluon strings between two quarks moving away from each other.

The increase of α_s towards lower energies has a good and a bad side: the good one is that three quarks can form stable nucleons, hence nuclei, matter, and finally our universe including physicists who can wonder about its origin. The bad one is that perturbation theory fails since there is no convergence for $\alpha_s = o(1)$. This fact has so far prevented QCD from calculating low energy hadronic processes, and in particular the proton wave function. In the absence of an exact theory to calculate "soft" hadronic properties , various phenomenological models have been proposed, e.g. Regge trajectories, vector dominance, duality, or chiral symmetry, but I will not go into a discussion of their respective merits. In the following I want to describe the theoretical models which are most commonly used to describe the results obtained by experiments studying antiproton-nucleon reactions.

Nucleon-nucleon interaction and meson exchange

Historically, strong interaction theory started with the description of nucleon-nucleon (NN) forces by the exchange of pions (in analogy to the QED electron-photon vertex) . The physical picture is a "bare" nucleon - today it is called the "core of valence quarks" - surrounded by a cloud of (virtual) pions. The exchange of the (pseudoscalar) pion leads in the non-relativistic approximation [2] to the potential shown in fig. 1. In practical calculations, both the nucleon and the pion are treated as pointlike particles, but since the tensor part diverges like r^{-3}, a cut-off is needed at small distances. A natural cut-off would be the structure of the nucleon and the pion, which are however not well known at these low energies. The divergence is therefore avoided by a regularization procedure. The one-pion exchange model is nevertheless very successful: it explains the size of nuclei (by the effective pion range $1/m_\pi \approx$ 1.4 fm) and the bulk of low energy nucleon-nucleon scattering data.

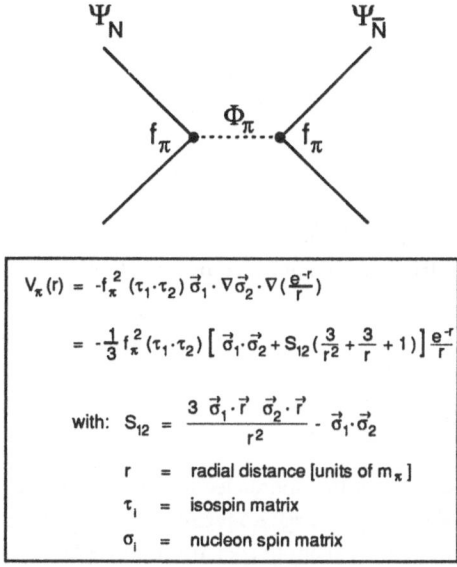

$$V_\pi(r) = -f_\pi^2 (\tau_1 \cdot \tau_2)\, \vec{\sigma}_1 \cdot \nabla \vec{\sigma}_2 \cdot \nabla\left(\frac{e^{-r}}{r}\right)$$

$$= -\frac{1}{3} f_\pi^2 (\tau_1 \cdot \tau_2) \left[\vec{\sigma}_1 \cdot \vec{\sigma}_2 + S_{12}\left(\frac{3}{r^2} + \frac{3}{r} + 1\right) \right] \frac{e^{-r}}{r}$$

$$\text{with:} \quad S_{12} = \frac{3\, \vec{\sigma}_1 \cdot \vec{r}\;\, \vec{\sigma}_2 \cdot \vec{r}}{r^2} - \vec{\sigma}_1 \cdot \vec{\sigma}_2$$

r = radial distance [units of m_π]

τ_i = isospin matrix

σ_i = nucleon spin matrix

Fig. 1 One-pion exchange potential

The potential between two nucleons must also have a repulsive part to keep protons and neutrons apart. This is obvious since the volume of nuclei is approximately equal to the sum of the volume of their constituent nucleons. The repulsive part of the potential is also needed to explain the properties of the deuteron and the NN scattering data. Encouraged by the success of the one-pion exchange potential (OPEP), the exchange of heavier mesons like the η, ρ, and ω was postulated [3]. The ω exchange leads to a repulsive NN interaction of short range ($1/m_\omega \approx 0.25$ fm) explaining the "hard core" of nucleons. The contribution of heavier mesons also allows a better adjustment to more precise experimental data, which showed discrepancies with the simple OPEP picture. Today, most models of NN interactions are based on meson exchange potentials (MEP) including the exchange of the η, ρ, and ω mesons, and - additionally - the exchange of 2 pions or of scalar mesons to parametrize exchange forces with scalar properties.

However, the physical reality of heavy meson exchange is questionable. According to recent work based on the MIT bag model and chiral symmetry [4] the proton consists of a core of quarks (r ≈ 0.7 - 0.9 fm) surrounded by a cloud of virtual pions, adding 0.1 - 0.3 fm to the core radius. Heavy particles like the ρ, ω, and the scalar mesons have interaction ranges of 0.25 fm and less, well inside the region where the quark structure of the nucleon must play an important role. If two nucleons are close enough to allow the exchange of these heavy mesons, their quark cores must significantly overlap, and the limits of the two "bags" are no longer well defined. How should one imagine the exchange of a heavy meson inside a small region filled with 6 quarks? It seems more likely that the repulsive part of the NN interaction is a consequence of the Pauli principle on the quark level, not allowing the presence of identical fermions at the same point in space.

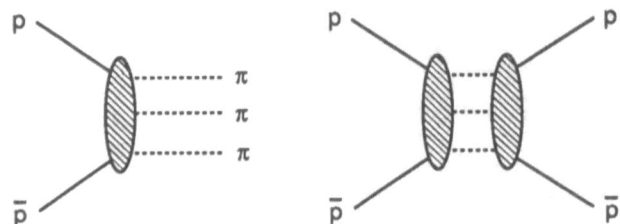

Fig. 2. Graphs contributing to the absorptive (left graph) and the real part (right graph) of the complex potential

Antinucleon-nucleon interaction

In the meson-exchange picture, the antinucleon-nucleon potential is linked to the nucleon-nucleon potential by a simple symmetry. Charge conjugation shows that the antinucleon-pion coupling constant is just (-1) times the nucleon-pion coupling. Hence, if we replace the nucleon by an antinucleon, the amplitude for any n-pion exchange process changes by $(-1)^n$. This symmetry also applies to exchanging other mesons with well-defined G-parity ("mesonic charge") [5], and the proportionality factor is $(-1)^G$. The potential generated by the exchange of particles with odd G-parity (π, ω) changes sign, while the G-even part (2π, ρ) remains unchanged. The experimental comparison of the two systems is of considerable interest for meson-exchange models since forces which cancel in the nucleon-nucleon system act coherently in the antinucleon-nucleon interaction.

A complication arises from the possibility of annihilation, which is not present in the NN system. The annihilation "potential" is of short range (about the size of the nucleon) and can lead to the transition into a mesonic final state, or to a re-scattering into the same antinucleon-nucleon final state (fig. 2). It is therefore described by a complex ("optical") phenomenological potential with an absorptive (imaginary) and an elastic (real) part. Since many annihilation channels contribute, the overall spin- or isospin-dependence is assumed to be negligible. For calculations, the Schroedinger equation is solved numerically, and the phase shifts of the partial waves are used to determine the cross-sections.

The first calculations along these lines (using the WKB approximation, since computers were not yet available) were carried out by Ball and Chew [6] in 1957. They could reproduce the observed large ratio of 2:1 for the annihilation to the elastic scattering cross-sections. The annihilation was parametrized by an absorptive boundary condition at a radius near 0.4 fm. The Ball-Chew model was improved in 1968 by Bryan and Phillips [7]. Their approach was based on the Bryan-Scott potential which successfully described NN scattering. It included the exchange of the π, η, ρ, ω and scalar mesons "σ_0", "σ_1" to parametrize non-resonant exchanges with the scalar quantum numbers and isospin 0 or 1. The annihilation potential was of short-range ($r \approx 0.2$ fm) and purely absorptive; its real part was put in the heavy meson exchange potential. Their work was further refined by Dover and Richard [8]. Dalkarov and Myhrer assumed meson exchanges at large distances and a black sphere absorption at short distances ($r \approx 0.5$ fm) [9].

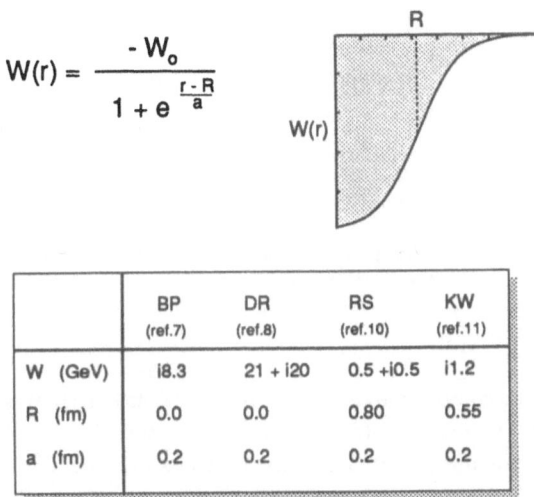

$$W(r) = \frac{-W_o}{1 + e^{\frac{r-R}{a}}}$$

	BP (ref.7)	DR (ref.8)	RS (ref.10)	KW (ref.11)
W (GeV)	i8.3	21 + i20	0.5 + i0.5	i1.2
R (fm)	0.0	0.0	0.80	0.55
a (fm)	0.2	0.2	0.2	0.2

Fig. 3. Comparison of different parametrizations of the annihilation potential

More recent work [10, 11] has concentrated on a better understanding of the influence of the various parts of the potential on the predictions. Fig. 3 shows a comparison of the different parametrizations of the annihilation potential. Since the wave function is strongly suppressed for $r < 0.5$ fm, different parametrizations in the region of smaller radii do not lead to significantly different results. In models motivated by the baryon-exchange picture of the annihilation, the annihilation has a very short range of $1/2m_p \approx 0.1$ fm [12]. Therefore, the annihilation potential has to be very deep [10 ... 20 GeV] to reproduce the data. Other models start from a slowly varying annihilation potential [300 ... 500 MeV deep] inside the nucleon ($r \approx 0.8$ fm), decreasing exponentially with the slope parameter of 0.2 fm for $r > 0.8$ fm.

The physical meaning of the annihilation range as a uniquely defined quantity has been questioned by Richard and co-workers [13]. They argue that the range depends on the considered reaction: in exclusive reactions like $\bar{p}p \rightarrow \Phi\Phi$, where all initial $\bar{q}q$ pairs annihilate, the annihilation occurs at a much shorter separation than for the reaction $\bar{p}p \rightarrow 3\pi$, which can be generated by the rearrangement of the three initial $\bar{q}q$ pairs (fig. 4). The annihilation radius as defined by a geometrical parameter where a certain percentage of the wave function is absorbed [14] would then only be an average over many different radii with a weight proportional to the relative branching ratio of the specific channel.

"Complete" annihilation: very short range

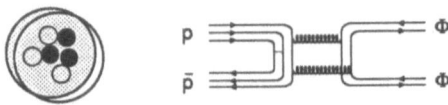

"Rearrangement" : long range

Fig.4. Dependence of the annihilation radius on the annihilation channel

Finally, an interesting comparison has been done by Shibata [15] showing that elastic scattering and charge exchange data at low momenta are fairly well described by one-pion exchange alone, together with an optical annihilation potential which is only slightly different from those used in the other models.

II   SHORT HISTORY OF EXPERIMENTS WITH ANTIPROTONS

The beginning

The history of antiproton physics would fill several books, and the following chapter only highlights some of the important episodes and discoveries associated with antiproton experiments. In the absence of powerful particle accelerators, elementary particle physics in the late 1940's and 1950's was dominated by cosmic ray experiments. These experiments were often done in scenic locations, e.g. on top of Swiss or French mountains, so that the experimenters did not mind waiting for the rare occasion when events with sufficiently high energy produced new "elementary" particles. Using cloud chambers or photographic emulsions, most of the strange particle "zoo" was discovered in this way between 1950 to 1955, namely the Λ°, the Σ^+, the Ξ^-, the K°, and five different decay modes of the K^+. However, as far as the antiproton was concerned, no sufficient evidence for a positive identification was obtained.

The Cosmotron (3 GeV/c protons, Brookhaven, 1952) and, in particular, the Bevatron (6.2 GeV/c protons, Berkeley, 1954) ended the great times of cosmic ray physics. The antiproton was finally discovered in autumn 1955 by Chamberlain, Segrè, Wiegandt and Ypsilantis [16] . They steered the 6.2 GeV/c proton beam of the Bevatron on a carbon target, and antiprotons were produced in the reaction $p + C \rightarrow p + (p + \bar{p}) + X$, where X denotes the debris of the carbon nucleus. The mass spectrograph used to identify the antiprotons is shown in fig. 5. It consisted of several scintillators, Cerenkov counters, dipole and quadrupole magnets. The apparatus was used to measure the mass , the charge and a lower limit of the antiproton life time (> 100 ns), all agreeing with expectation within the experimental error. The original apparatus delivered an antiproton "beam" of one transmitted antiproton of momentum 1.19 GeV/c per $1.6 \cdot 10^{12}$ protons impinging on the target, accompanied by a very heavy background flux of pions (500,000 π : 1 $\bar{p}$).

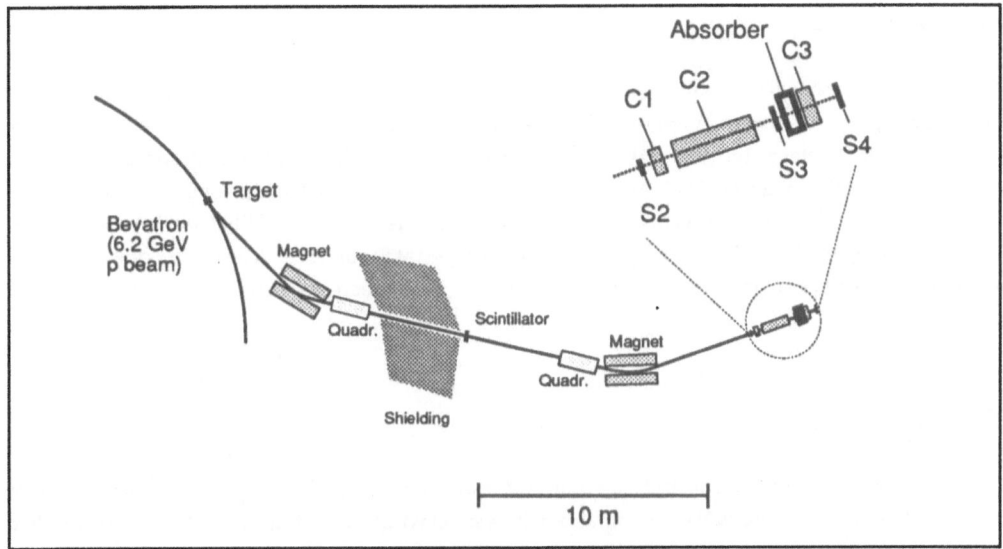

Fig. 5.   Mass Spectrometer of Segrè et al. used for
the discovery of the antiproton

In the following years, the cross-sections for elastic, inelastic, charge exchange scattering and annihilation were measured at fixed angles for $\bar{p}$ momenta between 500 and 1150 MeV/c. The experimental error was about 10-50 %, and simple arrangements of Cerenkov counters and scintillators were used. The salient fact emerged that antiproton reactions have much larger cross-sections than the corresponding proton reactions, and that the annihilation cross-section exceeded the elastic cross-section by a factor 2. However, the most interesting new property of antiprotons was annihilation. By 1958, about 220 annihilation stars had been observed and analyzed, mainly in emulsions, and a lot of useful information on the antiproton-nucleus interactions was derived, among them the average multiplicity of pions produced in antiproton-proton annihilation (5.3 ± 0.4), and the low yield of charged K mesons (3.5 ± 1.5 %). In 1959, the $\bar{\Lambda}$ particle was discovered using a hydrogen bubble chamber to stop and to annihilate antiprotons. For more information on the childhood of antiproton physics, I recommend the review by E. Segrè [17].

The bubble chamber era

In 1960, two new powerful proton accelerators were commissioned in Brookhaven (AGS, 30 GeV) and at CERN (PS, 25 GeV). Both laboratories decided to launch their own antiproton programme. Due to the higher production momentum, the yield of antiproton production and the purity of the beams was significantly improved, and antiproton beams in the momentum range 0.4-20 GeV/c became available.

Two different types of experiments were done: high statistics counter experiments measuring cross-sections with a relatively simple setup, and bubble chamber experiments reconstructing charged particle tracks and measuring their momentum from the curvature in a magnetic field. Bubble chamber experiments gave precise informations about the final state of annihilation, but the tedious scanning and reconstruction procedure (several minutes per picture) limited the available statistics. In the time between 1960 and 1972, only about 100,000 events were completely reconstructed and analyzed. 1.5 million pictures were scanned for visible $K_s \rightarrow \pi^+\pi^-$ decays to obtain a data sample of about 80,000 events with kaonic final states.

In spite of the tedious analysis procedure, bubble chamber results gave important contributions to the development of elementary particle physics. Among these were the discovery of the ω meson and the E(1420) [see chapter 5], the precise determination of the width of the Φ and the ω meson, and the spin-parity determination of the D(1285). An interesting episode was the clarification of the vector nature of the K*(890), which was the subject of a controversy between the groups at CERN (spin 1) and Berkeley (spin 0). In 1962, the charged Σ antihyperons were discovered at Brookhaven. The review of Armenteros and French [18] gives an overview about the experimental and theoretical development between 1958 and 1968.

The hunt for baryonium

In 1966, an experiment studying pion scattering on nuclei [19] had found evidence for a new meson state not far from the NN threshold, but with a small width of less than 30 MeV. In 1970, another experiment [20] reported even more narrow peaks (fig. 6) in the same mass region. This caused some excitement, since strongly bound antinucleon-nucleon states ("baryonium") were predicted owing to the attractive short-range part of the antinucleon-nucleon potential. This prediction was supported by "duality", a model based on the quark-picture of the annihilation (fig. 7) and associating the exchange of a resonant state in the t-channel with a bound or resonant state in the s-channel. For more details on baryonium and duality, see ref. 21.

The narrowness of the new states remained however puzzling, since the large annihilation rate should lead to widths of several hundreds of MeV. Therefore, new models were constructed explaining the narrow widths e.g. by internal selection rules originating from the exotic quark configuration of these objects.

In the 1970's, the hunt for narrow baryonium states became fashionable. There were two main strategies: a) to measure the total and differential elastic, charge exchange and annihilation cross-sections, or b) to search for narrow peaks in inclusive π° and γ spectra, indicating the transition from the initial state into a baryonium state by the emission of a π° or γ. The initial state of enthusiasm about the

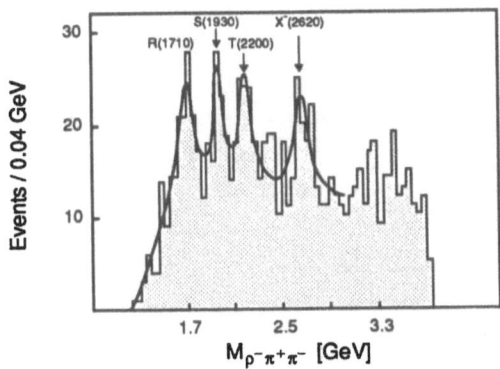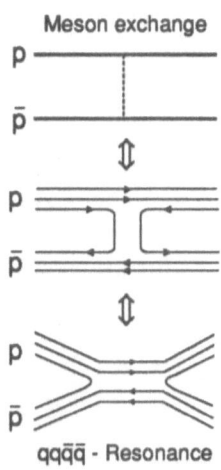

Meson exchange

$qq\bar{q}\bar{q}$ - Resonance

Fig. 6. Evidence for "narrow baryonium"
(from ref. 20)

Fig. 7. "Duality" argument for the existence of 4-
quark resonances

the multitude of "observed" states (baptized R, S, T, U, ...) was soon followed by a more cautious approach, since many of the initial observations were not confirmed by subsequent experiments. The states were usually "observed" with a statistical significance of 3-4σ, also because the quality of the available antiproton beams (intensity, momentum spread, purity) was insufficient to accumulate large enough data samples to settle this question once and for all. Two statements from review talks on antiproton-nucleon conferences illustrate the slow progress. In 1972 : "There probably exist several objects in the S region [≈1940 MeV]... This agrees with some theoretical ideas suggesting multipole resonances on the A_2 trajectory or quasi-nuclear structures at about M = 2 m_p." [22]. 10 years later, in 1982: "The existence of a narrow $\bar{p}p$ state at M = 1935 MeV seen in some formation and production experiments has not been confirmed by later ones. However, the situation concerning the existence of this state is still unsettled... Additional information is clearly still needed " [23].

LEAR

By 1980, high energy physics had developed far beyond the 1-5 GeV region and had crossed the charm and the beauty threshold. A new project at CERN started in 1977 with the aim to collide antiproton and proton beams inside the SPS, which had been constructed as a 400 GeV/c proton accelerator. The "S$\bar{p}$pS" collider would reach center-of-mass energies of 620 GeV for $\bar{p}$p collisions, providing for the first time sufficiently high energies to produce the elusive W and Z bosons, the carriers of the weak interaction and massive "brothers" of the photon. As we know today, the idea was very successful.

A necessary condition for the realization of this project was the invention of "stochastic cooling". Since this technique is very useful to provide high intensity antiproton beams at high and at low energies, I will give a short introduction to stochastic cooling in the following paragraph.

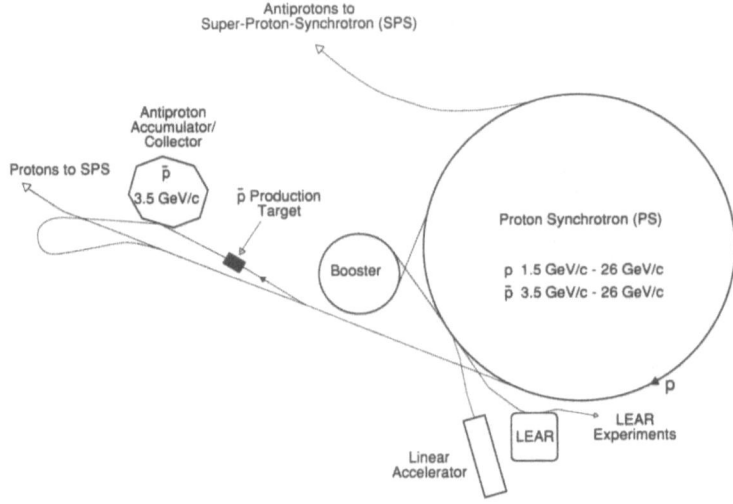

Fig. 8. Antiproton complex at CERN

The layout of the CERN antiproton complex is shown in fig. 8. Protons are accelerated to 26 GeV/c in the PS and are used to produce antiprotons in a copper target. The antiproton accumulator (AA) ring accepts a batch of these with momenta around 3.5 GeV/c every 2.4 seconds. The antiprotons are produced in a broad angular and broad momentum range ("large phase space volume"). Therefore the AA must catch the antiprotons coming from the target with a large acceptance, much larger than that of the SPS ring, where the antiprotons are finally stored. Before a transfer of the antiproton beam between the AA and the SPS is possible, the large oscillations of the particle trajectories around the central orbit in the AA must be reduced ("cooled"), i.e. the particle density in the central region of phase space must be increased.

The cooling of a single particle circulating in a ring is achieved as follows (fig. 9). Under the influence of the focusing fields, the particle executes betatron oscillations around its central orbit. At each passage of the particle a so-called differential pickup provides a short pulse signal that is proportional to the distance of the particle from the central orbit. This is amplified and applied to the kicker, which will deflect the particle. If the distance between pick-up and kicker contains an odd number of quarter

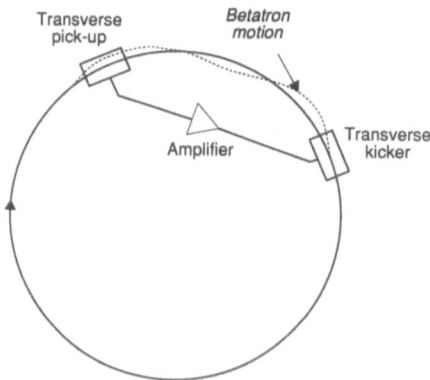

Fig. 9. Principle of (transverse) stochastic cooling

betatron wavelengths, and if the gain is chosen correctly, any oscillation will be cancelled. The signal should arrive at the kicker at the same time as the particle: because of delays in the cabling and amplifiers, the signal path must cut off a bend in the particle's trajectory. In practice, antiprotons circulate in bunches each containing a very large number of particles. Nevertheless, each particle's individual signal will still contribute to the cooling, but the correction now takes care of the average position of the bunch. Therefore the technique is called "stochastic" cooling, since the cooling process acts on the ensemble of the particles and leads to a decrease of the "temperature" of the beam instead of correcting each single orbit separately. For more details I refer to the article of Simon van der Meer [24], who received the Nobel Prize in 1984 for the development of this technique.

The " Low Energy Antiproton Ring (LEAR) " (fig. 10) was proposed in 1979 by a community of enthusiastic physicists envisaging a broad range of experiments with $\bar{p}$ beams of previously unknown quality [25, 26] . The AA would be capable of accumulating 10^{11} $\bar{p}$ per day at a momentum of 3.5 GeV/ c. The LEAR project involved the construction of a small stretcher ring (78.5 m circumference) with three main features: a stochastic cooling system to cool the antiproton beam at low momenta, a RF system to vary the $\bar{p}$ momentum between 100 and 2000 MeV/c, and an "ultraslow" ejection system allowing the extraction of typically $3 \cdot 10^9$ antiprotons over 1 hour. The improvement of the beam quality compared to previous antiproton beams was enormous: a beam purity of 100%, since all other particles had decayed long before leaving LEAR, a very small emittance, a momentum spread $\Delta p/p$ of 10^{-3} instead of 10^{-2} , and an intensity of 10^6 s^{-1} instead of less than 10^3 s^{-1}. In addition, owing to the achievable low momenta it became possible to stop antiprotons in targets of low densities (see chapter 5), allowing the development of new experimental techniques. More details about the initial performance of the LEAR ring can be found in [27].

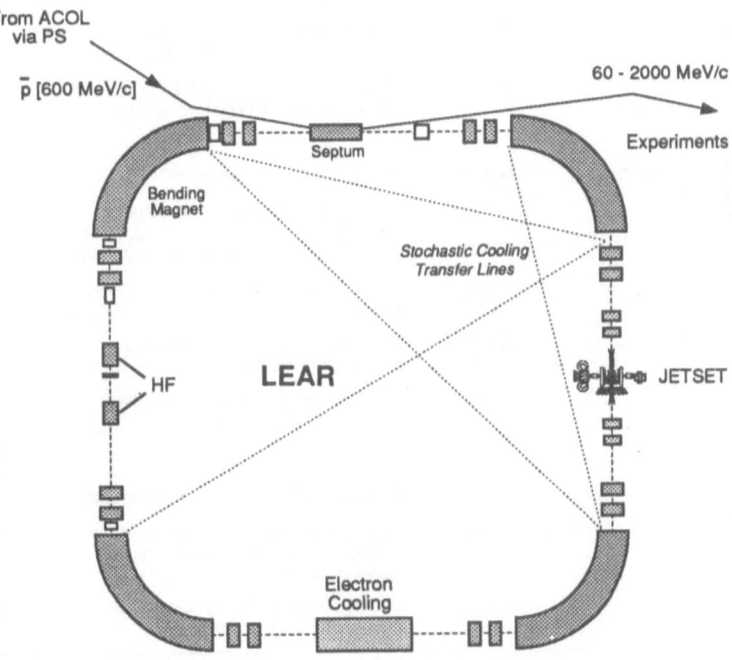

Fig. 10.   Schematic design of the LEAR machine (Status
1990, including the installation of the Jetset detector)

The physics motivation for LEAR was very broad. Clearly, the still open question of the existence of baryonium was one of the main arguments . However, many other experiments were proposed, ranging from meson spectroscopy, exclusive two-body annihilations and the measurement of the electromagnetic form factor of the proton in the time-like region to atomic and nuclear physics experiments, like the search for K X-rays from $\bar{p}p$ atoms, or the study of antiproton-induced fission. A list of the 15 (!) experiments approved at LEAR between 1983 and 1987 is given in fig. 11. In the following chapters, I want to illustrate the objectives, the detectors and the results of the initial LEAR programme by describing three experiments in more detail. These experiment stand representative for similar experiments, and the selection represents a personal choice. For more comprehensive reviews, I recommend the references 14 and 28–32.

Field	Experiment Code	Experiment Title
Particle Physics $\bar{N}N$ Interaction	PS 170 (APPLE)	Precision measurement of the proton electro-magnetic form factors in the time-like region and vector meson spectroscopy
	PS 171 (ASTERIX)	Study of $\bar{p}p$ interactions at rest in a hydrogen gas target
	PS 172 (SING)	$\bar{p}p$ total cross-sections and spin effects in $\bar{p}p \rightarrow K^+K^-$, $\pi^+\pi^-$ above 200 MeV/c
	PS 173	Measurement of $\bar{p}p$ cross-sections at low $\bar{p}$ momenta
	PS 182	Investigations on baryonium and other rare $\bar{p}p$ annihilation modes using high-resolution π° spectrometers
	PS 183	Search for bound $\bar{N}N$ states using a precision gamma and charged pion spectrometer
	PS 185	Study of threshold production of hyperon-antihyperon pairs in $\bar{p}p$ interactions
Nuclear Physics	PS 177	Search for heavy hypernuclei
	PS 179	Study of the interaction of low-energy antiprotons with Deuterium, He-3, He-4 and Neon nuclei using a streamer chamber
	PS 184	Study of antiproton-nucleus interaction with a high-resolution magnetic spectrometer
	PS 186	Nuclear excitations by antiprotons and antiprotonic atoms
	PS 187	A good statistics study of antiproton interactions with nuclei
Atomic Physics	PS 174	Precision survey of X-rays from $\bar{p}p$ ($\bar{p}d$) atoms
	PS 175	Measurement of the antiprotonic Lyman and Balmer X-rays of $\bar{p}H$ and $\bar{p}D$ atoms at very low target pressures
	PS 176	Study of X-ray and γ-ray spectra from antiprotonic atoms

Fig.11. List of approved LEAR experiments taking data between 1983 and 1987

The experiment PS 173 was designed to measure with high precision the elastic, charge exchange and annihilation cross-sections for antiproton-proton scattering at momenta between 180 and 600 MeV/c. Before LEAR, only few data with low statistics existed below 400 MeV/c. In particular, a search for small bumps in the region of the so-called "S meson" (E_{cm} = 1938 MeV, $p_{\bar{p}}$ = 500 MeV/c) was proposed. Some of these objectives were also pursued by experiment PS 172, and the interested reader is referred to ref. 33. In addition, PS 172 used a polarized target to study the spin-dependence of the $\bar{p}p$ interactions, in particular by measuring the analyzing power for specific two-body annihilations. The very interesting results have recently been reviewed [34].

Experimental considerations

In the following paragraph, some basic considerations for the design of the PS 173 detector are outlined. The detector consisted of a liquid hydrogen target surrounded by scintillators, a forward hodoscope, multi-wire proportional chambers and antineutron counters (fig. 12). The event trigger (and hence the normalization with respect to the incoming flux of antiprotons) was provided by a coincidence of two scintillators before the entrance to the target. At 500 MeV/c, the total cross-section for $\bar{p}p$ reactions is 174 mb, giving an interaction probability (n·σ) of 0.36 % per centimeter of liquid hydrogen. A target size of 2 cm was used for momenta above 220 MeV/c. Since the cross-sections increase with decreasing momentum (approximately: σ [mb] ≈ 54/p_{lab} [GeV/c] + 66 mb), the target size was reduced to 0.7 cm for p < 220 MeV/c to reduce multiple Coulomb ("Molière") scattering which smears out the differential cross-section at small angles.

For elastic scattering, two cases have to be considered. If the antiproton is scattered in the forward direction, the recoil proton usually stops in the target, and the event signature is a single antiproton identified by TOF and its energy loss information from the forward hodoscope. For large angle scattering of the antiproton, the proton is knocked in the forward direction and identified by TOF and dE/dx, while the antiproton annihilates in the target. The annihilation produces several pions which are detected in the plastic scintillators surrounding the target.

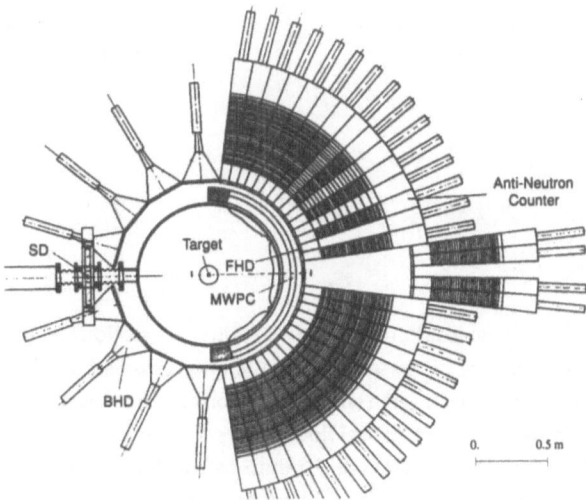

Fig. 12. Detector of experiment PS 173

In the charge-exchange reaction (CEX), the final state consists of an antineutron-neutron pair, which give no signal when traversing the scintillator box around the target. However, the antineutron can annihilate in the antineutron counters, and by appropriate corrections for solid angle and detection efficiency the CEX cross-section is determined. In the case of annihilation, several charged pions are produced and detected in the scintillator box, and no heavy charged particle (p or $\bar{p}$) is in the final state.

For the measurement of the differential cross-section, the direction of the beam axis has to be compared with the frequency of antiproton scattering under a certain angle. The beam axis was determined by a careful geometrical alignment of the beam line and the measurement of its position by a cylindrical MWPC behind the target. The scattering angle was determined from the coordinates of the hit in the cylindrical MWPC and the geometrical center of the target. The angular resolution was better than 1 degree.

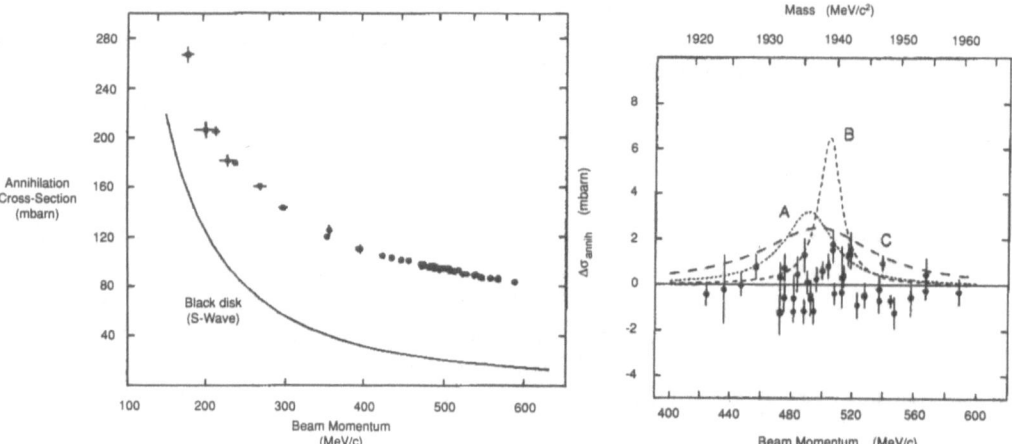

Fig. 13. Antiproton-proton annihilation cross-section at low momentum

Fig. 14. Scan of the "S meson" region and comparison with previous results

Results

In the following section the results on the annihilation cross-section and the total and differential elastic-scattering cross-section are discussed. The annihilation cross-section σ_{ann} was measured with high statistics [35]. Fig. 13 shows that σ_{ann} is a smooth function of momentum, and no sign of a resonance is observed within the limits given by statistical fluctuations. The region between 400 to 600 MeV/c was scanned in very small momentum steps to look for the S meson, but no structure was found and only an upper limit of 5 mb·MeV was given. It is of historical interest to compare the results of previous experiments claiming evidence for the S meson (fig. 14). With the beginning of LEAR operation, the last and "best established" of the now infamous narrow baryonium resonances had to be buried.

The annihilation cross-section as a function of momentum can be compared with the simple "black disk" model, describing the $\bar{p}p$ interaction by a totally absorbing black disk with radius R, for which the inelastic and the elastic cross-sections are equal:

$$\sigma_{el} = \sigma_{inel} = (\pi/k^2) \, \Sigma \, (2l+1)$$

As shown in fig. 13, the annihilation cross-section exceeds by far the limit of $\pi(1/k)^2$ for S-wave annihilation alone. Therefore a strong P-wave contribution must already exist at 180 MeV/c. This is different from nucleon-nucleon scattering where the P-wave contribution to the total cross-section is less than 10 % at 300 MeV/c.

The measurement of the elastic cross-section as a function of momentum [36] yields two more important informations: the partial wave composition (mainly S- and P-wave) obtained from the angular distribution, and the "ρ parameter", derived from the interference of Coulomb and nuclear scattering (see below). The angular distributions for elastic scattering at three different momenta are shown in fig. 15. The solid, dashed and dash-dotted curves represent model calculations [37, 38, 39]. It is interesting to note that a simple one-pion exchange potential plus an appropriately chosen optical annihilation potential [15] yields comparable results. The most outstanding feature of the data is the strong forward enhancement even at small momenta (which must however not be confused with the enormous peak at $\Theta < 10°$ due to Coulomb scattering). This indicates a large contribution of $l \geq 1$ partial waves (in agreement with the annihilation data) since a pure S-wave should yield an isotropic distribution. A partial wave analysis shows indeed a strong P wave contribution down to low momenta, while D-wave amplitudes are small below 300 MeV/c.

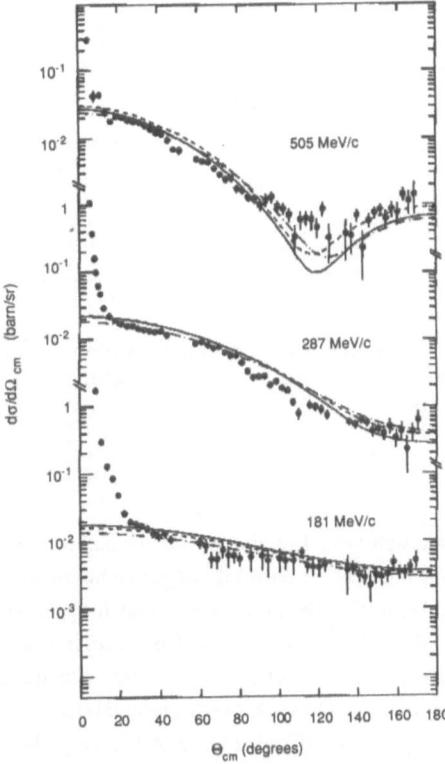

Fig.15. Differential elastic $\bar{p}p$ scattering cross-section at
three different momenta (PS 173)

The ρ parameter is defined as the real-to-imaginary ratio of the forward elastic scattering amplitude $[\rho = \mathrm{Re}\, f_N\, (t=0)\, /\, \mathrm{Im}\, f_N\, (t=0)]$ and is sensitive to resonances in the $\bar{p}p$ system. The cross-sections are proportional to the square of the scattering amplitudes, and the ρ parameter has to be extracted from an interference term. Since Coulomb scattering dominates for very small angles ($\sim 1/\sin^4\theta/2$) and nuclear scattering for large angles, the Coulomb-nuclear interference term is extracted from the angular region between 10 .. 30°. Fig. 16a shows the cross-section at 181 MeV/c together with three fits corresponding to a ρ parameter of +1, -1, and 0.1, the latter being the best fit to the data. The formulae in fig. 16b list the different contributions from nuclear and Coulomb scattering which enter in the calculation of the elastic cross-section.

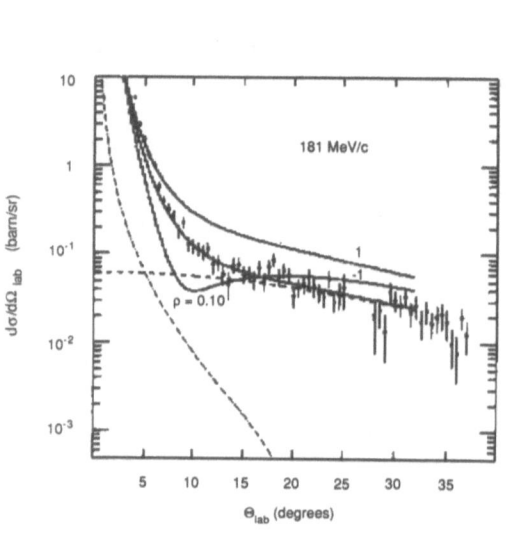

$$\frac{d\sigma}{dt} = |f_c + f_N|^2 = \frac{d\sigma_c}{dt} + \frac{d\sigma_N}{dt} + \frac{d\sigma_I}{dt}$$

$$\frac{d\sigma_c}{dt} = \frac{1}{\pi}\left(\frac{\alpha\, hc}{bt}\right)^2 F(t)^2 \sim \frac{1}{\sin^4\frac{\theta}{2}} \quad \text{[Rutherford]}$$

$$\frac{d\sigma_N}{dt} = \frac{\pi}{4}\left(\frac{\sigma_{tot}}{hc}\right)(1+\rho^2)\, e^{-bt}$$

$$\frac{d\sigma_I}{dt} = \frac{\sigma_{tot}}{\beta t}\, F(t)\, e^{-\frac{b}{2}t}(\rho\cos\delta_c - \sin\delta_c)$$

$$\rho \equiv \frac{\mathrm{Re}\, f_N\, (t=0)}{\mathrm{Im}\, f_N\, (t=0)}$$

δ_c = Coulomb scattering phase shift
$$= -\frac{\alpha}{\beta}\{\ln[9.5t] + 0.57\} \approx 0.1$$

$F(t)$ = electromagnetic form factor of proton
$$= \frac{1}{(1+\frac{t}{0.71})^4} \approx 1 \quad (t \leq 0.01\ \mathrm{GeV}^2)$$

Fig. 16a. Differential elastic cross-section at 181 MeV/c and comparison with fit for ρ = -1., +1., and +0.10 (best fit)

Fig. 16b. Contributions from Coulomb and nuclear scattering to the elastic cross-section

The behaviour of the ρ parameter as a function of the momentum is peculiar [fig. 17]. At high momentum, ρ is slightly greater than zero, becoming negative below 400 MeV/c, but rising again towards (and beyond?) zero at 200 MeV/c. The asymptotic value for p=0 is known from the measurement of the the shift and width of the 1s level of protonium (see chapter 5): $\rho = -1.08 \pm 0.09$! This "oscillation" of the ρ parameter has led several authors to the conclusion that a strongly bound $\bar{p}p$ state exists below threshold [40,41]. More data below 100 MeV/c would be of great importance to clarify the question, but measurements become very difficult owing to the very short range of antiprotons in scintillators and other detector materials at such low momenta .

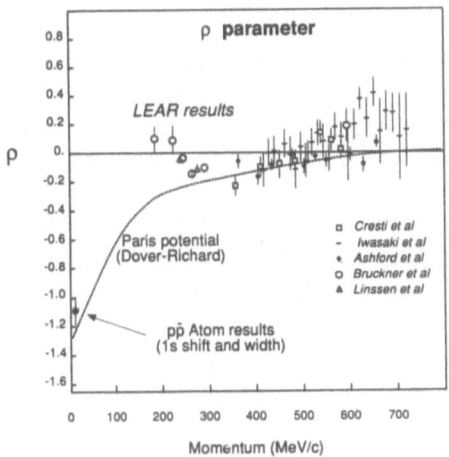

Fig. 17. Measurements of the ρ parameter for p̄p
elastic scattering at low momentum

IV HYPERON-PAIR PRODUCTION AT THRESHOLD

The aim of experiment PS 185 was the study of hyperon-antihyperon production at threshold, making use of the low emittance and the high intensity of the LEAR beam. Before describing the detector in more detail, I want to give an overview about the physics motivation.

Physics motivation

The hyperons (Λ, Σ, Ξ) belong to the same SU(3) octet with spin 1/2 and positive parity as the proton and the neutron, but the hyperons contain one or two strange quarks (fig. 18). The Λ is a (uds) state, and the graph in fig. 19 illustrates the production of $\bar{\Lambda}\Lambda$ pairs from p̄p annihilation. It should be remembered that the quark-line graph is only an illustration, but not a prescription for calculating the transition amplitude of the depicted process (unlike Feynman graphs). The reaction involves the creation of a strange quark pair inside a hadronic "bag", and a systematic study of hyperon pair production could improve our understanding of the basic operators involved in quark-pair production and annihilation.

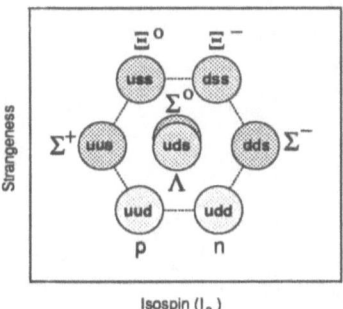

Fig. 18. The 1/2$^+$ baryon octet with the
nucleon and the hyperon family

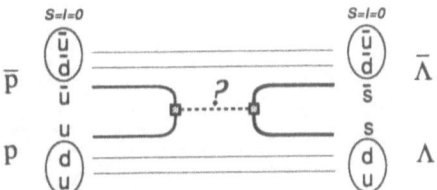

Fig. 19. Annihilation graph contributing to $\bar{\Lambda}\Lambda$ pair
production in p̄p annihilation

In a simple approximation, called the "diquark spectator model", the up- and down-quark pairs are assumed to be only spectators. During the reaction, they remain in their spin and isospin zero state required by SU(3) symmetry. The quantum numbers of the final $\bar{\Lambda}\Lambda$ state are then determined by those of the $s\bar{s}$ quark pair. If the effective $\bar{q}q$ creation operator has vacuum quantum numbers ("$^3P_0$ model" corresponding to a multi-gluon intermediate state), the quark-pair and hence $\bar{\Lambda}\Lambda$ is expected to be in a relative P-wave (L=1), while the one-gluon exchange model predicts a relative S-wave (L=0, "$^3S_1$ model"). The production cross-section for a two-body final state close to threshold is determined by phase space. For pure S-wave production, the cross-section rises proportional $\varepsilon^{1/2}$ (ε = excess energy = $E_{cm} - 2m_\Lambda$), while for pure P-wave production ~ $\varepsilon^{3/2}$. The correlation of the Λ and the $\bar{\Lambda}$ spin reflect the $s\bar{s}$ spin correlation. The spin-correlation is measurable, since the $\Lambda \to p\pi$ decay is self-analyzing: owing to parity violation in weak interaction, the proton tends to go into the direction of the Λ spin, so that the direction of the decay proton relative to the Λ production plane correlates with the spin direction.

Experimental considerations

The PS 185 detector (fig. 20) accounted for the reaction kinematics, the low production cross-section and the need for a momentum scan around the threshold in small steps. A thorough study of the threshold region requires sensitivities in the 1 µb region, and a comparison with the total cross-section (≈100 mb at the threshold momentum of 1435 MeV/c), shows that a selectivity of 1:10^5 is required.

The Λ decay length is between 2 and 10 cm, depending on the production momentum and the angular distribution. Since the target had a smaller dimension, the first level trigger required that no charged particles leave the target region, reducing the event rate by one order of magnitude. Both Λ and $\bar{\Lambda}$ decays into charged final states were selected, therefore four charged particles had to be detected in the forward hodoscope. Since most of the neutral channels (except K_sK_s, which is rare) do not decay into a four charged particle final state, this trigger reduced the background by another two orders of magnitude. The four charged particles were then tracked in a stack of 10 MWPC and 13 DC planes, allowing the determination of their directions and the reconstruction of the secondary and the primary decay vertices. At the downstream end of the detector, three drift chambers inside a solenoidal magnetic field were used to determine the charge of the particles by the curvature of their tracks.

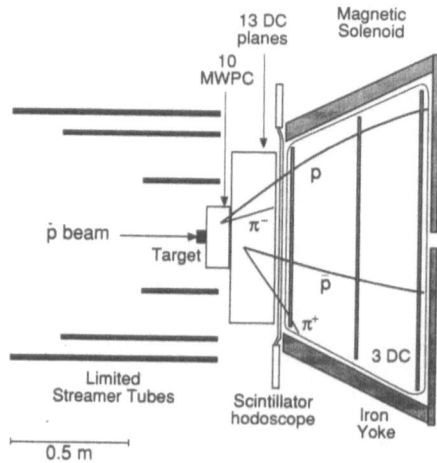

Fig. 20 Detector of experiment PS 185

The momentum fine-scan was achieved in a very elegant way: the target (poly-ethylene) was divided into four segments, each surrounded by a box of scintillator material (fig. 21). The momentum loss of the beam in each target segment was 0.8 MeV/c, and by reconstructing the primary production vertex, the incoming momentum of the antiproton for this event was known. In this way, four different momenta could be studied at the same time using one LEAR momentum setting.

<u>Results</u>

Fig. 22 shows the $\Lambda\bar{\Lambda}$ production cross-section as a function of energy in the threshold region. The dotted and the dash-dotted lines indicate the respective S- and P-wave contributions. Surprisingly, the data require a P-wave contribution already for $\varepsilon \geq 0.8$ MeV. The strong P-wave contribution near to threshold is confirmed by the forward peak in the differential cross-sections (fig. 23), already appearing for $\varepsilon < 0.8$ MeV. Pure S-wave distributions are flat. The forward peak becomes more pronounced at higher energies, showing that the relative P-wave contribution increases.

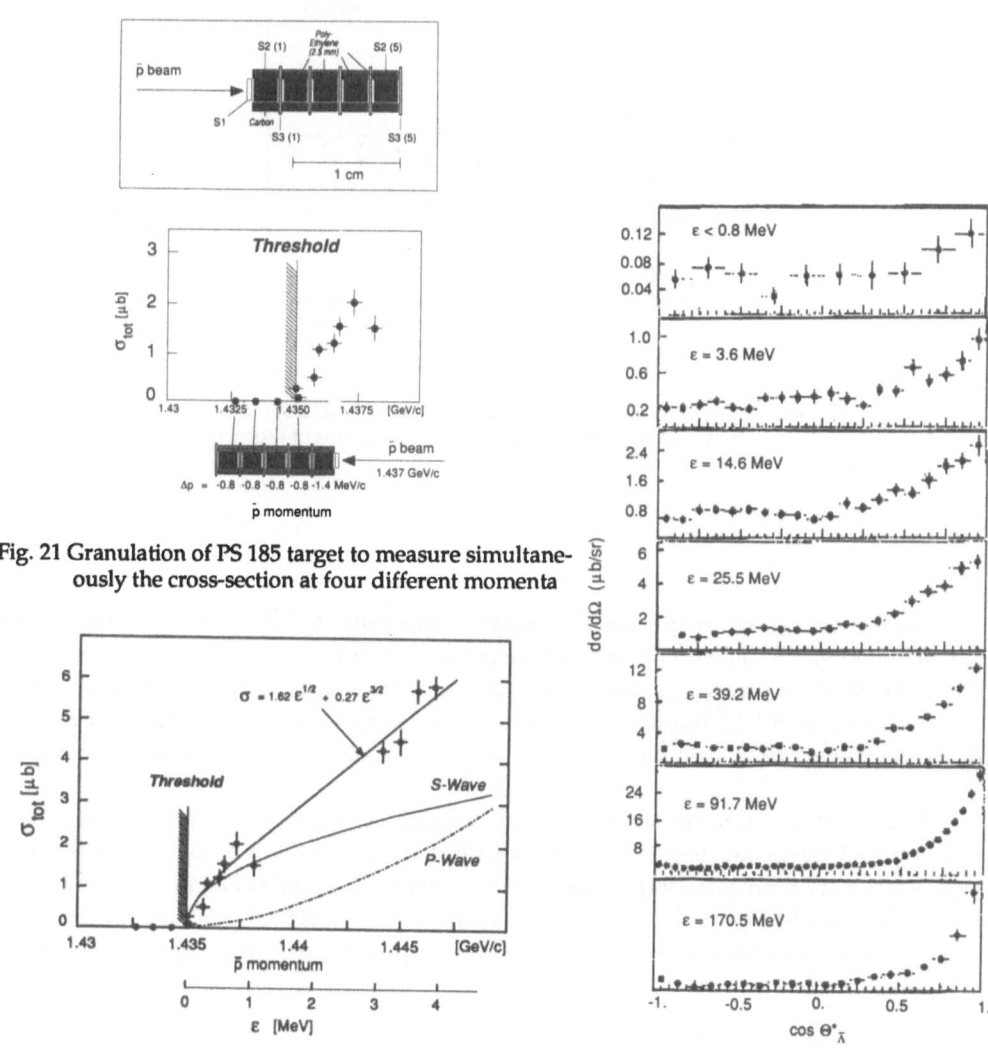

Fig. 21 Granulation of PS 185 target to measure simultaneously the cross-section at four different momenta

Fig. 22 Cross-section for $\bar{p}p \to \bar{\Lambda}\Lambda$ close to threshold

Fig. 23 Differential cross-section for $\bar{p}p \to \bar{\Lambda}\Lambda$ from threshold to $\varepsilon = 170$ MeV

137

The correlation between the two hyperon spins is described by the "singlet fraction":

$$F_s = 1/4 \ [\ 1 - (\sigma_\Lambda \cdot \sigma_{\bar{\Lambda}})\ ] = \begin{matrix} 0 & \text{for pure triplet} \\ 1 & \text{for pure singlet} \\ 1/4 & \text{for uncorrelated spins} \end{matrix}$$

The singlet fraction is consistent with zero at all measured momentum transfers. Fig. 24 shows the results for p = 1546 and 1695 MeV/c, leading to the conclusion that $\Lambda\bar{\Lambda}$ pairs are produced in a pure triplet state, independent of the c.m. energy.

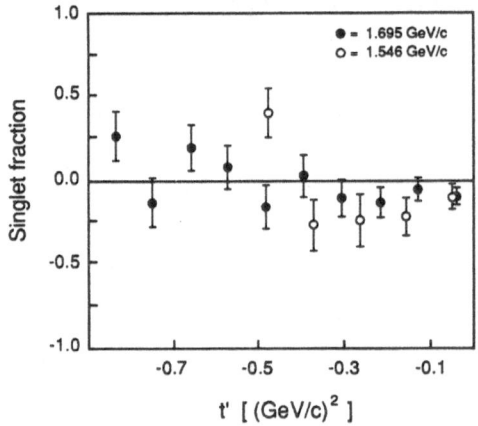

Fig. 24.   Singlet fraction of the $\bar{\Lambda}\Lambda$ spin correlation as a function of the reduced momentum transfer, at two different momenta

The high quality of the data imposes strong constraints on theoretical models describing $\Lambda\bar{\Lambda}$ production. Unfortunately, microscopic models incorporate too many unknowns to give reliable results [42]. Most of the theoretical work still relies on MEP models based on the exchange of the kaon and its excited states [43,44,45,46]. However, since the K and K* exchange have typical ranges of 0.4 and 0.22 fm, the uncertainties due to short-range effects (annihilation potential, regularization) are considerable.

It is interesting to note that like in $\bar{p}p$ elastic scattering, a strong P-wave contribution is present close to the $\Lambda\bar{\Lambda}$ threshold. Shapiro and co-workers [47] claim - on the basis of coupled channel calculations - that this provides clear evidence for the existence of a strongly bound antibaryon-baryon state with L=1. It will be very interesting to see how the future results of PS 185 on $\Lambda\bar{\Sigma}$ and $\Sigma\bar{\Sigma}$ production compare to those from $\Lambda\bar{\Lambda}$, since the spin and isospin composition of the "spectator" diquarks will be different.

The experiment PS 171 ("ASTERIX") involved various aspects of atomic physics, annihilation dynamics and meson spectroscopy. The detector was therefore a multi-purpose instrument incorporating a gaseous hydrogen target at NTP, an X-ray Drift Chamber, and a magnetic spectrometer to measure the momentum of final state mesons (fig. 25). A detailed description can be found in ref. 48.

ASTERIX SPECTROMETER
(PS 171 at LEAR)

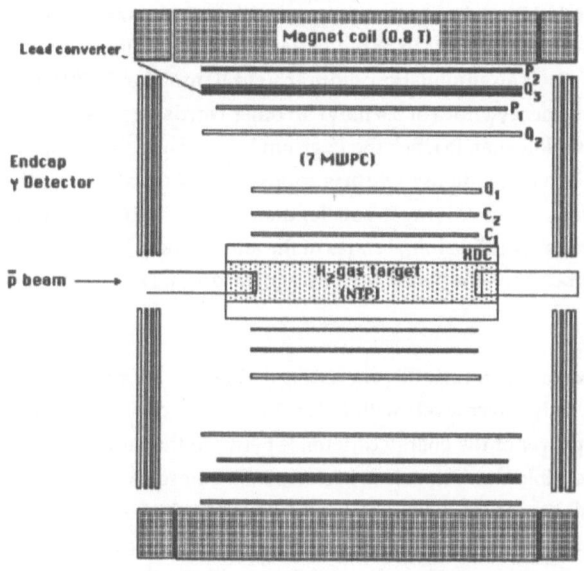

Charged particles : $\frac{\sigma_p}{p}$ = 2.1 % at 500 MeV/c , $\frac{\Omega}{4\pi}$ ≈ 0.6

γ Detection   : Position only (≈ 25 % acc.)

X-rays   : 1 - 20 keV, 10 - 40 % det. eff.

Fig. 25 Detector for experiment PS 171 (ASTERIX)

The organization of this chapter follows the phenomenology of $\bar{p}p$ annihilation at rest. It starts with the capture and the atomic cascade process, followed by atomic X-ray transitions into the 2P and 1S level, from where annihilation takes place. The different branching ratios of the various two- and three-body annihilation channels and their dependence on the angular momentum and the isospin of the initial states are linked to the dynamics of the annihilation process. The search for new mesons or exotic states produced by the annihilation is discussed in the last section on meson spectroscopy.

Atomic cascade of protonium and S- versus P-wave annihilation

Annihilation at rest proceeds via the formation of a $\bar{p}p$ atom ("protonium"). It is formed by the emission of an electron when an antiproton is sufficiently close to a hydrogen molecule. The capture probability is highest when the antiproton is captured into the orbit of the ejected electron, implying $n_{capt} \approx (m_p / 2m_e)^{1/2} \approx 30$, where the factor 2 is due to the reduced masses. The initial distribution among levels with different angular momentum l is not known, but one often assumes a capture probability proportional to the statistical weight (2l+1) of an $|n,l\rangle$ state.

Annihilation takes place when the wave functions of the proton and the antiproton overlap. This is the case for L=0 (S-wave) orbitals. For P-wave orbitals (L=1) there is no overlap at r=0. However, since the nucleon has a finite radius, there is a considerable probability for P-wave annihilation. As a rule of thumb, the annihilation probability decreases by four orders of magnitude per unit of angular momentum. Typically, the annihilation width of the 1S level is 1 keV, scaling with $1/n^3$ for higher nS levels. One would therefore expect an annihilation width around 10 meV for the 2P level, to be compared with a much smaller radiative decay width of 0.4 meV! In other words, protonium in the 2P level has a much higher chance to annihilate than to reach the 1S ground state. This is also true for higher P states, since the scaling of annihilation and radiative width is approximately equal. The annihilation width of D- and higher angular momentum states is much smaller than their radiative width. Therefore we only have to consider S- or P-states as possible initial states of the annihilation, but the question which type of annihilation dominates depends on the density of the target medium. This is explained in more detail in the following section.

We consider first deexcitation in vacuum proceeding exclusively by electromagnetic transitions. Since the photon has spin 1, only lower levels with l' = l ± 1 can be reached. The transition probability is proportional to the third power of the energy difference between the initial and the final state, and transitions into the lowest possible atomic level are preferred. Therefore, the atomic cascade in vacuum leads to an increasing population of the "circular" levels (l = n-1), which have the maximum orbital angular momentum for a given energy level n (fig. 26). After a series of X-ray emissions, the protonium will finally arrive in the 2P level, from where it annihilates with 99 % probability. In vacuum, we therefore have practically pure P wave annihilation.

In a dense medium like liquid hydrogen, an alternative and faster deexcitation mechanism is provided by collisions of the protonium with neighbouring hydrogen molecules (fig. 27). Protonium is a neutral system which penetrates into the electron cloud of colliding hydrogen molecules. Inside the electron cloud, protonium can deexcite by electron emission ("external" Auger effect), or its atomic levels with same n, but different orbital angular momentum l are mixed ("Stark effect", see below). The influence of these collisional effects on the atomic cascade depends on the target pressure.

The proportion of S- versus P-wave annihilation is determined by the frequency of the Stark mixing collisions. The orbital angular momentum l is a conserved quantum number for any spherically symmetric problem, in particular for the radial electric field of an atomic nucleus. However, it looses its meaning as a conserved quantity in an external electric field. In other words, the atomic levels of protonium with same n, but different l, becomes mixed during a collision. Due to the admixture of S-states, such collisions lead to rapid annihilation: the S-wave annihilation rate is of the order of 10^{15} s^{-1} (for n=10), which is much faster than the duration of a collision. The mechanism of S-wave annihilation from high nS protonium states was first proposed in 1960 [49]. Modern cascade calculations are now able to reproduce the absolute X-ray intensities, cascade times and S/P-wave annihilation with reasonable accuracy [50,51] .

Now we are able to understand the pressure dependence of S- versus P-wave annihilation. In liquid hydrogen, the density and therefore the collision rate is high. This leads to S-wave admixture and to a large fraction (≈ 90 %) of S-wave annihilation. In hydrogen gas at 1 atm pressure and room temperature (NTP), the density is 800 times lower than in liquid hydrogen. Therefore, the atom has a higher chance to reach lower levels, in particular circular levels with n=l-1. The cascade calculations predict about equal amounts of S- and P-wave annihilation, in agreement with the experimental results (see below). An elegant experimental trick to enhance P-wave annihilation is to detect L X-rays

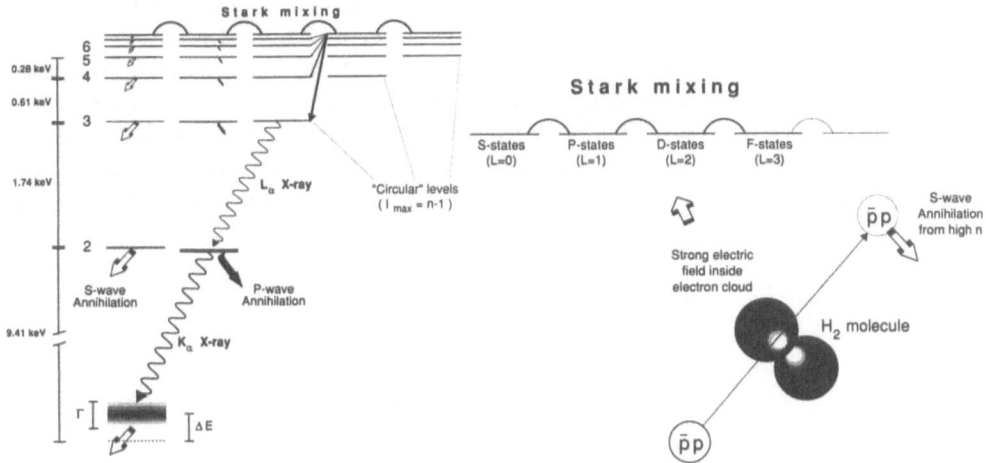

Fig. 26 Atomic energy levels of protonium with indications for Stark mixing, radiative transitions and annihilation.

Fig. 27 Stark mixing due to the electric field experienced by the protonium atom during collisions

signalling the transition into the 2P state (fig. 26), yielding almost exclusive P-wave annihilation. However, a natural X-ray background exists in $\bar{p}p$ annihilation owing to inner bremsstrahlung associated to the emission of charged particles in the final state. This X-ray background with a continuous energy spectrum (~ 1/E) and a yield of a few percent leads to a "dilution" of the cleanliness of the L X-ray trigger. Hence, requiring an "L X-ray" in coincidence tags P wave annihilation with approximately 90 % probability.

Shift and width of the 1S level of protonium

The energy levels of protonium are mainly determined by the electromagnetic interaction, but the real and the imaginary part of the strong interaction $\bar{p}p$ potential lead to a shift and broadening of the atomic S- and the P-states. The shift and width of the 1S ground state is related to the complex S-wave $\bar{p}p$ scattering length by:

$$a_s(0) = M/2\pi \ | \psi_{1s}(0) |^2 \ (\Delta E_{1s} + i/2 \ \Gamma_{1s})$$

where $\psi_{1s}(0)$ is the atomic wave function at the origin. The ρ parameter at rest is approximately

$$\rho \approx 2 \cdot \Delta E_{1s} / \Gamma_{1s}$$

141

A correction of about 10 % is needed to account for the effect of the Coulomb interaction [30]. The shift and width of the 1S ground state are obtained by measuring the energy of K X-ray transitions from nP-states to the 1S state, and then comparing the value with the expectation from QED (9.41 keV for the K_α transition). As discussed in the previous section, K X-ray transitions are rare in liquid hydrogen due to Stark mixing and subsequent annihilation from high nS states, but by lowering the target density, the Stark mixing can be reduced.

The experiment PS 171 used a cylindrical H_2 gas target at NTP with an effective length of 76 cm and a radius of 8 cm. The stop distribution of antiprotons at 105 MeV/c after traversing an appropriately chosen moderator is shown in fig. 28a. Protonium X-rays were detected by their conversion in an X-ray Drift Chamber (XDC) [52], surrounding the gas target, from which it was separated by a 6 μm thin mylar foil. The XDC (fig. 28b) covered 90 % of the solid angle and had a detection efficiency of about 30 % for L X-rays and 20 % for K X-rays. The energy resolution was typical for a proportional chamber (σ = 10 % at 5.5 keV). The L X-ray yield in hydrogen gas at NTP is 12 %, and together with a detection efficiency of 30 %, a trigger on L X-rays (signalling transitions into the 2P state) selected 3.6 % of all antiproton annihilations. This technique allowed to record a large number of events annihilating from the 2P state.

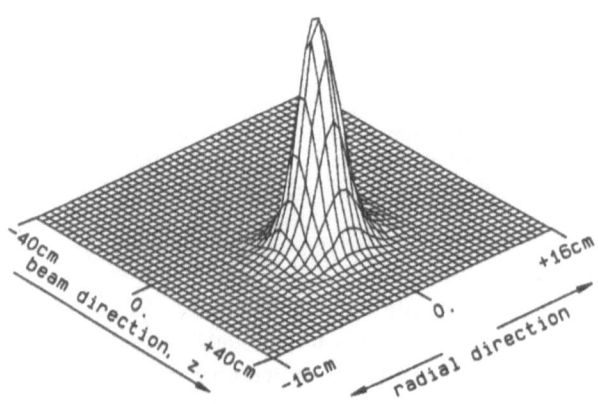

Fig. 28a Stop distribution of 105 MeV/c antiprotons in a H_2 gas target at NTP

The inclusive X-ray spectrum of events with only neutral particles in the final state is shown in fig. 29a. Those events were preferred in the X-ray analysis due to the absence of charged particles giving rise to a background from inner bremsstrahlung. However, only a broad structure comprising the unresolved K line series is observed. Fig. 29b shows the X-ray energy spectrum where a second X-ray is required in coincidence with an L X-ray, and a clear peak from the K_α transition (which is singled out by the L X-ray coincidence) is observed. From a fit to the K_α peak, the shift and the width of the 1S ground state were determined: $\Delta E_{1S} = -0.70 \pm 0.15$ keV, $\Gamma_{1S} = 1.60 \pm 0.4$ keV [53].

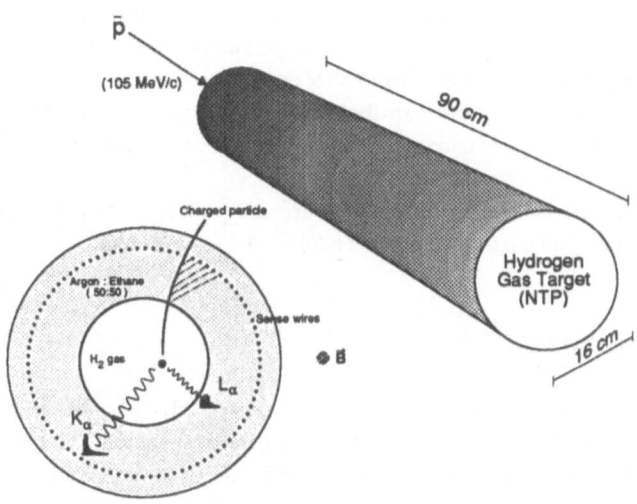

Fig. 28b   Hydrogen gas target and X-Ray Drift Chamber (XDC)

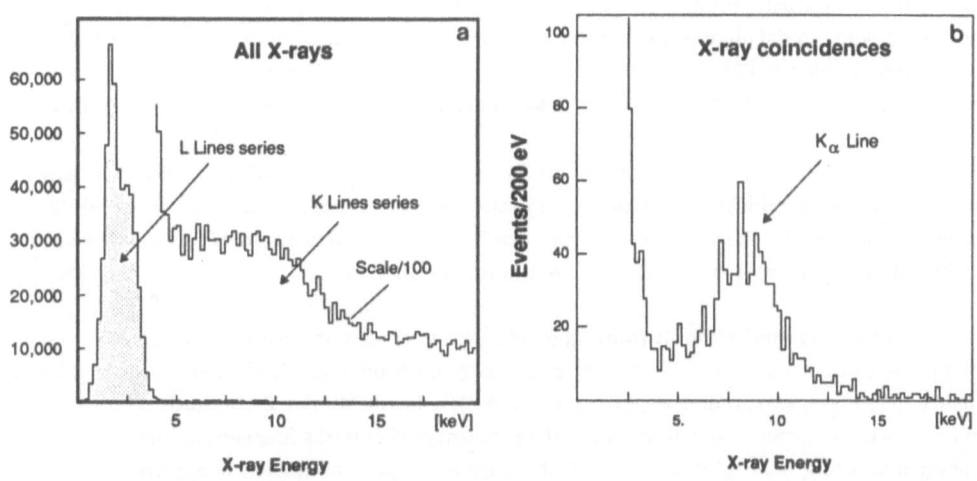

Fig. 29   Protonium X-ray energy spectrum (PS 171) a) for all neutral events, b) for neutral events
with two coincident X-rays

Two other experiments at LEAR also observed protonium K X-rays using Si(Li) detectors with a much better resolution, but smaller solid angle coverage. PS 174 varied the target density by using a gas target in which the pressure could be reduced (density $\rho < \rho_{NTP}$) or the gas cooled ($\rho > \rho_{NTP}$). X-rays were observed either by a Si(Li) detector or, in a later stage of the experiment, by two gas scintillation proportional detectors [54]. PS 175 used a clever trick to stop antiprotons in a target at very low densities (30 mbar): antiprotons enter a cylinder inside a homogeneous magnetic field (called "cyclotron trap"); the anticyclotron is filled with gaseous hydrogen, and the kinetic energy of the antiprotons is slowly decreased by energy loss in the target gas; the radius of the cyclotron orbits become smaller, and a weak focussing magnetic field guides the particles to the center of the magnet, where they are finally captured to form protonium atoms; the X-ray energy is then measured by Si(Li) detectors [55].

The average values from all three experiments are : $\Delta E_{1s} = -0.72 \pm 0.04$ keV, $\Gamma_{1s} = 1.11 \pm 0.07$ keV, yielding a Coulomb corrected ρ parameter at rest $\rho (0) = -1.08 \pm 0.09$ [30].

Models of the annihilation process

The most general description of annihilation at rest is given by the transition amplitude $<f | T_{ann} | i>$, where $| i >$ stands for the wave function of the initial $\bar{p}p$ atom, $| f >$ for the mesonic final state, and T_{ann} for a "transition operator" containing the details of the annihilation process. Unfortunately, only little is known about the dynamics and the symmetry of this transition operator.

The initial protonium wave function depends on the orbital angular momentum, the spin and the isospin of the two nucleons. While the $\bar{p}n$ system is pure isospin one (I=1), the $\bar{p}p$ wave function is a mixture of I=0 and I=1, and annihilation can take place from either of the two isospin states. The knowledge of the initial angular momentum (L=0 or L=1) is very helpful in reducing the number of possible initial states and facilitates the partial wave analysis of the final state. However, all possible spin-singlet or spin-triplet substates must be taken into account (2 or S-wave, 4 for P- and higher waves).

The final state of $\bar{p}p$ annihilation at rest contains in the average about 5 pions. Nevertheless, the direct (incoherent) production of 5 pions is very rare, and most pions are due to secondary decays of heavier mesons, e.g. $\bar{p}p \rightarrow \pi + X$, $X \rightarrow \pi + \omega$, $\omega \rightarrow \pi\pi\pi$. For broad intermediate resonances, the pion spectrum is difficult to disentangle from incoherent continuum production.

The most important quantity from a theoretical point of view is the transition operator. There is no hope that the structure of this operator can be derived from QCD in the near future: while the interaction between two gluons or between two infinitely heavy quarks can be studied by numerical simulations on a discrete space-time lattice, the problem of six quarks in an environment filled with strongly interacting "soft" gluons is outside the reach of the present theoretical approaches.

Several models of $\bar{p}p$ annihilation exist, and extensive reviews have been written [28, 56]. The following comments are only a guideline for further reading.

The "threshold dominance" model [57,58] is simple, but surprisingly successful. It assumes that the annihilation proceeds via two-meson "doorway" states, and that the intermediate two-meson threshold closest to the actual center-of-mass energy dominates. The model describes correctly the pion multiplicities and their relative cross-sections up to 3.5 GeV/c. However, the suppression of kaonic final states (contributing about 7 % of all annihilations) has to be put in by hand, and the model fails in describing the known exclusive branching ratios correctly.

The "quark-line rule" model is based on the conservation of SU(3) flavour symmetry [59,60] and describes a wide range of absolute and relative branching ratios fairly accurate. The model assumes that the transition operator is unity and the initial state interactions are negligible. Since the branching ratios of many interesting final states, in particular involving those with several neutral mesons, are only known with poor accuracy or not at all, a challenging test of this model has to wait for the new results from the Crystal Barrel experiment (see chapter 6).

More ambitious approaches are based on microscopic models describing annihilation by a sequence of $\bar{q}q$-pair annihilations and/or re-arrangement. Several types of quark-line diagrams can contribute (fig. 30), but the relative importance of these graphs is difficult to calculate [61–64]. Again, more experimental guidance is desperately needed.

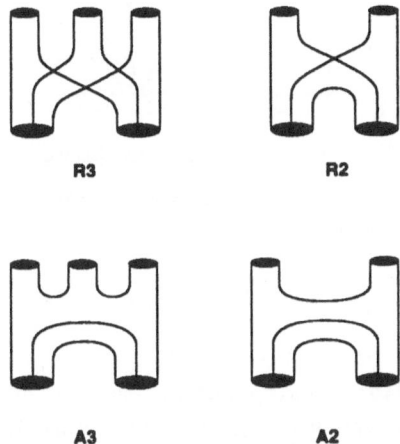

Fig. 30 Quark-line diagrams contributing in in lowest order to $\bar{p}p$ annihilation

Annihilation at rest into $\pi\pi$ and $K\bar{K}$ from atomic S- and P-states

The initial states of $\bar{p}p$ annihilation at rest are either atomic S- or P-states. In S-states, the two spins may be parallel (triplet) or antiparallel (singlet). The parity of a state is determined by the product of the inner parity (negative for fermion-antifermion pairs) and the angular momentum factor $(-1)^l$, which is positive for even and negative for odd angular momentum. Hence, the parity of the protonium S-states is negative. The C parity $(-1)^{l+s}$ is negative for the triplet and positive for the singlet state. Therefore, the S-states have the quantum numbers $J^{PC} = 0^{-+}$ (singlet) and 1^{--} (triplet), respectively. Similar considerations for the P states give $J^{PC} = 0^{++}$, 1^{++}, 2^{++} (triplet) and 1^{+-} (singlet).

The L X-ray tagging technique allowed PS 171 to carry out - for the first time - a systematic study of the dependence of two-body annihilation branching ratios on the initial angular momentum. The simplest final states are $\pi^+\pi^-$ and K^+K^-, identified by requiring two collinear tracks of opposite charge in the detector. Fig. 31 shows the momentum spectra for events with both tracks fitting to a common helix. The left spectrum is taken from liquid hydrogen bubble chamber data [18], while the middle and right spectrum are from PS 171 in H_2 gas without and with L X-ray trigger [65]. The two main observations are that the BR of $\pi^+\pi^-$ increases with increasing P wave annihilation, while the K^+K^- BR decreases by more than a factor of 3.

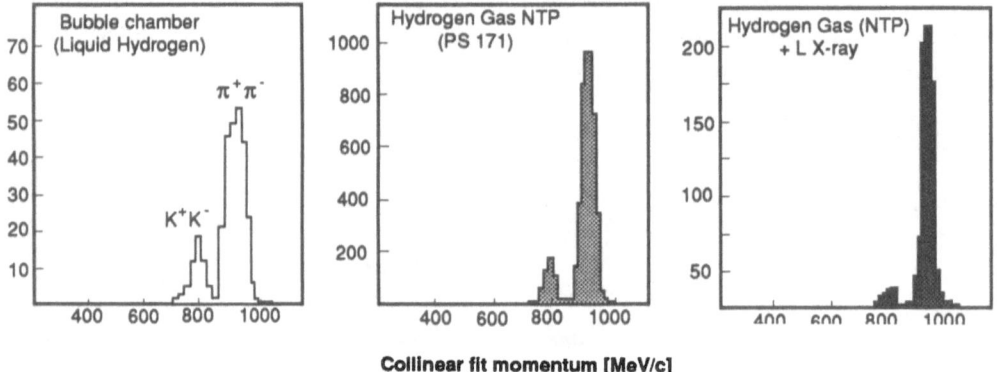

Fig. 31.   Momentum spectra for collinear events in bubble chambers, hydrogen gas target at NTP, and with L X-ray coincidences

Annihilation into neutral kaon pairs (K_sK_s and K_sK_L) shows a similar dependence. It follows from Bose-Einstein statistics and conservation of C-parity that K_sK_s can only come from P-states, and K_sK_L only from S-states. In P-wave annihilation, the branching ratio of K_sK_s, and hence $K^0\bar{K}^0$, is a factor 10 lower than of K_sK_L in S-wave annihilation [66]. Fig. 32 shows the dependence of $\pi\pi$ and $K\bar{K}$ branching ratios on the proportion of S- to P-wave annihilation.

At present, there is no unique explanation of these effects. A simple explanation of $K\bar{K}$ suppression from P states comes from the $^3P_0$ model of $\bar{q}q$ pair creation. The lowest order quark-line graph contributing to $K\bar{K}$ production is the annihilation graph shown in fig. 33. The initial P-states decaying into $K\bar{K}$ have quantum numbers 0^{++} and 2^{++}, i.e. positive parity. Kaons are pseudoscalar particles with spin 0, and their relative orbital angular momentum is determined by the angular momentum between the $s\bar{s}$ quark pair created during the annihilation. In the $^3P_0$ model, this $s\bar{s}$ pair is created with angular momentum L=1, so that the $K\bar{K}$ pair must be in an L=1 state with negative parity! This is forbidden by parity conservation in strong interaction, and could explain the suppression of $K\bar{K}$ from P-states.

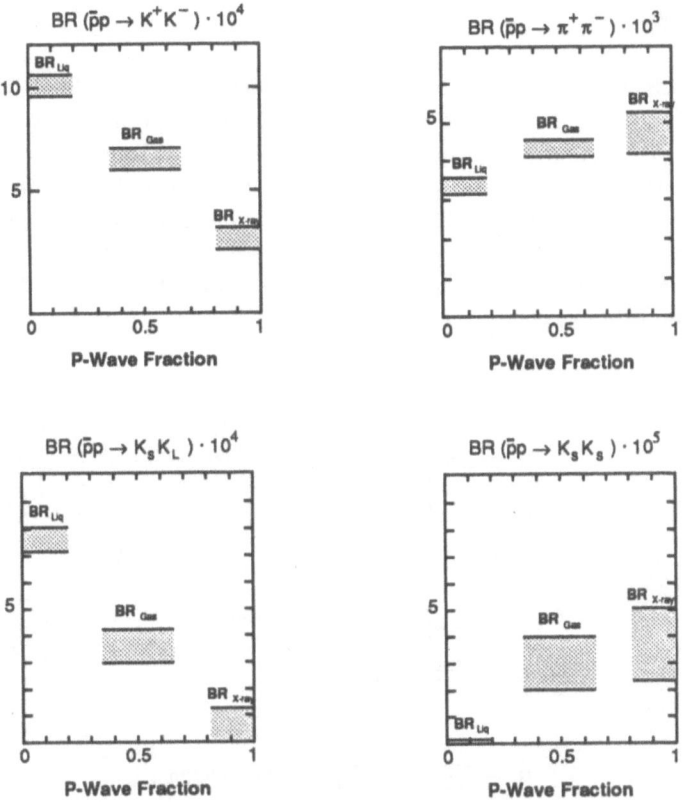

Fig. 32 Dependence of the $\pi\pi$ and $K\bar{K}$ branching ratios in $\bar{p}p$ annihilation at rest on the relative P-wave contribution

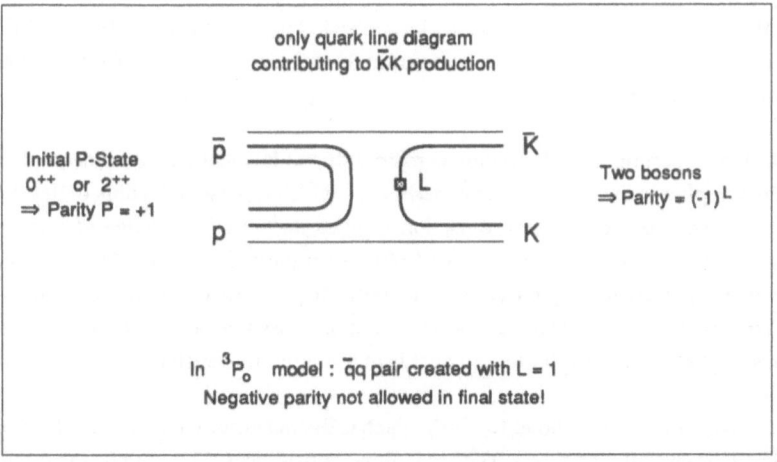

Fig. 33 Explanation of the $K\bar{K}$ suppression in P-wave annihilation by the ^3P_o model

Meson spectroscopy

The emphasis of meson spectroscopy has moved from the study of the SU(3) flavour nonets to the search for non-$\bar{q}q$ states. These exotic states obey the QCD dogma of colour-neutrality, but they have four ($\bar{q}\bar{q}q\,q$), five ($\bar{q}qqqq$) or six ($qqqqqq$, $\bar{q}\bar{q}\bar{q}qqq$) constituent quarks, or - as a manifestation of the gluon-gluon self-interaction - gluons as valence particles (glueballs, hybrids). It seems plausible that an extended object of quarks and gluons (like protonium) is a good source of either gluon-enriched or purely gluonic states (see fig. 34), and they could be produced in two-body reactions - like most $\bar{q}q$ mesons - with branching ratios of order 10^{-3}. A systematic study at the level of BRs of 10^{-4} should therefore reveal the existence of these states.

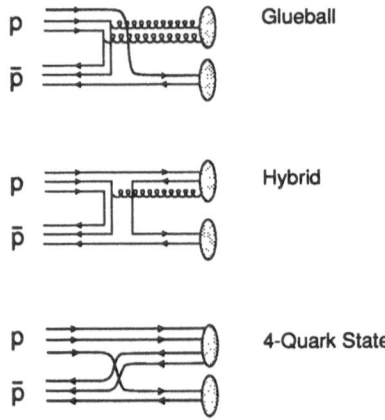

Fig. 34   Possible production of exotic resonances in $\bar{p}p$ annihilation

I see three possible reasons why exotics have resisted their discovery until now: incomplete experimental information, search in the wrong mass range, or the difficulty of separating exotics from the $\bar{q}q$ spectrum ("camouflage").

The lack of experimental information is mainly due to low statistics and incomplete final state detection. Both applies to bubble chamber experiments, which represent the bulk of data before the PS 171 experiment. The absence of γ-ray detection implies that our knowledge of final states of $\bar{p}p$ annihilation with two or more neutral particles ($\geq 60\ \%$ of annihilations) in the final state has remained marginal. However, glueballs might decay flavour-blind or prefer to decay into $\eta\eta$ or $\eta\eta'$ states. In both cases, fully reconstructed neutral and kaonic final states are essential, and both the new Crystal Barrel and the Obelix detector (see chapter 6) are capable of accumulating such data with high statistics.

Exotics may have a mass above 1.7 GeV, which is the maximum mass for particles produced by pion emission in $\bar{p}p$ annihilation at rest. The predictions from lattice QCD suffer from a highly uncertain absolute mass scale. While several calculations agree that the scalar glueball should have approximately 2/3 of the tensor glueball mass, the absolute value for $M(0^{++})$ is predicted between 1 and 2 GeV. Hybrid

2/3 of the tensor glueball mass, the absolute value for $M(0^{++})$ is predicted between 1 and 2 GeV. Hybrid states ($\bar{q}qg$) are usually placed into the 1.5 - 2.5 GeV region. Therefore Crystal Barrel and Obelix will also measure annihilation in flight (p = 2 GeV/c) to extend the mass region up to 2.2 GeV. The Jetset experiment studies $\bar{p}p \rightarrow \Phi\Phi$ in a formation experiment, where the maximum mass obtainable at LEAR is 2.4 GeV.

The "camouflage" problem is the most difficult one. Firstly, non-$\bar{q}q$ states with quantum numbers accessible to $\bar{q}q$ can mix with $\bar{q}q$ states. The clearest evidence for a non-$\bar{q}q$ state would therefore come from states with "exotic" quantum numbers like 0^{+-}, 0^{--}, 1^{-+}, 2^{+-}, etc. There is only one candidate for a 1^{-+} state documented in literature [67], but it needs confirmation. Otherwise, exotics with "ordinary" quantum number can only be found by a thorough understanding of all $\bar{q}q$ states and by identifying "superfluous" members of an already completed SU(3) nonet. Secondly, if the exotic state has a very large width, there is little chance to find it by a partial wave analysis or even as a bump in the mass spectrum. However, the production probability of this state may depend on the initial angular momentum or the isospin. Then, a systematic comparison of the same spectra obtained for different initial states could reveal even broad new resonances.

Since PS 171 has pioneered the technique of comparing production of mesons from S- and P-states, I will illustrate meson spectroscopy "beyond bubble chambers" by two results from PS 171. The next section describes the discovery of a new resonance in the 1.5 GeV mass region, originally baptized "Ax", which is now an established tensor particle "$f_2(1515)$". The other result concern the E(1420) meson, originally discovered in $\bar{p}p$ annihilation and since its re-discovery in radiative J/ψ decays a hot glueball candidate.

Annihilation into three pions and the discovery of the Ax/f_2

In a detector without γ-ray detection, the final state is $\pi^+\pi^-\pi^0$ identified by requiring two charged particles recoiling against the missing mass of a neutral pion. The $\pi^+\pi^-\pi^0$ final state has been studied in detail using bubble chambers (S wave annihilation) [68]. In the corresponding Dalitz plot, strong bands from the production of ρ^+, ρ^- and ρ^0 are visible, but there is no clear sign of any other resonance. The $\pi^+\pi^-$ invariant mass plot (fig. 35a) only shows ρ^0 production. With the L X-ray tagging technique of PS 171, two different data samples (with and without L X-ray coincidence) have been obtained. From the knowledge of the respective P-wave contributions (derived from the $\pi\pi$ and $K\bar{K}$ data), an extrapolation to the "pure" S- and P-wave spectra is possible. Fig. 35b and 35c show the extrapolated mass spectra for S- and P-wave annihilation, and the S-wave plot of PS 171 agrees indeed very nicely with the bubble chamber data. However, the $f_2(1270)$ and a new resonance, the Ax, are the prominent structures in the P-wave plot.

The properties of the "Ax" resonance were extracted from a complete partial wave analysis of the corresponding Dalitz plot. The mass and the width are M = 1565 ± 20 MeV and Γ = 160 ± 40 MeV, and spin 2 is preferred over spin 0. It has positive parity and C-parity (from its decay into $\pi^+\pi^-$, the absence of charged modes [I=0] and G-parity conservation). The absolute branching ratio is about $4 \cdot 10^{-3}$ - or larger, if other decay channels (e.g. $\rho\rho$, πa_2, $\eta\eta$, $\omega\omega$) exist. The discovery of a new resonance with a branching ratio of $4 \cdot 10^{-3}$ in a channel which looks "boring" in S-wave production illustrates impressively how efficient the new technique can be in finding new resonances.

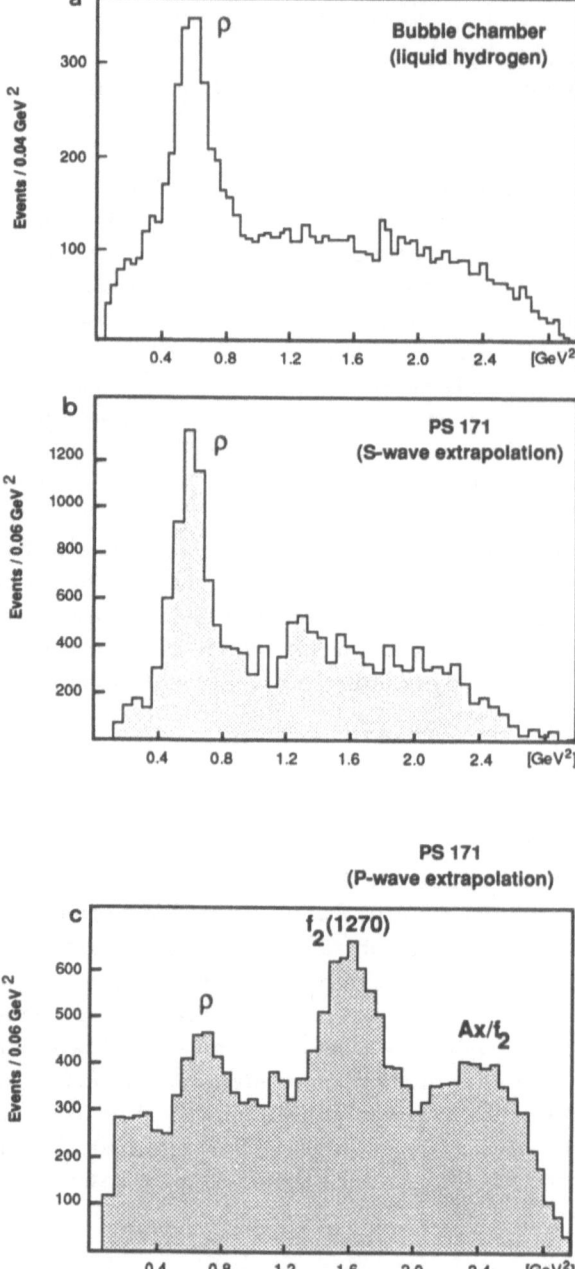

Fig. 35. Comparison of $\pi^+\pi^-$ invariant mass spectra in the $\pi^+\pi^-\pi^0$ final state.

    a) For annihilations in liquid hydrogen (bubble chamber data), mainly S-wave
    b) Extrapolated spectrum for S-wave annihilation (PS 171)
    c) Extrapolated spectrum for P-wave annihilation (PS 171)

In the meantime, the Crystal Barrel experiment has cleanly identified final states of $\bar{p}p$ annihilation with several neutral particles (e.g. $\pi^\circ\pi^\circ$, $\pi^\circ\pi^\circ\pi^\circ$, $\eta\pi^\circ\pi^\circ$, $\eta\eta\pi^\circ$, $\pi^+\pi^-\pi^\circ\pi^\circ$, $\pi^+\pi^-\pi^\circ\eta$). Its first physics result is the confirmation of the Ax, observed as a $(\pi^\circ\pi^\circ)$ mass peak in the $3\pi^\circ$ final state [70]. The Ax is produced in the same reaction ($\bar{p}p \rightarrow \pi^\circ Ax$), but it decays into $\pi^\circ\pi^\circ$, which is expected on the basis of its isospin. The mass is 1520 ± 10 MeV and the width 120 ± 20 MeV. The mass shift is probably due to the non-uniform background in the ASTERIX ($\pi^+\pi^-\pi^\circ$) Dalitz plot due to the $\rho^+\rho^-$ interference in the Ax/f_2 mass region.

What is the Ax? It does not fit into the conventional tensor meson nonet, since the nonet is already filled with well-established candidates. In particular, it cannot be identical with the $f_2(1525)$, which has a very strong coupling to $K\bar{K}$: the observed branching ratio into $\pi\pi$ would then require more than 50 % of "Ax" production in $\bar{p}p$ annihilation at rest, which is excluded by the data. It is probably not a radial excitation of the $f_2(1270)$: there is already a 2^3P_2 candidate around 1810 MeV , and the calculations of Godfrey and Isgur [71] predict that the 2^3P_2 state should be at a mass around 1800 MeV. Is it a glueball? It has not been observed in other reactions which are supposed to be "gluon-rich" (radiative J/ψ decay, central $\pi p/pp$ collisions). There are several possible interpretations: a meson-meson molecule, a 4-quark state or a quasi-nuclear bound state, which have been predicted in this mass range [12, 72]. At present, no final conclusion can be drawn. More experimental information on the Ax/f_2 decaying into different final states would be of great interest.

Annihilation into $\pi^+\pi^-K\bar{K}\pi$ and the E/$\iota(1420)$

The E(1420) was discovered in 1963 in $\bar{p}p$ annihilation at rest in a hydrogen bubble chamber [73,74]. In spite of its age, considerable confusion persists about its nature, and the number of reviews about the E meson exceeds the number of experimental results by far. The most controversial points are: Is the E(1420) a single resonance, or are there two or more mass degenerate objects? In particular, is the $\iota(1440)$, observed in radiative J/ψ decays [75,76], identical with the E(1420)? Are the quantum numbers 0^{-+} or 1^{++}? Does the decay of the E(1420) proceed via E$\rightarrow a_0(980)\pi\rightarrow K\bar{K}\pi$ or via E$\rightarrow K^*\bar{K}\rightarrow K\bar{K}\pi$? If there is a decay chain via the $a_0(980)$, and since the $a_0(980)$ decays more frequently into $\eta\pi$, why is there only one observation of the decay E $\rightarrow \eta\pi\pi$ [77]? Is the E(1420) mainly an $s\bar{s}$ member of the 1^{++} SU(3) nonet, as indicated by its strong kaonic decay modes? If yes, why is it produced in πp scattering [78] but not in K⁻p scattering [79]?

The tentative interpretation of the 0^{-+} object as a glueball candidate is mainly due to its observation in radiative J/ψ decays and the difficulty to associate it with a $\bar{q}q$ nonet. The motivation of a further study of the $\pi\pi(K\bar{K}\pi)$ channel in $\bar{p}p$ annihilation at rest by experiment PS 171 was an independent study of the quantum numbers of the E(1420), in particular by comparing its production in S- versus P-wave annihilation.

The production and the decay of about 400 E meson events in the channel $\bar{p}p \rightarrow \pi^+\pi^-$[E $\rightarrow$ K$^\circ$K$^+\pi^-$ + c.c.] was observed [80]. Fig. 36 shows the neutral $K\bar{K}\pi$ mass distribution, where the E meson appears with a mass $M=1413 \pm 8$ MeV and a width $\Gamma = 62 \quad \pm 16$ MeV. A peak with a significance of 3σ is observed at the mass of the $f_1(1285)$ [the former D meson], a well-established resonance with $J^{PC}=1^{++}$ known to decay into $K\bar{K}\pi$. The decay Dalitz plot for events with M($K\bar{K}\pi$) = 1370 ... 1480 MeV is shown in fig. 37. The distribution shows an accumulation of events in the region of low $K\bar{K}$ masses (right upper corner), but no clear K* bands. This agrees well with the Dalitz plot from the previous bubble chamber experiment, but is strikingly different from the plot obtained in pp central production (fig. 38) [81]. The partial wave analysis of the PS 171 data exclude a dominant $K^*\bar{K}$ decay mode, and give a strong contribution from the low $K\bar{K}$ mass enhancement ("$a_o(980)$"). However, there is uncertainty about this structure: is the $a_o(980)$ really a $\bar{q}q$ resonance just below the $K\bar{K}$ threshold, a $K\bar{K}$ final state interaction, or a $K\bar{K}$ molecule [82]? Different parametrizations of the $a_o(980)$ influence the angular distribution stronger than the difference stemming from a $J^{PC}=0^+$ or 1^{++} assignment of the E(1420). Therefore, no clear preference can be given for either of the two possibilities.

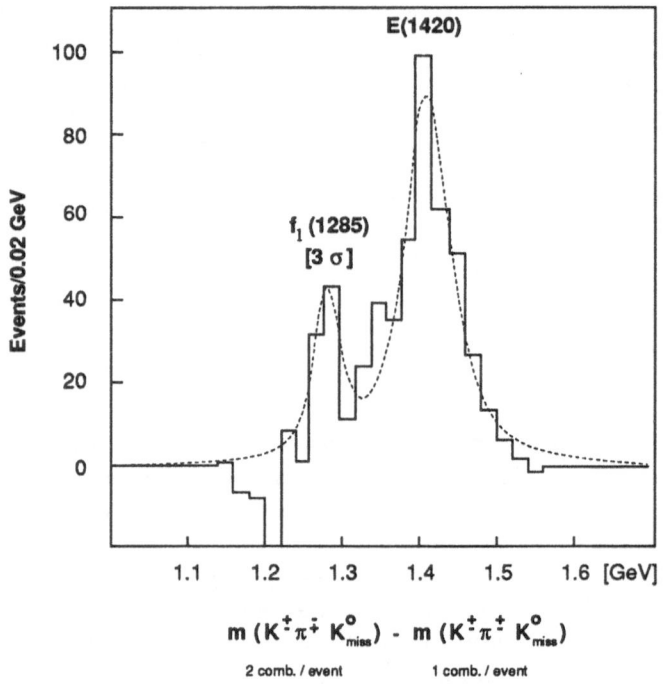

Fig. 36 Observation of the E(1420) in $\bar{p}p$ annihilation at rest (PS 171) decaying into $K\bar{K}\pi$

The study of the production $\bar{p}p \to \pi^+\pi^- E$ is more conclusive. The data show that $\pi\pi E$ production from P wave is compatible with zero: the S-wave branching ratio is $7 \cdot 10^{-4}$, while the P-wave branching ratio is less than $1 \cdot 10^{-4}$. This suppression can be explained in the following way: The $(\pi\pi)$ system recoiling against the E meson is most likely in a 0^{++} state (since $p_{\pi\pi} \leq 180$ MeV/c). The relative angular momentum L between the di-pion and the E is L=0 because of the limited phase space available [80]. For L=0, a pseudoscalar state (recoiling against a 0^{++} $\pi\pi$ pair) is only produced from S-wave (^1S_o), while a 1^{++} state can only come from P wave ($^3P_1$). The $f_1(1285)$ meson is a 1^{++} particle and follows this prediction. The E meson is not produced from P states, and hence the 0^+ assignment is the most probable. The PS 171 result therefore confirms the 0^+ assignment of the E (1420) observed in $\bar{p}p$ annihilation. The difference between the E(1420) decay Dalitz plot in $\bar{p}p$ annihilation and in pp central production suggests, however, that there are two mass-degenerate resonances, one with 0^+ and one with 1^{++}.

The first generation of LEAR experiments has finished data taking in 1987. Very interesting results on many different physics topics have emerged from the various experiments, e.g. on the electromagnetic form factor of the proton in the time-like region, the spin dependence of antinucleon-nucleon interaction, K X-rays in protonium, hypernuclei, antiproton-nuclear cross-sections, X-ray spectra in heavy antiprotonic atoms, and on antiproton induced fission, which could not be discussed for lack of space. For a good overview, I recommend the proceedings of the 1990 conference on antinucleon-nucleon physics in Stockholm [83].

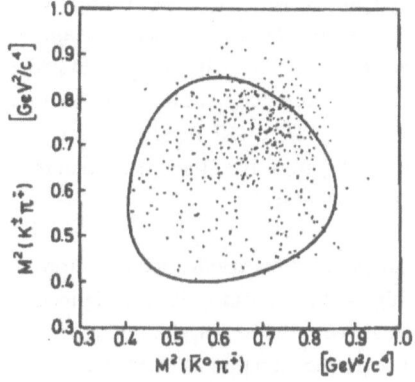 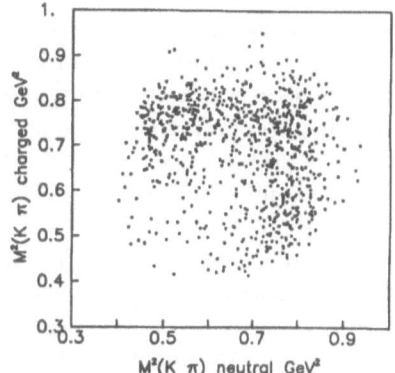

Fig. 37   Dalitz plot for E(1420) decay into $K\bar{K}\pi$ (produced in $\bar{p}p$ annihilation)

Fig. 38   Dalitz plot for E(1420) decay into $K\bar{K}\pi$ (produced in pp central production)

For particle physics, the major results were the discovery of the Ax/f2, and the refutation of the "narrow baryonium" states. However, the "oscillatory" behaviour of the ρ parameter in elastic $\bar{p}p$ scattering at low energies, the strong P-wave contribution in $\Lambda\bar{\Lambda}$ production near threshold, and the strong coupling of the Ax/f2 resonance to antinucleon-nucleon states may all be indications that nuclear bound states of antibaryon-baryon pairs with typical hadronic widths exist, and some of them very close to their respective thresholds. The PS 171 method of P-wave selection by L X-ray tagging has proven to be an important new tool in the search for new resonances and the determination of quantum numbers.

The LEAR programme after 1987 [84] differs in some respects from the first round of experiments. The detectors used for meson spectroscopy are more complete, and the technique of comparing data from different initial states (liquid hydrogen versus gas to study the S/P-wave annihilation, hydrogen versus deuterium to study the isospin dependence, annihilation at rest versus in flight to study the phase space dependence) is now established as a tool for searching exotic states. Both PS 197 (Crystal Barrel) and PS 201 (Obelix) feature neutral particle detection over almost the whole solid angle, and charged particle momentum measurement over $\approx 2/3$ of 4π. Crystal Barrel obtains a very good γ-ray energy resolution (≈ 2.5 % at 1 GeV) by means of 1380 CsI crystals, and is able to completely reconstruct final states with very high γ multiplicities. Obelix measures the position of γ-ray conversion very precisely, but is (because of the fair energy resolution) limited to $N_\gamma \leq 4$. On the other hand, its TOF measurement system allows the online identification of charged kaons, and one can hope for a significant increase of annihilation data involving kaonic final states. The Jetset detector studies $\Phi\Phi$ formation in a gas jet type experiment with a very selective trigger on 4 charged kaon final states.

The spin-dependence of the antinucleon-nucleon interaction is studied in elastic and charge-exchange scattering of antiprotons on a polarized hydrogen target (PS 198, PS 199). A significant progress would come from polarized antiproton beams, but the practicality of the two propositions to polarize the antiproton beam circulating in LEAR [85,86] has not yet been demonstrated.

Antiprotons are used to test fundamental symmetries, like conservation of CP (see below), or CPT. The CPT symmetry implies the equality of fundamental properties of particle and antiparticle, like the inertial mass, the charge, the magnetic moment, and the life time. Two experiments are aiming at a measurement of the inertial antiproton mass with a precision of order 10^{-9}. The principle in both cases is the measurement of the cyclotron frequency in a very homogeneous magnetic field. One experiment uses antiprotons stored in a Penning trap (PS 196), the other one uses an RF mass spectrometer with antiprotons of 20 MeV/c momentum (PS 189). PS 196 has recently improved the precision of the mass comparison to $2 \cdot 10^{-8}$ [87] but has not found any difference in mass (yet?).

The study of CP violation in the interference of K° and $\bar{K}^\circ$ is a new application of antiproton annihilation. The conventional approach is to measure with high statistical and systematical precision the CP-violating decays of K_L into $\pi^\circ\pi^\circ$ and $\pi^+\pi^-$, and to compare with the corresponding K_s decays. The new approach of the PS 195 (CP-LEAR) experiment is to use the reaction $\bar{p}p \rightarrow K^\circ K^- \pi^+$ and c.c. as a CP-symmetrical source of neutral kaons and antikaons, since the strong production process conserves CP-symmetry. Since a negative kaon is associated with a neutral kaon, and a positive kaon with a neutral antikaon, it is possible to study the decay of neutral kaons and antikaons separately as a function of their eigentime. The detector is capable of identifying the interesting $K\bar{K}\pi$ final state (BR $\approx 4 \cdot 10^{-3}$) and the charge of the kaon within less than 1 μs. The subsequent decay of the neutral kaon into two charged or neutral pions, into 3 pions, or into the semileptonic decay channel is then recorded on tape. With the goal of observing $4 \cdot 10^9$ neutral kaon decays over a period of three years, the experiment aims at improving our knowledge on CP violation in the 3π and semi-leptonic decay channels as well as contributing a new measurement of ε'/ε with a precision of about $1.5 \cdot 10^{-3}$.

VII   CONCLUSION AND OUTLOOK

Experiments with antiprotons have a long and rich history, and their results have contributed to the progress of elementary particle physics. Elementary particle physics has now moved to energies of hundreds of GeV. There, protons are more a loosely bound system of 3 quarks than a classical nucleon, and the observations are the results of quark-antiquark or quark-gluon scattering, with soft hadronic effects like the production of jets as a disturbing, but unavoidable background.

The "classical" antiproton physics (scattering, two-body reactions) studying the nucleon as an entity and trying to understand its interaction at low energy has come to a turning-point. Theoretical models based on light and heavy meson-exchange describe the measurements quite accurately, but these models are based on a doubtful phenomenological picture and contain enough free parameters to adapt quickly to new experimental results, hence they are difficult to discard. On the other hand, we are still light years away from a full non-perturbative QCD calculation of the nucleon, and even simpler models "inspired" by QCD but containing simple approximations still encounter enormous difficulties. New experimental results would be most welcome to help in finding better approximations on the long way to an understanding of hadronic phenomena.

The main motivation to continue antiproton physics comes from the annihilation as a source of exotic meson states. The present LEAR experiments will hopefully shed more light on the question if glueballs and hybrids exist and what there properties are, and this would certainly be a pivot for attempts of non-perturbative QCD calculations. But whatever new state will be found in antiproton-proton annihilations, the comparison with different production mechanisms (central production in πp or pp scattering, Kp reactions, $\gamma\gamma$ collisions, radiative J/ψ decays) is essential to understand the nature of this state.

Going beyond the present LEAR programme, I believe that experiments with antiprotons at higher energies are very important. At LEAR, the mass range for the production of resonances associated to the emission of one or two pions is limited to 1.6 - 1.7 GeV. A facility like SuperLEAR (a LEAR-type ring with supraconducting bending magnets, cooling, antiproton momentum range 2-12 GeV/c, two gas jet targets for luminosities of 10^{32} cm^{-2}s^{-1} and an ultraslow extraction) would allow the study of the 2-3 GeV mass range, where the identification of exotic states (with low angular momentum) could be much easier. In addition, it would give access to the charmonium range. The experiments R704 at the ISR (CERN) and E760 at Fermilab have shown that the direct production of $c\bar{c}$ states can be studied with high precision in antiproton-proton formation experiments. A similar experiment doing a systematic scan of the 3-5 GeV mass range looking for exotic charmonium states ($c\bar{c}g$ hybrids) or glueballs could be the first to announce the discovery of an exotic meson beyond reasonable doubt: since the charmonium spectrum is very well understood, any additional state has to be of a different nature.

REFERENCES

[1]   F.E. Close, in: Quarks and Nuclear Forces (Springer Verlag, 1982)
[2]   J. Iwadare et al., Suppl. to the Progr. Theor. Phys. 3, 1956 (Japan)
[3]   R. A. Bryan and B.L. Scott, Phys. Rev. B135 (1964) 434
[4]   A.W. Thomas, Adv. Nucl. Phys. 13 (1983) 1
[5]   T.D. Lee and C.N.Yang, Nuov. Cim. 3 (1956) 749
[6]   J.S. Ball and G.F.Chew, Phys. Rev. 109 (1958) 1385
[7]   R.A. Bryan and R.J.N. Phillips, Nucl. Phys. B5 (1968) 201
[8]   C.B. Dover and J.M. Richard, Phys. Rev. C21 (1980) 1466
[9]   O.D. Dalkarov and F. Myhrer, Nuov. Cim. A40 (1977) 152
[10]   J.M. Richard and M.E. Sainio, Phys. Lett. B110 (1982) 349
[11]   M. Kohno and W. Weise, Nucl. Phys. A454 (1986) 429
[12]   I.S. Shapiro, Phys. Rep. 35C (1978) 129
[13]   J. Carbonell, G. Ihle and J. M. Richard, Z. Phys. A334 (1989) 329
[14]   T. Walcher, Ann. Rev. Nucl. Part. Sci. 38 (1988) 67
[15]   T.A. Shibata, Phys. Lett. B189 (1987) 232
[16]   O. Chamberlain et al., Phys. Rev. 100 (1955) 947
[17]   E. Segré, Ann. Rev. of Nucl. Science 8 (1958) 127

[18]   R. Armenteros and B. French, Antinucleon-Nucleon Interactions, in: "High Energy Physics", Vol IV, ed. E.H.S. Burhop, London, Acadamic (1969); p. 237-417

[19]   G. Chikovani et al., Phys. Lett. **22** (1966) 233

[20]   G. Caso et al., Lett. Nuov. Cim. **3** (1970) 707

[21]   L. Montanet, G.C. Rossi and G. Veneziano, Phys. Rep. **63** (1980) 149

[22]   M. Laloum, in Proc. of the Symposium on Nucleon-Antinucleon Annihilations, Chexbres, 1972, p.1-5; ed. L. Montanet, CERN 72-10, Geneva, 1972

[23]   S.U. Chung, in Proc. of the 6th European Symposium on Nucleon-Antinucleon and Quark-Antiquark Interactions, Santiago de Compostela, Spain, 1982, p.49; ed. J. Campos, Ann. di Fisica, Serie A, Vol. 79, Num. 1, 1983, Madrid, Spain

[24]   S. van der Meer, Stochastic Cooling and the Accumulation of Antiprotons, Science 230 (1985) 900

[25]   U. Gastaldi, K. Kilian and G. Plass, A low energy antiproton facility at CERN, Report to the PSCC; CERN/PSCC/79-17

[26]   Design Study of a Facility for Experiments with Low Energy Antiprotons (LEAR), CERN/PS/DL 80-7, ed. G. Plass

[27]   P. Lefevre, in Third LEAR Workshop, Physics with Antiprotons at LEAR in the ACOL era, Tignes, 1985, p. 33; ed. by U. Gastaldi et al., Editions Frontières, Gif-sur-Yvette, France, 1985

[28]   C. Amsler and F. Myhrer, CERN/PPE/91-29, to be published in Ann. Rev. Nucl. Sc. (1991), and ref. therein

[29]   R. Landua, Conference Summary, in Proc. of the First Biennial Conference on Low Energy Antiproton Physics, Stockholm, Sweden, July 1990, p. 525; eds. P. Carlson et al., World Scientific, Singapore, 1991

[30]   C.J. Batty, Rep. Prog. Phys. **52** (1989) 1165

[31]   B.O. Kerbikov et al., Sov. Phs. Usp. **32** (1989) 739

[32]   J. Sedlak and V. Simak, Sov. J. Part. Nucl. **19** (1988) 191

[33]   L. Linssen et al., Nucl. Phys. **A469** (1987) 726; P. Schiavon et al., Nucl. Phys. **A505** (1989) 595; and references therein

[34]   F. Bradamante,in Proc. of the First Biennial Conference on Low Energy Antiproton Physics, Stockholm, Sweden, July 1990, p. 219; eds. P. Carlson et al., World Scientific, Singapore, 1991

[35]   W. Brückner et al., Z. Phys. **A335** (1990) 217

[36]   W. Brückner et al., CERN-PPE/91-41, submitted to Z. Phys. A

[37]   O.D. Dalkarov and F. Myhrer, N. Cim. **40A** (1977) 152

[38]   J. Coté et al., Phys. Rev. Lett. **48** (1982) 1319

[39]   P.H. Timmers et al., Phys. Rev. **D29** (1984) 1928

[40]   P. Kroll and W. Schweiger, Nucl. Phys. **A503** (1989) 865

[41]   I.S. Shapiro, private communication, and J. Carbonell et al., in preparation

[42]   M.A. Alberg et al., Nucl. Phys. **A508** (1990) 323c

[43]   F. Tabakin and R.A. Eisenstein, Phys. Rev. **C31** (1985) 1857

[44]   M. Kohno and W. Weise, Nucl. Phys. **A479** (1988) 433c

[45]   S. Furui and A. Faessler, Nucl. Phys. **A508** (1987) 669

[46]   P. Lafrance et al., Phys. Lett. **214B** (1988) 317

[47]   O.D. Dalkarov, K.V. Protasov, and I.S. Shapiro, Preprint 37, Moscow, FIAN, 1988

[48]   S. Ahmad et al., Nucl. Instr. and Meth. **A286** (1990) 76

[49]   T.B. Day, G.A. Snow, and J. Sucher, Phys. Rev. **3** (1960) 864

[50]   E. Borie and M. Leon, Phys. Rev. **A21** (1980) 1460

[51]   G. Reifenröther and E. Klempt, Nucl. Phys. **A503** (1989) 885

[52]   U. Gastaldi, Nucl. Instr. and Meth. **156** (1978) 257; Nucl. Instr. and Meth. **157** (1978) 441; Nucl. Instr. and Meth. **176** (1980) 99; and Nucl. Instr. and Meth. **188** (1981) 459

[53]   M. Ziegler et al., Phys. Lett. **B206** (1988) 151

[54]   C.A. Baker et al., Nucl. Phys. **A483** (1988) 631; and C.W.E. van Eijk et al., Nucl. Phys. **A486** (1988) 604

[55]   R. Bacher et al., in Proc. of the First Biennial Conference on Low Energy Antiproton Physics, Stockholm, Sweden, July 1990, p. 373; eds. P. Carlson et al., World Scientific, Singapore, 1991

[56]   A.M. Green and J.A. Niskanen, Prog. Part. Nucl. Phys. **18** (1987) 93

[57]   G.F. Chew, in "Fundamental Problems in Elementary Particle Physics", Proc. 14th Conference on Physics at Univ. Brussel, p. 72, London, Wiley (1968)

[58]   J. Vandermeulen, Z. Phys. **C37** (1988) 563

[59]   H. Genz, M. Martinis and S. Tatur, Z. Phys. **A335** (1990) 87

[60]   U. Hartmann, E. Klempt and J. Körner, Z. Phys. **A331** (1988) 217

[61]   S. Furui, Nucl. Phys. (Proc. Suppl.) **B8** (1989) 231

[62]   T. Gutsche, M. Maruyama and A. Faessler, Nucl. Phys. **A503** (1989) 737

[63]   Yu.S. Kalashnikova and V.P. Yurov, Phys. Lett. **B231** (1989) 341

[64]   E.M. Henley, T. Oka and J. Vergados, Phys. Lett. **B166** (1986) 274

[65]   ASTERIX collaboration, M. Doser et al., Nucl. Phys. **A486** (1988) 493

[66]   ASTERIX collaboration, M. Doser et al., Phys. Lett. **B215** (1988) 792

[67]   GAMS collaboration, D. Alde et al., Phys. Lett. **B205** (1988) 397

[68]   M. Foster et al., Nucl. Phys. **B6** (1968) 107

[69]   ASTERIX collaboration, B. May et al., Z. Phys. **C46** (1990) 191

[70]   Crystal Barrel collaboration, K. Peters et al., $\bar{p}p$ annihilation at rest in liquid hydrogen into three neutral pseudoscalars, in Proc. of the First Biennial Conference on Low Energy Antiproton Physics, Stockholm, Sweden, July 1990, p. 161; eds. P. Carlson et al., World Scientific, Singapore, 1991

[71]   S. Godfrey and N. Isgur, Phys. Rev. **D32** (1985) 189

[72]   W.E. Buck, C.B. Dover, and J.M. Richard, Ann. Phys. **121** (1979) 47; C.B. Dover and J.M. Richard, Ann. Phys. **121** (1979) 70

[73]   R. Armenteros et al., Proc. Sienna Int. Conf Elem. Part., Vol I, eds. G. Bernadini and G. Puppi, Soc. Italiana di Fisica, Bologona, 1963; p. 287

[74]   P. Baillon et al., Nuov. Cim. **50A** (1967) 393

[75]   D.L. Scharre et al., Phys. Lett. **B97** (1980) 329

[76]   C. Edwards et al., Phys. Rev. Lett. **49** (1982) 259

[77]   A. Ando et al., Phys. Rev. Lett. **57** (1986) 1296

[78]   C. Dionisi et al., Nucl. Phys. **B169** (1980) 1

[79]   A. Gurtu et al., Nucl. Phys. **B151** (1979) 181

[80]   ASTERIX collaboration, K.D. Duch et al., Z. Phys. **C45** (1989) 223

[81]   T.A. Armstrong et al., Z. Phys. **C34** (1987) 23

[82]   J. Weinstein and N. Isgur, Phys. Rev. **D41** (1990) 2236

[83]   LEAP '90, First Biennial Conference on Low Energy Antiproton Physics, Stockholm, Sweden, July 1990; eds. P. Carlson, A. Kerek and S. Szilagyi, World Scientific, Singapore, 1991

[84]   R. Landua, Proc. First Workshop an Antimatter Physics at Low Energy (Fermilab, Batavia, Illinois, USA, March 1986), p. 35, and CERN-EP/86-136

[85]   H. Döbbeling et al., FILTEX proposal, P92, CERN/PSCC/85-80

[86]   N. Akchurin et al., The Spin Splitter, Proposal P120, CERN/PSCC/89-22

[87]   G. Gabrielse et al., Phys. Rev. Lett. **65** (1990) 1317

PS185: STRANGENESS PHYSICS AT LEAR

Staffan Ohlsson

Dept. of Radiation Sciences
Uppsala University, Sweden

representing the PS185 Collaboration:

P.D. Barnes[9], P. Birien[2], B.E. Bonner[6], W.H. Breunlich[8], G. Diebold[9],
W. Dutty[2], R.A. Eisenstein[4], G. Ericsson[7], W. Eyrich[3],
R. von Frankenberg[3], G. Franklin[9], J. Franz[2], N.Hamann[2],
D. Hertzog[4], A. Hoffmann[3], T. Johansson[7], K. Kilian[5],
C. Maher[9], D. Malz[3], W. Oelert[5], S. Ohlsson[7], P. Pawlek[8],
B. Quinn[9], E. Rössle[2], H. Schledermann[2], H. Schmitt[2],
J. Seydoux[9] and F. Stinzig[3]

[1] CERN, Geneva, Switzerland
[2] University of Freiburg, Freiburg, Fed. Rep. Germany
[3] University of Erlangen – Nürnberg, Erlangen, Fed. Rep. Germany
[4] University of Illinois, Urbana – Champaign (IL), USA
[5] IKP Forschungszentrum Jülich, Jülich, Fed. Rep. Germany
[6] Los Alamos National Laboratory, Los Alamos (NM), USA
[7] University of Uppsala, Uppsala, Sweden
[8] Institut für mittelenergiephysik der ÖAW, Vienna, Austria
[9] Carnegie Mellon University, Pittsburg (PA), USA

Introduction

The aim of PS185 is to study the antiproton – proton into antihyperon – hyperon channel near reaction threshold. This channel is suited for study of antistrange – strange ($\bar{s}s$) quark – pair creation. In the $\bar{p}p \rightarrow \bar{\Lambda}\Lambda$ case the additive quark model tells us that we can study the spin properties of the $\bar{s}s$ by observing them in the $\bar{\Lambda}\Lambda$ final state, since the spin of the Λ is that of its strange quark. The Λ decay into $p\pi^-$ is self analyzing , which means that one can get the polarization of the Λ by measuring the proton emission direction.

So far, $\bar{p}p \rightarrow \bar{\Lambda}\Lambda$ differential cross sections and polarizations have been published for several incoming momenta [1]. Spin correlation coefficients were extracted only for some high statistics data takings [1]. The $\bar{\Lambda}\Sigma^0$ + c.c. have been measured and published at incoming momenta of 1.695 GeV/c [2]. Polarization and differential cross section were extracted. The experiment has also taken data for the channel $\bar{p}p \rightarrow K_S K_S$ while scanning the region around the resonance X(2220) (former ξ) [1].

Medium-Energy Antiprotons and the Quark–Gluon Structure of Hadrons
Edited by R. Landua *et al.*, Plenum Press, New York, 1991

159

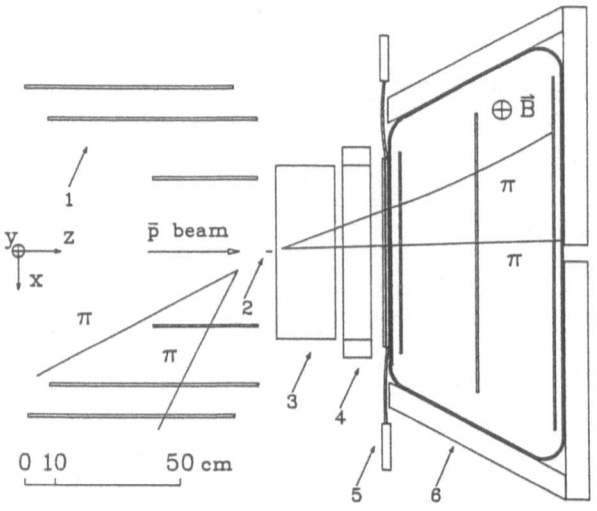

Fig. 1   The Setup of PS185
     1 = Streamer tubes; 2 = Target; 3 = Multiwire proportional chamber
stack; 4 = Drift chamber stack; 5 = Hodoscope; 6 = Solenoid magnet.

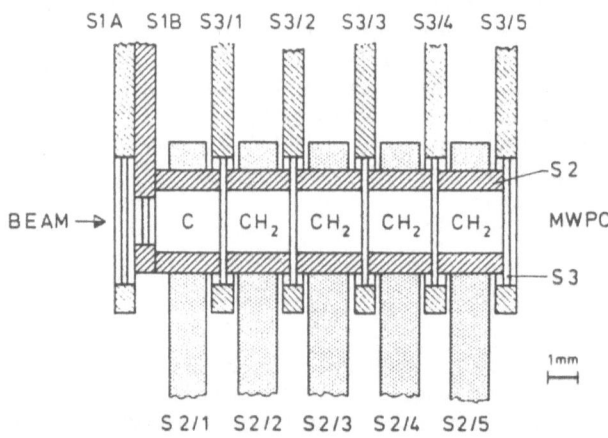

Fig. 2   Sandwiched target of PS185
     S1, S2 and S3's are scintillators surrounding the cells.

160

Among the interesting results the group has found, two results should be mentioned. The first is that PS185 found a strong $\ell \neq 0$ partial wave very close to threshold. The other result is that the polarization displays a very regular behaviour, being positive at small t′ and switching sign at the same t′ (-0.15 GeV^2) for all measured energies.

The experiment

The reaction $\bar{p}p \rightarrow \bar{\Lambda}\Lambda$ is the easiest to detect. The Λ decays in 64% of the time into $\bar{p}\pi^-$ with a $c\tau$ of 8 cm, which make the Lambda appear as a charged 'V' in our detector. The kinematics makes the V's go forward, and we need only to equip the forward region with detectors. Our setup is shown in Fig. 1. The multiwire proportional chamber stack and the driftchamber stack are used for tracking of the 2 V's. The reconstruction is purely geometrical, and the only thing needed as extra information is the charge of the decay particles, to distinguish the hyperons from the antihyperons. This charge is gotten from our solenoid magnet with 3 driftchamber planes, which determine the bending in the field.

Since the cross section of $\bar{p}p \rightarrow \bar{\Lambda}\Lambda$ near threshold is around 1 μb we need an efficient trigger. The sandwiched target arrangement shown in Fig. 2 helps us. Scintillators surround each target cell, and we take only events where we have a charged particle coming into the cell and a neutral system leaving it. To further sharpen the trigger we demand that we have charged particles in our hodoscope. This will give us events of the form: Charged – neutral – charged. The hodoscope also register multiplicity. The target – cell arrangement will give us an interaction point on – line and is also used as a degrader for measurements close to threshold.

For the other channels the detection and reconstruction is a little bit harder. The $\bar{p}p \rightarrow \bar{\Lambda}\Sigma^0$ + c.c. is detected as two V's as well, because Σ^0 decays in 100% of the time into $\Lambda\gamma$. The photon escapes detection, but there is enough information in the 2 V's to fully reconstruct the events. The $\bar{p}p \rightarrow K_S K_S$ is not restricted to the forward hemisphere as the other channels. That is why we have a streamertube arrangement to detect its backward going decay products. The $K_S K_S$ decays into charged particles ($\pi^+\pi^-$) in 68% of the time, with a $c\tau$ of 2.68 cm, as to give a 2 V pattern.

Results

The published cross – sections for $\bar{\Lambda}\Lambda$ and $\bar{\Lambda}\Sigma^0$ + c.c. are shown in Fig. 3. Here we can see the onset of the cross – section. Looking at the cross section closer to the threshold for $\bar{\Lambda}\Lambda$, which is done in Fig. 4, as a function of excess energy ($\varepsilon = \sqrt{s} - 2M_\Lambda$), one sees that we do not have a pure S – wave behaviour, but there is a P – wave admixture [3]. Studying the differential cross section near threshold (see Fig. 5) we can see that we do not have a flat distribution, which is expected for a pure S – wave. Even at our lowest data point at 1.436 GeV/c (excess energy 300 KeV) there is a forward peaked distribution [4].

Looking at the $\bar{\Lambda}\Sigma^0$ + c.c. and overlaying it with $\bar{\Lambda}\Lambda$ data at corresponding excess energy, we can see that both are displaying a similar structure, which might indicate that the reaction mechanisms for the two channels are related [2].

Mark III has seen a bump in the $K_S K_S$ invariant mass, when studying $J/\Psi \rightarrow \gamma K_S K_S$ [5]. This bump is interpreted as a resonance ($X(2220)$). Our data, shown in Fig. 6, does not show a bump at the invariant mass of 2.22 GeV.

Outlook

Since we have taken data since 1984, there is a lot of data on $\bar{p}p \rightarrow \bar{Y}Y$. More $\bar{\Lambda}\Lambda$ data will be published. We have taken data up to incoming momentum of 1.922 GeV/c. $\bar{\Lambda}\Sigma^0$ + c.c. data, where we enhanced our detector with a calorimeter, to detect the γ, have been taken. There is an ongoing

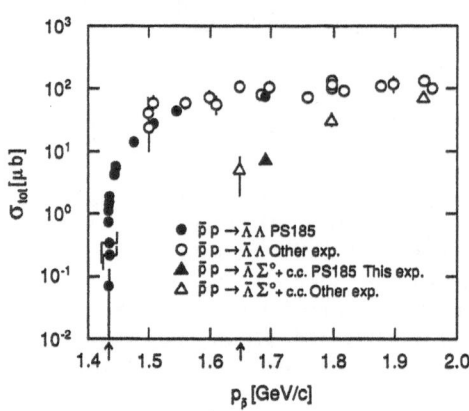

Fig. 3. Cross sections of $\bar{p}p \rightarrow \bar{\Lambda}\Lambda$ and $\bar{p}p \rightarrow \bar{\Lambda}\Sigma^0$ + c.c.

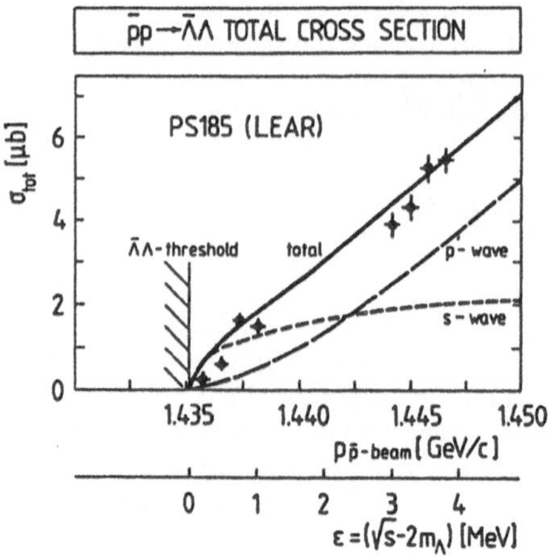

Fig. 4. Cross section of $\bar{p}p \rightarrow \bar{\Lambda}\Lambda$ near threshold as a function of excess energy. The expected behaviour for pure S- and P-wave are shown.

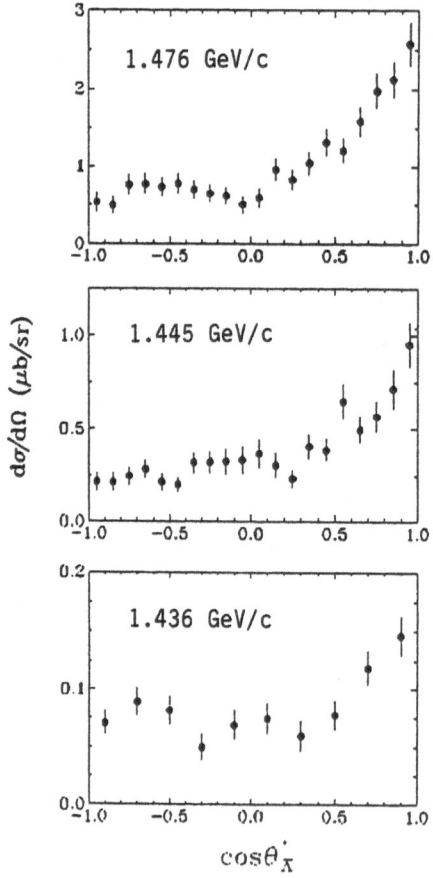

Fig. 5. Some differential cross sections near threshold for the reaction $\bar{p}p \rightarrow \bar{\Lambda}\Lambda$.

162

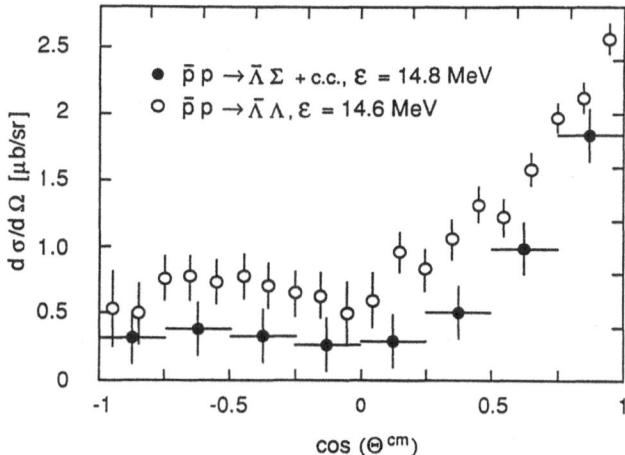

Fig. 6. Overlay of $\bar{\Lambda}\Sigma^0$ + c.c. and $\bar{\Lambda}\Lambda$ differential cross section at similar excess energies.

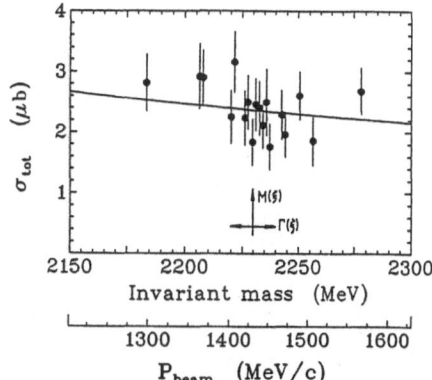

Fig. 7. Cross section for $\bar{p}p \to K_S K_S$ for incoming momenta corresponding to invariant mass around the X(2220) (former ξ).

analysis of these data. A test run on $\bar{\Sigma}^+\Sigma^+$ and $\bar{\Sigma}^-\Sigma^-$ have been done, using microstrip planes to detect the kinks due to the charged Σ decaying into Nπ. There has also been a measurement of $\bar{\Lambda}$p scattering. In this case we used a scintillator arrangement as an active scatterer. Our plans is to take $\bar{\Lambda}\Sigma$ + c.c. data with a fully equipped calorimeter, around the threshold. Moreover a micro−vertex detector will be installed, to be able to take precision data on $\bar{\Sigma}^+\Sigma^+$ and $\bar{\Sigma}^-\Sigma^-$.

References

1.   P.D. Barnes et al., Nucl. Phys. B(Proc. Suppl.) 8 (1989) 162.

2.   P.D. Barnes et al., CERN/EP − 90 − 40, Submitted to Phys. ett. B.

3.   K. Kilian, Nucl. Phys. Λ479 (1988) 425c.

4.   P.D. Barnes et al., Phys. Lett. B229 (1989) 432.

5.   R.M Baltrusaitis et al., Phys. Rev. Lett. 56 (1986) 10.

IN-FLIGHT ANNIHILATION p̄p→φφ AND p̄p→K̄K WITH JETSET AT LEAR

Nikolaus Hamann

CERN, EP Division
CH − 1211 Geneva 23, Switzerland

representing the JETSET Collaboration:

R. Armenteros[1], D. Bassi[3], P. Birien[2], R.K. Böck[1], A. Buzzo[3], E. Chesi[1],
P.T. Debevec[4], R. Dobinson[1], R.A. Eisenstein[4], T. Fearnley[1], M. Ferro − Luzzi[1],
J. Franz[2], M.A. Graham[4], N. Hamann[1], R. Harfield[1], P. Harris[4], D. Hertzog[4],
S. Hughes[4], T. Johansson[7], R. Jones[1], K. Kilian[5], K. Kirsebom[3], A. Klett[2],
H. Korsmo[6], M. Lovetere[3], A. Lundby[6], M. Macri[3], M. Marinelli[3], P. Martinengo[3],
L. Mattera[3], B. Mouellic[1], W. Oelert[5], S. Ohlsson[7], B. Osculati[3], J. − M. Perreau[1],
M.G. Pia[3], M. Price[1], P. Reimer[4], E. Rössle[2], A. Santroni[3], A. Scalisi[3],
H. Schmitt[2], B. Stugu[6], R. Tayloe[4], S. Terreni[3], H. − J. Urban[2] and H. Zipse[2]

[1] CERN, Geneva, Switzerland
[2] University of Freiburg, Freiburg, Fed. Rep. Germany
[3] University of Genova and INFN, Genova, Italy
[4] University of Illinois, Urbana − Champaign (IL), USA
[5] IKP Forschungszentrum Jülich, Jülich, Fed. Rep. Germany
[6] University of Oslo, Oslo, Norway
[7] University of Uppsala, Uppsala, Sweden

Abstract

The JETSET experiment (PS202) at CERN−LEAR uses an internal gas-jet target surrounded by a compact general-purpose detector. The initial physics programme involves the spectroscopy of hadronic states, including glueballs, hybrids and multi-quark states, in the mass range up to 2.4 GeV. Emphasis is put on the reactions p̄p→φφ, ωφ and ωω, and also on p̄p→K̄K, K̄K* and K̄*K*, using antiprotons up to 2 GeV/c momentum. This report gives the physics motivation and objectives of the experiment, and it describes the experimental technique chosen. A detailed account is given of data around and above 2 GeV invariant mass, which were obtained by previous experiments using hadronic reactions and heavy quarkonium decays. Future possiblities include threshold studies in multi-kaon final states, as well as high-statistics studies of exclusive hyperon−antihyperon production.

Medium-Energy Antiprotons and the Quark–Gluon Structure of Hadrons
Edited by R. Landua *et al.*, Plenum Press, New York, 1991

165

Introduction

According to the present understanding of particle physics, Quantum Chromodynamics (QCD) is the candidate theory underlying hadronic interactions. The theory describes hadrons as bound states of quarks that interact through colour forces by exchange of gluons. However, the equations of motion in QCD have yet to be solved in an exact manner, and many experimentally observed phenomena in hadronic interactions have yet to be understood. One prominent example for such a phenomenon is called 'colour confinement'. It points to the fact that the coloured quarks and gluons appear to exist not as free particles but only as colour-neutral bound hadronic states. Speaking in general terms, it is still one of the fundamental questions in QCD what the physical states of the theory are.

Lacking a more elegant description, our theoretical understanding of the hadronic spectrum is largely based on the Constituent Quark Model (QM). This model treats the 'conventional' hadrons. Baryons are colour-neutral triplets of quarks, (qqq), and mesons are colour-neutral quark−antiquark pairs, (q$\bar{\text{q}}$). But the conventional hadrons are only one class of physical objects that QCD is dealing with. The other class, which does not exist in the Constituent Quark Model, is characterized by the fact that, in addition to quarks, gluons too can play a dynamical role in hadronic states. Since gluons are coloured they are subject to self-coupling, and so gluons themselves can be constituents of hadronic matter. This self-coupling of the mediators of the strong interaction is a peculiarity in Quantum Chromodynamics. The situation in Quantum Electrodynamics (QED) is quite different, since there the mediating photons are electrically neutral.

For those states in QCD, which cannot be explained in terms of conventional hadrons, we may adopt the name 'non-quark-model' states (NQM). There are three types of such NQM states. Glueballs are quark-less states consisting of two or more real gluons, (gg) or (ggg). Hybrids − alternatively called meiktons − are composed of a quark−antiquark pair and a real gluon, (q$\bar{\text{q}}$g). Multi-quark states are colour-neutral combinations of four or more quarks and/or antiquarks, for instance (qq$\bar{\text{q}}\bar{\text{q}}$) or (qqqq$\bar{\text{q}}$). Since QCD allows the existence of such states, without giving arguments against it, their clear-cut experimental discovery would be of greatest importance for the verification of QCD in the low-energy regime. On the other hand, the proof of non-existence of NQM states could be a genuine problem with far-reaching consequences for our current understanding of hadronic interactions. Experimentally, the difficulty lies in the fact that there is almost no clear signature which would allow NQM states and conventional hadrons to be distinguished from each other. Therefore, some candidates have been established based mainly on the lack of consistency with our picture of the conventional hadronic spectrum.

The JETSET experiment (PS202) at CERN−LEAR aimes at the spectroscopy of hadronic states produced in the annihilation of in-flight antiprotons with protons at rest [1, 2]. The mass range covered in formation studies extends up to 2.43 GeV. The experiment uses a molecular hydrogen-cluster jet target inserted in the LEAR ring. A compact general-purpose detector surrounds the interaction region. One advantage of this technique is the possibility to perform fine-tuned momentum scanning of the circulating antiproton beam. This fully exploits the excellent beam-momentum resolution of LEAR, and it results in a correspondingly good mass resolution for states that may be observed in $\bar{\text{p}}$p interactions. Another advantage is the high luminosity. It is achieved by making most efficient use of the antiprotons, and it is unprecedented in extracted-beam experiments with in-flight antiprotons.

The experiment initially focusses on the investigation of $\bar{\text{p}}$p annihilations into $\phi\phi$, $K^+K^-\phi$ and $K^+K^-K^+K^-$ from the respective reaction thresholds up to the 2.0 GeV/c beam momentum. This is to be complemented by studies of $\bar{\text{p}}$p annihilations into $\pi^0\phi$, $\eta\phi$, $\omega\phi$ and $\omega\omega$, which involve the detection both of charged particles and of photons, and also by measurements of other 'strange' channels, such as $\bar{\text{K}}$K, $\bar{\text{K}}$K* and $\bar{\text{K}}^*$K*. In order to have an access to hadronic states with quantum numbers not available in $\bar{\text{p}}$p formation experiments, three-particle reactions such as $\bar{\text{p}}$p→$\pi^0\phi\phi$ are to be measured using the highest possible beam momenta at LEAR. In the following chapter we describe the experimental technique chosen. Then we discuss the physics objectives of the JETSET experiment and review previously obtained data relevant to the mass region around and above 2 GeV. We conclude with an outlook into the future in terms of a substantial improvement of the experimental boundary conditions and the subsequent widening of the physics programme.[1]

Experimental technique

LEAR and the internal jet target

A central feature of the JETSET apparatus (see figure 1) is the molecular hydrogen-cluster jet target. The system is installed in the straight section SL2 of the LEAR ring. The basic principle is the expansion of molecular hydrogen kept at low temperature and high pressure, which creates an intense supersonic flow of clusters of H_2 molecules. The jet is oriented in the horizontal direction and intersects the antiproton beam perpendicularly. In the interaction region the jet provides a low-mass pure gaseous hydrogen target. The jet has a density $\rho = 5 \times 10^{13}$ atoms/cm^2 and a diameter of 8 mm (FWHM). For the later stage of the experiment the operation of a polarized jet target is foreseen, which would make observables such as spin-transfer parameters accessible.

Some considerations for the operation of LEAR with the jet target are discussed in refs. [4] and [5]. At a momentum of 1.5 GeV/c the antiprotons stored in the machine have a revolution frequency f = 3.2 MHz. With an unbunched beam initially consisting of $N_0 = 6 \times 10^{10}$ antiprotons, the peak luminosity is thus $L_0 = \rho f N_0 = 10^{31}$ cm$^{-2} \cdot$sec^{-1}. The hadronic $\bar{p}p$ total cross-section being $\sigma \approx 100$ mb in the relevant energy range, we have a hadronic interaction rate of up to $\sigma L_0 = 10^6$ sec^{-1}. This is one of several factors which cause the luminosity to decrease exponentially in time, $L = L_0 \exp(-t/\tau)$, and which thus influence the beam lifetime τ in the machine. The angular acceptance of the machine for scattering from the hydrogen jet is limited by the size of the vacuum chamber. Without a special low-beta insertion the acceptance is of the order of a few mrad. A fraction of the Coulomb-scattered antiprotons can thus be recuperated. However, the presence of the hydrogen jet causes a local pressure bump in the machine, which would lead within some minutes to an unacceptably large blow-up of the beam. This must be compensated by continuous transverse stochastic cooling, so that the transverse dimensions of the beam are always kept smaller than those of the intersecting jet. The energy loss of up to a few keV/sec must be compensated by continuous longitudinal stochastic cooling, so as to maintain a good momentum resolution and to keep the beam on its nominal orbit. In view of a possible upgrade of the JETSET apparatus with a surrounding magnetic field, schemes for the compensation of such a field and the induced machine resonances have already been studied. The calculations seem to indicate that, except for very low beam momenta, LEAR can be safely operated with a solenoidal field up to about 1 T·m.

In the region where the JETSET detector system is located around the interaction volume, the LEAR ring is equipped with a special vacuum chamber. The size of the chamber is dictated by the machine parameters: it is an oval with horizontal and vertical half-axes of 78 mm and 38 mm, respectively. This immediately limits the geometrical acceptance of the detector to angles $\theta > 10°$. It also means that particles have to travel some distance and must traverse the chamber wall before they can reach the nearest detector. At present the best solution is a corrugated chamber made of 0.3 mm Inconal, the hope being that future developments will lead to a reduced amount of material or perhaps even to a smaller-sized vacuum chamber.

---

[1] Throughout this report some particle names [3] are shortened as follows: $\rho = \rho(770)$, $\omega = \omega(783)$, $K^* = K^*(892)$, $\eta' = \eta'(958)$, $f_0 = f_0(975)$, $a_0 = a_0(980)$, $\phi = \phi(1020)$, $K_2^* (1430) = K_2^*$, $\eta_c = \eta_c(2980)$ and $J/\psi = J/\psi(3097)$.

One may assume that the stochastic cooling systems work perfectly well and that they always keep the beam emittance below the machine acceptance. If, for reasons of simplicity, one further assumes that there are no losses in the machine other than due to hadronic interactions and (multiple) Coulomb scattering in the hydrogen jet, one can estimate the beam lifetime. The number of particles in the machine is then given by $N(t) = N_0 \exp(-\rho f \Sigma t)$, where Σ denotes the total cross-section for 'absorption' or 'off-acceptance scattering' of antiprotons. For $\Sigma \approx 120$ mb, the 1/e lifetime would be $\tau = (\rho f \Sigma)^{-1} = 14.5$ hr. The longitudinal stochastic cooling of the beam also allows the antiproton momentum to be continuously scanned across a band in $\sqrt{s}$, for instance a hadronic resonance that one may hope to find. The momentum resolution of the beam, $\Delta p/p \approx 10^{-3}$, and the low mass of the hydrogen jet would then translate into an experimental invariant-mass resolution of about 500 keV for such a state.

Basic event signatures

For the task of on-line triggering, the $\bar{p}p$ reactions in the prime interest of the JETSET experiment can be grouped into three categories. The decay modes assumed for this are $\phi \to K^+ K^-$, $\omega \to \gamma \pi^0$, $K^0(K_S) \to \pi^+ \pi^-$, and $K^* \to \pi^0 K^0(K_S) \to \pi^0 \pi^+ \pi^-$.

- Channels $\phi\phi$, $K^+K^-\phi$, $K^+K^-K^+K^-$, and $\pi^0\phi\phi$:
  - final state with four charged K and with no or one π^0.
- Channels $\omega\phi$ and $\omega\omega$:
  - final state with one $\gamma\pi^0$ and two charged K, or with two $\gamma\pi^0$ and no charged particles.
- Channels $\bar{K}K^*$, $\bar{K}K^*$ and $\bar{K}^*K^*$:
  - final state with four charged π, featuring a $2V^0$ topology of two delayed decays, and with no or one or two π^0.

The fast on-line triggering schemes are thus based on the multiplicity of charged particles, on the discrimination of π / K / p, and on the multiplicity of photons.

Detection of charged-particle multiplicities

The vacuum chamber of the LEAR machine is immediately surrounded by 2 mm thick scintillator strips running along the beam direction (Z). One set of 40 strips, each 10 mm wide, covers polar angles $\theta \leq 45°$. Another set of 20 strips, each 20 mm wide, gives the complementary coverage for larger angles. These 'inner' scintillators with their fine azimuthal segmentation provide a fast trigger on charged-particle multiplicities and they also give the time reference for $\bar{p}p$ interactions. The information obtained from them is not much distorted by particle decays, secondary interactions or time-of-flight effects. However, such effects can become large as one goes to low beam momenta or approaches a reaction threshold, so that the inner trigger counters then play a key role in keeping the acceptance at a reasonable level.

Additional sets of 5 mm thick segmented 'outer' scintillation counters are positioned 40 to 60 cm away from the interaction region. Charged-particle multiplicity information in the forward part is provided by 48 pie-shaped segments, each covering $\Delta\phi = 7.5°$. This plane is overlayed by two other planes, each of which contains 24 segments curved with opposite sense. The three layers form a fine grid that can provide fast kinematical information on the events, for instance hit coordinates and track angles, which are to be used in the higher-level trigger. Similarly, the barrel part comprises a layer of 12 straight segments running along the Z direction, each covering $\Delta\phi = 30°$. This is interwoven with two layers of 12 helical pieces each, which are wound with opposite screw sense.

The combined information from the inner and the outer scintillation counters is used to establish a 'charged multiplicity step' trigger on particles that have relatively long decay lengths, such as K_S or Λ. More generally speaking, one wants to discriminate particles pointing back to the interaction volume against those showing a finite impact parameter relative to that region.

168

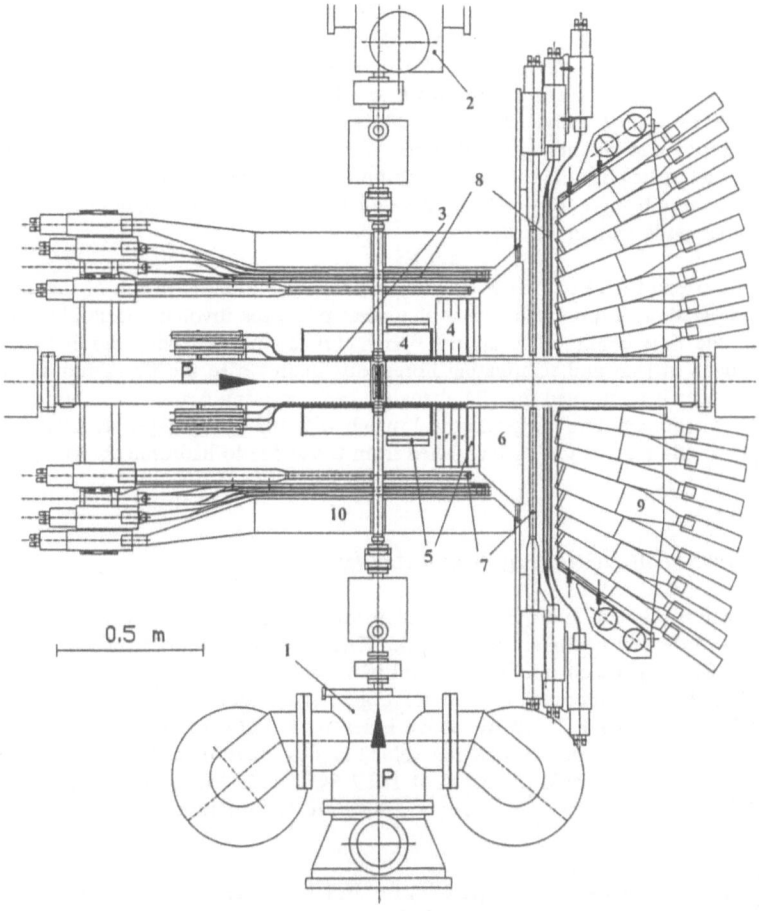

Figure 1. Experimental set-up of the JETSET (PS202) apparatus at LEAR. The elements are: 1 = jet-target source; 2 = jet-target sink; 3 = inner trigger scintillators; 4 = barrel and forward drift-tube chambers; 5 = silicon pad detectors; 6 = ring-imaging Cherenkov counters; 7 = threshold Cherenkov counters; 8 = outer trigger scintillators; 9 = forward electromagnetic calorimeter; 10 = barrel γ-veto counters.

Gamma detection

In the forward region the JETSET apparatus is equipped with an electromagnetic calorimeter. Its main purpose is to detect γ-rays coming from the decays of neutral mesons, such as π^0, η or ω, which may be produced in $\bar{p}p$ interactions. For measurements of the exclusive production of $\phi\phi$ or $\bar{K}K$, it provides an efficient veto on 'unwanted' events involving photons in the final state ($\phi{\rightarrow}K^+K^-$; $K_S{\rightarrow}\pi^+\pi^-$; $K^*{\rightarrow}\pi K$). But studies of other channels, such as $\pi^0\phi$, $\eta\phi$, $\omega\phi$, $\omega\omega$ or $\pi^0\phi\phi$, require that the direction and energy of the photons be measured well ($\pi^0{\rightarrow}\gamma\gamma$; $\eta{\rightarrow}\gamma\gamma$, $\pi^0\pi^0\pi^0$, $\pi^+\pi^-\pi^0$; $\omega{\rightarrow}\pi^+\pi^-\pi^0$, $\gamma\pi^0$). The calorimeter is made in a technique employing lead and scintillating fibres. It consists of 300 individual modules ('towers') grouped in eight rings. There are 12, 24 or 48 modules per ring in the azimuthal direction, and the modules have nearly constant front-face areas. Each of these modules contains plastic fibres of 1 mm diameter, which are embedded between sheets of corrugated lead-alloy glued on top of each other. The filling by volume is 50 % fibres, 35 % lead-alloy and 15 % optical epoxy glue. The blocks are 12.5 radiation lengths deep. They are individually machined and oriented such that the fibres running along their depths approximately point to the interaction region. The energy resolution was measured over a wide range of electron energies, a typical value being $\sigma/E = 0.063/\sqrt{E}$ (with E in GeV). The angular resolution for γ-rays coming from the target is around 7 mrad.

A project to extend the electromagnetic calorimeter into the barrel region of the apparatus is in progress. This upgrade is important in view of several processes involving multiple photons in the final state. For the time being, however, the barrel part is equipped with simple γ-veto counters. They are also made of lead and scintillating fibres, but in this case the fibres run along the beam direction. The device is segmented into 24 elements, each covering $\Delta\phi = 15°$. For the calorimeter as well as for the γ-veto counters the geometrical match with the outer trigger scintillators makes sure that showers due to photons can be distinguished from those due to hadronic charged particles.

Charged-particle identification by β and dE/dx

For the reaction $\bar{p}p{\rightarrow}\phi\phi$ and others, sets of threshold Cherenkov counters provide a fast and efficient rejection of the charged-pion background. There are 24 pie-shaped detector segments in the forward part and 24 straight elements in the barrel part, so that each of the pieces covers $\Delta\phi = 15°$ and thus matches the segmentation of the outer trigger scintillators. The 2 cm thick counters are filled either with liquid freon $FC-72$ (refractive index n = 1.26) or with water (n = 1.33). The detectors are thus sensitive to particles having velocities $\beta > 0.79$ or $\beta > 0.75$, respectively. This corresponds to π / K / p threshold momenta of 182 / 645 / 1225 MeV/c in the first case and 159 / 563 / 1070 MeV/c in the second case. These values are well-matched in particular to studies of $K^+K^-K^+K^-$ final states.

A ring-imaging Cherenkov counter (RICH) for JETSET is under development. It will provide additional information about the momenta of charged particles by measuring their β values. The RICH is conceived to have a 1 cm thick quartz radiator (n = 1.46). This is followed by a 6 cm deep drift space that allows the Cherenkov light cone to broaden before detection. The detector is sensitive to particles having velocities $\beta > 0.68$, which corresponds to π / K / p threshold momenta of 131 / 464 / 882 MeV/c. The photons are detected by means of photon$-$electron conversion in TMAE gas that is carried in a helium$-$ethane mixture at room temperature. The detector employs a modular structure, the basic element being a miniature wire chamber with $8{\times}8$ mm^2 cell size and a single wire running perpendicular to the radiator surface. A detector module is formed by a matrix of 64 such cells. The pattern of struck cells provides the information necessary to determine the β value. One can estimate that under these conditions the measurement error in β will be around 10 %.

Silicon pad counters measure the specific energy loss dE/dx of charged particles and thus greatly help to distinguish $K^+K^-K^+K^-$ final states from backgrounds such as $\bar{p}p\pi^+\pi^-$. Their information is evaluated during the off-line event reconstruction. The silicon detector is a natural complement to the threshold Cherenkov counters and the RICH, as its best information is obtained at low particle energies. The 280 μm thick counters are arranged in two planes in the forward region and two layers in the lower-angle part of the barrel region. They consist of elements 2.0×2.4 cm^2 in area, each of which contains four individual pads. The dE/dx signal resolution obtained from two measurements is 10 to 20 % for particles having $\beta < 0.8$. This would then allow the particle momentum to be 'predicted', typical precisions for kaons being 10 % at 300 MeV/c and 30 % at 600 MeV/c. In addition to dE/dx measurements, the fine segmentation of the silicon counters also provides geometrical information on charged-particle tracks.

Tracking of charged particles

The 'heart' of the detection system and the central element for the off-line track reconstruction of charged-particles is a wire chamber made of individual drift tubes. This device is divided into a forward part covering polar angles $\theta < 50°$ and a barrel part being sensitive to tracks with $\theta > 40°$.

The barrel tracking chamber is composed of about 1500 individual drift tubes ('straws') running parallel to the beam direction. The tubes are glued together in two self-supporting units which constitute the top and the bottom halves of the barrel chamber. The horizontal split facilitates the installation of the detector and allows for the entry of the jet target. The 300 tubes in the horizontal midplane of the jet are to be mounted at a later stage. Each of the barrel straws consists of an extruded aluminium tube with 60 μm wall thickness and 8 mm diameter. The tubes are 436 mm long. Each tube is equipped with a stainless steel wire of 30 μm diameter running down its centre. The wires are kept in position by precision plastic endpieces. Low-mass sandwich structures serve as endplates and gas manifolds, and they also provide the electrical connections to the wires. Such a construction reduces the amount of support material needed at the forward end of the assembly, and it thus reduces the multiple scattering of particles leaving the barrel and entering the forward tracking chamber. For this reason all connections to gas supply, high voltage and readout systems are made at the barrel rear endplate. Each wire is equipped separately with front-end electronics consisting of a charge-sensitive preamplifier, a postamplifier with differential analog drivers, and a discriminator with differential ECL drivers. The electrons' drift-times are read from each wire in order to measure the radial distance of impact (rϕ). For the measurement of the longitudinal coordinate (Z) by means of charge division, two wires are connected to each other 'back-to-back' using resistors in SMD technology which are located at the forward end of the assembly. The combined information from TDC's (drift time) and ADC's (charge division) provides unambiguous three-dimensional information for track and vertex reconstruction. When operated with a gas mixture of 50 % Ar and 50 % CO_2 at atmospheric pressure, the average resolutions obtained are $\sigma(\perp) = 150$ μm from the drift-time information and $\sigma(\|) = 8$ mm from the charge-division information.

The forward tracking chamber uses basically the same drift tubes as described above, although the geometrical configuration and the endplate assembly are quite different. In this case the straws are directly glued head-on onto precision printed-circuit boards that fix the wire position and provide electrical connections. The straws are mounted in 12 planes perpendicular to the beam direction. The alternating x and y modules, arranged as (3x)(3y)(3x)(3y), employ a total number of about 1000 straws. Such an arrangement provides complete tracking information on the basis of drift-time readout only.

Physics objectives of the JETSET experiment

Survey of phenomenological and theoretical approaches

The questions that are addressed in phenomenological and theoretical approaches to non-quark-model hadrons, such as glueballs, hybrids and multi-quark states, are the masses, the ordering of spin states, and the most likely modes of production and decay. The reader is referred to several excellent recent reviews on this subject [6−8]. As mentioned in the introduction, the identification of an NQM state as such goes largely by 'excluding' that it be a conventional hadron. It thus requires a rather good understanding of the spectrum of $(q\bar{q})$ mesons. However, a problem arises from the possibility that the real physical states, which we observe, are mixtures of conventional and non-conventional hadrons. Another difficulty for experimenters is that some NQM states, in particular light glueballs, are expected to lie in mass regions which are already densely populated with $(q\bar{q})$ states.

A number of models have been developed to predict spectra of non-quark-model states, or they have been adopted from their original application to $(q\bar{q})$ mesons. In Bag Models [9, 10] the massless spin-1 gluons are confined in a spherical bag of suitable size. Two types of eigenmodes with angular momentum L are distinguished: transverse electric $(TE)_L$ with parity $P = (-1)^{L+1}$, and transverse magnetic $(TM)_L$ with parity $P = (-1)^L$. Correcting for gluon−gluon interactions in the bag, the two-gluon state with the lowest mass is of the type $(TE)_1(TE)_1$ with $J^{PC}=0^{++}$ and a mass of 1 to 1.3 GeV. In order of increasing mass, the spectrum is supposed to be $J^{PC} = 0^{++} < 0^{-+} < 2^{++} < 2^{-+}$ for two-gluon states, the highest mass being around 2.3 GeV. In Potential Models one takes gluons to be massive spin-1 particles, which interact through a long-range confining potential that is linearly increasing with distance. The assumed effective gluon mass is of the order of a few hundred MeV, which is similar to the effective masses of light quarks bound in hadrons. In QCD Sum Rule approaches one studies functions associated with a current carrying the same quantum numbers as a resonance. The extrapolation from the high-momentum regime down to the non-perturbative one allows the mass of such a resonance to be extracted. The Flux-Tube Model [11] is a QCD-based approach in which glueballs can be formed by removing the quarks from the colour-flux tubes and joining the end of those flux tubes together. The lightest glueball is predicted to have $J^{PC}=0^{++}$ with a mass around 1.5 GeV, and most of the other states are expected above 2 GeV. In Lattice Gauge calculations [12] methods of Monte−Carlo simulations are used to evaluate Feynman path integrals on a lattice and to study two-point correlation functions. This numerical type of approach can provide, in principle, the most accurate description of non-perturbative aspects of QCD. For reasons of computational feasibility, however, the lattice must be of finite spacing and of finite box size. The goal is to choose conditions such that a perturbative treatment is applicable, so that the resulting mass spectrum is not much affected by the finite lattice spacing. Typical values compatible with today's computing power are a lattice spacing of 0.1 fm and a box size of 2 fm. The 0^{++} scalar glueball again turns out to be the lightest one, and the 2^{++} tensor state is expected to be about 1.5 times heavier [8, 12]. However, in Lattice Gauge calculations as well as in the Flux-Tube Model there is a considerable uncertainty in the overall mass scale of hadrons containing gluonic degrees of freedom.

Whereas there is some consensus on the J^{PC} ordering of glueballs, not much can be said about their widths. One may assume that those processes, in which the initial and the final states are not connected by quark lines, are dominated by multi-gluon intermediate states. Then a prediction can be made [13] that the width of such a glueball is of the order of 10 MeV. There is, however, considerable uncertainty about the justification for the above assumption, largely due to the lack of understanding of the dynamics underlying quark-disconnected processes. Another naive expectation concerns the decay modes of glueballs. A flavourless, electrically neutral two-gluon state should have no preferred quark flavour or charge for its decays to $(q\bar{q})$ mesons. However, this flavour symmetry is broken due to the mass difference between the strange quark and the lighter u- and d-quarks. There are, in fact, arguments derived from perturbation theory which favour a particularly strong coupling of gluons to strange quarks, rather than to u- or d-quarks [14].

Glueballs have been searched for in various hadronic reaction channels, such as πp or $\bar{p}p$, and in the decay of heavy quarkonia, in particular radiative J/ψ decays. Proton−antiproton annihilations at rest or in flight can provide a rich source of hard gluons. The production of final states containing one or more $(s\bar{s})$ quark pairs is likely to proceed via the annihilation of one or more light $(q\bar{q})$ pairs. In general, however, both annihilation and rearrangement mechanisms contribute to the production or formation of hadrons in $\bar{p}p$ interactions. The outstanding feature of radiative decays, $J/\psi \rightarrow \gamma + X$, is the cleanness and the conceptual 'simplicity' of this process: the $(c\bar{c})$ quark pair must annihilate in order for hadrons to be created, and the photon provides an experimental tag for that. The decay is thought to be dominated by $J/\psi \to \gamma gg \to \gamma + $ hadrons, with $J^{PC} = 0^{++}$, 0^{-+} and 2^{++} being the most important partial waves contributing. In fact, the two most serious glueball candidates, $\eta(1430)$ with $J^{PC} = 0^{-+}$ and $f_2(1720)$ with $J^{PC} = 2^{++}$, were seen in radiative J/ψ decays. They seem to have no place in the spectrum of conventional $(q\bar{q})$ mesons, and their parameters are now often used in theoretical calculations in order to fix quantities such as the gluon self-energies. The question as to how 'glueish' a glueball candidate is cannot be answered easily in a quantitative way. For the comparison of different candidates with the same J^{PC}, however, Chanowitz [10] has defined the 'stickiness' of a state X as $S_X = [\Gamma(J/\psi \to \gamma X)/(\text{phase space})]/[\Gamma(X \to \gamma\gamma)/(\text{phase space})]$. This expresses the expectation that glueballs are copiously produced in radiative J/ψ decays and that they have, due to the lack of electric charge, very little coupling to photons.

Hybrids, being mixed states of quarks and gluons, have been studied theoretically in frameworks similar to those of glueball studies. Explicit calculations in the Flux-Tube Model have predicted a number of states around 2 GeV mass, which preferentially decay to two mesons with one of them being excited [15]. In the Bag Model the lightest hybrids arise from the coupling of a $(q\bar{q})$ pair with $L = 0$ and $S = 0$ or 1 to a $TE(J^P = 1^+)$ gluon [6, 7]. The resulting states, most of them below 2 GeV mass, are a spin singlet $J^{PC} = 1^{--}$ and a spin triplet $J^{PC} = 0^{-+}$, 1^{-+}, 2^{-+}. Excited states are due to a $TM(J^P = 1^-)$ gluon. This gives rise to a singlet $J^{PC} = 1^{+-}$ and a triplet $J^{PC} = 0^{++}$, 1^{++}, 2^{++}, with masses just below or well above 2 GeV. It should be noted that quantum numbers in the series $J^{PC} = (\text{odd})^{-+}$ or $(\text{even})^{+-}$ are not accessible for conventional $(q\bar{q})$ mesons. The identification of a state such as $J^{PC} = 1^{-+}$, a so-called J^{PC}-exotic, would thus be the most convincing evidence for the discovery of a non-conventional hadron.

A $(q'\bar{q}g)$ hybrid could, for instance, decay according to the following two-step scheme. The first step is $(q'\bar{q})_8 g \to (q'\bar{q})_8 (s\bar{s})_8$, where the subscript refers to colour octet. Two mesons can then be formed either by rearrangement, $(q'\bar{s})_1 (s\bar{q})_1$, or by gluon exchange, $(q'\bar{q})_1 (s\bar{s})_1$, the subscript denoting colour singlet here. The second possiblity is, due to disconnected quark lines, suppressed for the decay of any conventional meson, hence it may be taken as a decay signature of a hybrid. A speculative example for this is the state $\xi(2230)$ seen in $\bar{K}K$ final states. Its yet unobserved $\omega\phi$ decay mode would be an indication for the $(u\bar{u} + d\bar{d})g$ hypothesis. An 'ω-based' hybrid with $J^{PC} = 2^{++}$ is, in fact, predicted at 2.32 GeV mass, and it has therefore been suggested [14] to search for the decays of $\xi(2230)$ to $\omega\phi$ and $\bar{K}^*K^*$.

Multi-quark states too have been the subject of many theoretical investigations. A prominent example is the six-quark object called H, which has been predicted to be a stable strangeness -2 dibaryon about 80 MeV below the $\Lambda\Lambda$ invariant mass [16]. Other studies found that most $(q\bar{q}q\bar{q})$ configurations lead to two free mesons, rather than to bound four-quark states [17]. An exception are the so-called $\bar{K}K$ molecules, 'deuteron-like' systems, which have been predicted to exist as loosely bound states in the isospin 0 and 1 channels. Genuine multi-quark states, if they really exist and are observable, may have peculiar properties that can never be carried by ordinary $(q\bar{q})$ or (qqq) hadrons. Flavour-exotics, for instance, can be formed in $K^+ p \to X^*$, and the result would in this case be a doubly-positive charged object with strangeness $+1$.

A suggestion being of particular interest to the JETSET experiment has been made by Ioffe [18]. He expects hybrids and four-quark states made of s- and/or c-quarks (but without u- or d-quarks) to have rather small total widths of less than 50 MeV, which may lead to cleaner experimental signatures. At masses of (2.3 ± 0.1) GeV, hence close to the upper limit of LEAR, several $(s\bar{s}s\bar{s})$ states are predicted with $J^{PC} = 0^{++}$, 1^{+-} and 2^{++}. The main decay mode of the 2^{++} state at 2.3 GeV is $\phi\phi$, and that of the 1^{+-} state about 100 MeV lower is $\eta\phi$ or $\eta'\phi$. The cross-section for the formation of such states in $\bar{p}p$ annihilations is estimated to be only 10 to 100 nb, but the partial width for this channel would be well below 1 MeV.

The reactions $\bar{p}p \to \phi\phi$, $\bar{p}p \to \omega\phi$ and $\bar{p}p \to \omega\omega$

Before elaborating on possible mechanisms for the exclusive production of $\phi\phi$, $\omega\phi$ and $\omega\omega$ in $\bar{p}p$ annihilations, it is useful to examine the quantum numbers that are available both in the initial and in the final states. The quantum numbers [3] of ω and ϕ mesons are the same, $J^{PC}(I^G) = 1^{--}(0^-)$, hence the selection rules are the same for all three channels considered here. In the following we use capital letters, L and S, for the orbital and spin angular momentum quantum numbers of the final state; those of the $\bar{p}p$ state are denoted by lower case, ℓ and s. The $\phi\phi$ state restricts the quantum numbers of charge conjugation and of G-parity to be positive, C = +1 and G = +1, respectively. Since the charge conjugation is defined as $(-1)^{L+S}$ for $\phi\phi$ and $(-1)^{\ell+s}$ for $\bar{p}p$, it follows that both (L+S) and (ℓ+s) must be even numbers. The parity of $\phi\phi$ is $(-1)^L$, and it is $(-1)^{\ell+1}$ for $\bar{p}p$. Therefore, L must be odd when ℓ is even, and L must be even when ℓ is odd. The total spin of the $\phi\phi$ system can have the quantum numbers S = 0, 1 or 2; that of the $\bar{p}p$ initial state is restricted to s = 0 or 1. Finally, the only allowed values of the isospin quantum numbers are (I,I_3) = (0,0). The conservation of parity and total angular momentum (J = j) then leads to the following three sets of selection rules.

- ℓ even, s = 0; J = ℓ
  - L odd, S = 1; J = L$\pm$1 (J^{PC} = 0^{-+}, 2^{-+}, 4^{-+}, ...)
- ℓ odd, s = 1; J = ℓ
  - L even, S = 2; J = L$\pm$1 (J^{PC} = 1^{++}, 3^{++}, 5^{++}...)
- ℓ odd, s = 1; J = $\ell\pm$1
  - L even, S = 0; J = L (J^{PC} = 0^{++}, 2^{++}, 4^{++}, ...)
  - L even, S = 2; J = L (J^{PC} = 2^{++}, 4^{++}, ...)
  - L even, S = 2; J = L$\pm$2 (J^{PC} = 0^{++}, 2^{++}, 4^{++}, ...)

In total there are 8 kinds of amplitudes connecting the $\bar{p}p$ spin$-$parity initial state to the $\phi\phi$ (or $\omega\phi$, $\omega\omega$) spin$-$parity final state. We note that exotic quantum numbers, $J^{PC}=1^{-+}$ for instance, are in principle possible for states decaying to $\phi\phi$, but in $\bar{p}p$ annihilations they can be accessed only through reactions involving more particles, such as $\bar{p}p{\rightarrow}\pi^0\phi\phi$.

Physical states having the same J^{PC} and the same additive quantum numbers can mix because of SU(3) breaking. Examples where such mixing occurs [3] are given by pseudoscalar mesons, η and η', and by vector mesons, ϕ and ω. Each of these particles can be thought of as a linear combination of the corresponding SU(3) octet and singlet states. For the vector mesons the octet and singlet states, written in terms of unsymmetrized quark wave functions, are $\omega_8 = (u\bar{u} + d\bar{d} - 2s\bar{s})/\sqrt{6}$ and $\omega_1 = (u\bar{u} + d\bar{d} + s\bar{s})/\sqrt{3}$, respectively. The physical states ϕ and ω are then constructed from the SU(3) states involving a 'vector-mixing' angle: $\phi = \omega_8\cos\theta_V - \omega_1\sin\theta_V$ and $\omega = \omega_8\sin\theta_V + \omega_1\cos\theta_V$. The mixing is said to be 'ideal' if $\phi = (s\bar{s})$ and $\omega = (u\bar{u} + d\bar{d})/\sqrt{2}$. This is the case for $\tan\theta_V = 1/\sqrt{2}$, which corresponds to $\theta_V \approx 35.3°$. The θ_V value as determined from $\phi{\rightarrow}\pi\rho$ is about 39°. It can thus be assumed that the non-($s\bar{s}$) admixture in the ϕ-meson and the ($s\bar{s}$) admixture in the ω-meson are of the order of 1 %.

The immediate consequence for the hadronic reaction $\bar{p}p{\rightarrow}\phi\phi$ appears to be rather extreme. As the initial and the final states have no quark in common, the transition should proceed via an 'intermediate state' containing only mediators of the strong interaction, namely at least two gluons (see figure 2a). This would make the reaction an ideal hunting ground for glueballs. The cross-section to be expected for such a 'quark-disconnected' process is, however, small. But if so, it would be explained by the Okubo$-$Zweig$-$Iizuka (OZI) rule [19]. This empirical rule was established to account for the small ratio of branching fractions, BR($\phi{\rightarrow}\pi^+\pi^-\pi^0$)/BR($\omega{\rightarrow}\pi^+\pi^-\pi^0$). The OZI rule basically states that quark-disconnected processes are suppressed with respect to quark-connected ones. As a consequence, the occurrence of $\bar{p}p{\rightarrow}\phi\phi$ through a quark-disconnected diagram would have to be considered a violation of the OZI rule, or it could be a hint for 'new physics'. It should be noted, however, that the OZI rule itself is poorly understood in the context of the underlying QCD, and even more so are its limits of validity. The OZI suppression can be 'bypassed' if one considers that the reaction $\bar{p}p{\rightarrow}\phi\phi$ may proceed via the $\phi-\omega$ mixing. In this picture, the $\bar{p}p$ initial state would couple to the ($u\bar{u}$) and ($d\bar{d}$) components of the mixed $\phi-\omega$ system (mainly ω), whereas the final charged kaon state would arise through the ($s\bar{s}$) component of the $\phi-\omega$ system (mainly ϕ). The probability for such a mechanism may be small, but it can nevertheless turn out to be important if the quark-disconnected process is strongly suppressed by the OZI rule.

174

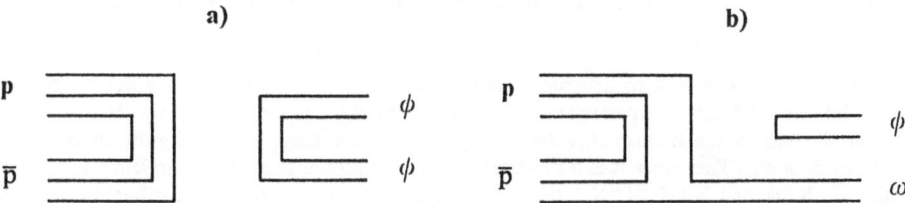

Figure 2.   Examples of quark-line diagrams. a) The reaction $\bar{p}p \rightarrow \phi\phi$, and b) the reaction $\bar{p}p \rightarrow \omega\phi$.

For the hadronic production of $\phi\phi$ states there are, in fact, several reaction mechanisms that seem to avoid the 'complication' of OZI suppression. The case due to $\phi - \omega$ mixing discussed above shows that the reaction $\bar{p}p \rightarrow \phi\phi$ may proceed in several steps each of which is quark-connected and thus not OZI-suppressed. Intermediate states relevant for the production of $\phi\phi$ are obviously those that contain a sizeable amount of strangeness. Donoghue [20] has identified $\eta\eta$ as an important intermediate state. As η mesons contain large components of all three quark types $(u\bar{u})$, $(d\bar{d})$ and $(s\bar{s})$, it is quite conceivable that the first reaction step, $\bar{p}p \rightarrow \eta\eta$, proceeds through the non-strange component of η and the second step, $\eta\eta \rightarrow \phi\phi$, through the strange one. Lipkin [21] has argued that in such a case there are additional contributions due to $\eta\eta'$ and $\eta'\eta'$, which may altogether tend to cancel. Similar considerations can be made for $\bar{K}K$ as another possible intermediate state between $\bar{p}p$ and $\phi\phi$. Again, additional contributions due to $\bar{K}K^*$ and $\bar{K}^*K^*$ may have a cancelling effect.

The reaction $\bar{p}p \rightarrow \phi\phi$ can also be seen to proceed via intermediate states consisting of two four-quark objects, $(qs\bar{q}\bar{s})(qs\bar{q}\bar{s})$, with yet unknown identity. This has been argued for by Dover and Fishbane [22] in an attempt to explain the rates observed experimentally for ϕ production in nucleon $-$ antinucleon annihilation. Roberts and Karl [23] have suggested a string-breaking mechanism for hadronic $\phi\phi$ production. After the creation of the first $(s\bar{s})$ pair, the second one is produced by breaking the QCD string that joins the first pair. The sequential process is then seen as $q\bar{q} \rightarrow s_1\bar{s}_1 \rightarrow s_1(s_2\bar{s}_2)\bar{s}_1 \rightarrow (s_1\bar{s}_2)(s_2\bar{s}_1) \rightarrow \phi\phi$. Recent data from deep-inelastic scattering of polarized muons on protons have revived interest in the question to what extent a sea of $(q\bar{q})$ pairs (q = u, d, s) and gluons may contribute to the proton wave-function, even when larger distances or small momentum transfers are involved. Ellis et al. [24] have demonstrated that the coupling of an $(s\bar{s})$ meson to a non-strange baryon via non-valence $(s\bar{s})$ quark pairs is indeed conceivable. For $\bar{p}p \rightarrow \phi\phi$ this constitutes another possible mechanism in which the otherwise expected OZI suppression is not in action.

We have briefly discussed six different reaction modes with which to produce $\phi\phi$ meson pairs in $\bar{p}p$ annihilations. Clearly, a measurement of this reaction alone will never be sufficient for unravelling the various mechanisms involved, nor will it be sufficient for the identification of possible NQM states as such. Therefore, it is indispensable that additional channels complementary to $\phi\phi$, in particular $K^+K^-\phi$ and $K^+K^-K^+K^-$ on one side, and $\omega\phi$ and $\omega\omega$ on the other, be investigated simultaneously by the experiment.

For the reaction $\bar{p}p \rightarrow \omega\omega$ the mechanism in the context of a simple quark picture is obvious, as there are no significant quark-disconnected contributions. But this means also that the ratio of resonant and non-resonant cross-sections may be very unfavourable. Ideally seen, however, measurements of the $\phi\phi$ and $\omega\omega$ channels together would allow the relative sizes of strange and non-strange components in a possible $(q\bar{q})$ resonance to be extracted. In the case of a glueball, the flavour (in-)dependence of its decay could be tested.

The reaction $\bar{p}p \rightarrow \omega\phi$ is OZI-suppressed for intermediate $(q\bar{q})$ states because of its quark-disconnected $(s\bar{s})$ vertex. However, the gluons associated with that vertex and the $(q\bar{q})$ nature of the ω seem to make the $\omega\phi$ channel very attractive for a hybrid search (see figure 2b). In a simple view the creation and decay of such a hybrid may be written as $\bar{p}p \approx (\overline{uud} + uud) \rightarrow (u\bar{u} + d\bar{d})(u\bar{u}) \rightarrow (u\bar{u} + d\bar{d})g \rightarrow (u\bar{u} + d\bar{d})(s\bar{s}) \approx \omega\phi$.

The reactions $\bar{p}p \rightarrow \bar{K}K$, $\bar{p}p \rightarrow \bar{K}K^*$ and $\bar{p}p \rightarrow \bar{K}^*K^*$

In ground-state kaons the spins of the quark $-$ antiquark pair are antiparallel, giving rise to $J^P(I) = 0^-(1/2)$. Therefore it is S = 0, J = L, and $P = C = (-1)^L$ for the $\bar{K}K$ systems. In the $\bar{p}p$ initial state, on the other hand, one has $P = (-1)^{\ell+1}$ and $C = (-1)^{\ell+s}$. The conservation of parity and charge conjugation is satisfied only for s = 1. In words this means that the annihilation $\bar{p}p \rightarrow \bar{K}K$ can proceed only from the spin-triplet state of the $\bar{p}p$ system. For the orbital angular momenta it must hold that $L = \ell \pm 1$. The isospin in the initial and the final states can have the values $(I, I_3) = (0,0)$ or $(1,0)$. In summary, the selection rules for $\bar{p}p \rightarrow \bar{K}K$ are the following.

- ℓ even, s = 1; J = $\ell\pm1$
  - L odd, S = 0; J = L ($J^{PC} = 1^{--}, 3^{--}, 5^{--}, ...$)
- ℓ odd, s = 1; J = $\ell\pm1$
  - L even, S = 0; J = L ($J^{PC} = 0^{++}, 2^{++}, 4^{++}, ...$)

For the K^+K^- system there are no further restrictions. However, when considering the neutral-kaon final state, $\bar{K}^0K^0$, the short- and the long-lived components of K^0 and $\bar{K}^0$ introduce more constraints. In the limit of CP conservation we can identify these components with the CP eigenstates K_1 and K_2: $K_S \approx K_1 = (K^0 + \bar{K}^0)/\sqrt{2}$ and $K_L \approx K_2 = (K^0 - \bar{K}^0)/\sqrt{2}$. For even L-values it is P = C = +1. The corresponding part of the $\bar{K}^0K^0$ wave-function is thus symmetric with respect to $K^0 - \bar{K}^0$ exchange and it consists of K_SK_S and K_LK_L components only. For odd L-values one has P = C = -1 with an antisymmetric wave-function containing only the K_SK_L component of $\bar{K}^0K^0$. If, as in the JETSET experiment, the detection of hadronically produced K^0 and $\bar{K}^0$ mesons proceeds via the two-pion decay of the short-lived component, the only accessible $\bar{K}^0K^0$ states are in the series $J^{PC} = (\text{even})^{++}$.

In K^* mesons the constituent quark−antiquark pair has the spins parallel, so that $J^P(I) = 1^-(1/2)$. In $\bar{K}K^*$ systems (and the charge-conjugated case $\bar{K}^*K$) it must therefore be S = 1. Again, parity conservation requires that L = $\ell\pm1$. The selection rules for $\bar{p}p \to \bar{K}K^*$ + c.c. are as follows.

- ℓ even, s = 0; J = ℓ
  - L odd, S = 1; J = $L\pm1$ ($J^{PC} = 0^{-+}, 2^{-+}, 4^{-+}, ...$)
- ℓ even, s = 1; J = ℓ
  - L odd, S = 1; J = $L\pm1$ ($J^{PC} = 2^{--}, 4^{--}, ...$)
- ℓ even, s = 1; J = $\ell\pm1$
  - L odd, S = 1; J = L ($J^{PC} = 1^{--}, 3^{--}, 5^{--}, ...$)
- ℓ odd, s = 0; J = ℓ
  - L even, S = 1; J = $L\pm1$ ($J^{PC} = 1^{+-}, 3^{+-}, 5^{+-}, ...$)
- ℓ odd, s = 1; J = ℓ
  - L even, S = 1; J = $L\pm1$ ($J^{PC} = 1^{++}, 3^{++}, 5^{++}, ...$)
- ℓ odd, s = 1; J = $\ell\pm1$
  - L even, S = 1; J = L ($J^{PC} = 0^{++}, 2^{++}, 4^{++}, ...$)

The reaction $\bar{p}p \to \bar{K}^*K^*$ involves even more partial waves because of the possibilities S = 0, 1 or 2 in the final state. From charge-conjugation invariance one has $(-1)^{\ell + s} = (-1)^{L+S}$. The selection rules are thus the following.

- ℓ even, s = 0; J = ℓ
  - L odd, S = 1; J = $L\pm1$ ($J^{PC} = 0^{-+}, 2^{-+}, 4^{-+}, ...$)
- ℓ even, s = 1; J = ℓ
  - L odd, S = 2; J = $L\pm1$ ($J^{PC} = 2^{--}, 4^{--}, ...$)
- ℓ even, s = 1; J = $\ell\pm1$
  - L odd, S = 0; J = L ($J^{PC} = 1^{--}, 3^{--}, 5^{--}, ...$)
  - L odd, S = 2; J = L ($J^{PC} = 1^{--}, 3^{--}, 5^{--}, ...$)
  - L odd, S = 2; J = $L\pm2$ ($J^{PC} = 1^{--}, 3^{--}, 5^{--}, ...$)
- ℓ odd, s = 0; J = ℓ
  - L even, S = 1; J = $L\pm1$ ($J^{PC} = 1^{+-}, 3^{+-}, 5^{+-}, ...$)
- ℓ odd, s = 1; J = ℓ
  - L even, S = 2; J = $L\pm1$ ($J^{PC} = 1^{++}, 3^{++}, 5^{++}, ...$)
- ℓ odd, s = 1; J = $\ell\pm1$
  - L even, S = 0; J = L ($J^{PC} = 0^{++}, 2^{++}, 4^{++}, ...$)
  - L even, S = 2; J = L ($J^{PC} = 0^{++}, 2^{++}, 4^{++}, ...$)
  - L even, S = 2; J = $L\pm2$ ($J^{PC} = 0^{++}, 2^{++}, 4^{++}, ...$)

The reactions $\bar{p}p \to \bar{K}K$, $\bar{p}p \to \bar{K}K^*$ + c.c. and $\bar{p}p \to \bar{K}^*K^*$ involve, in lowest order, the annihilation of two $(q\bar{q})$ pairs and the creation of one $(s\bar{s})$ quark pair. An alternative to this quark-connected process is the quark-disconnected one, in which all three $(q\bar{q})$ pairs of the $\bar{p}p$ initial state annihilate. The reactions are thus open for various NQM states as intermediate resonances. In particular, the simultaneous measurement of these channels can provide stringent tests for the quark—gluon structure of intermediate resonances. For glueballs, the various possibilities of decays to strange-meson pairs are an 'analyzer' with respect to the charge-conjugation quantum number [6]. A glueball having $C = +1$ can decay to $\bar{K}K$, $\bar{K}^*K^*$ or $\bar{K}K_2^*$, whereas a $C = -1$ glueball would select the decay channels $\bar{K}K^*$ or $\bar{K}^*K_2^*$. In contrast to this, a nonet of 'ordinary' quarkonium states has all these channels open, provided J^P is in the allowed series. In addition, the decays of a $(q\bar{q})$ state are somehow restricted by the OZI rule as discussed above. A 'smoking gun' glueball candidate can thus be identified as such if it would decay, for example, into both (s$\bar{s}$)-type channels such as $\phi\phi$ and non-(s$\bar{s}$) channels like $\omega\omega$ or $\pi\pi$, but into only one of the two classes of strange-meson pairs.

Previous data on $\phi\phi$, $K^+K^-\phi$ and $K^+K^-K^+K^-$ states

In this section we attempt to give an account of data that have been obtained in $\phi\phi$ and states related to this. The reaction $\bar{p}p\rightarrow\phi\phi$, which constitutes the main physics objective of the JETSET experiment, has been measured only once. In the CERN—ISR experiment R704 an energy scan was performed in order to detect $\bar{p}p\rightarrow\eta_c$ formation [25]. As a part of this, the reaction $\bar{p}p\rightarrow\phi\phi\rightarrow K^+K^-K^+K^-$ was also studied. On the basis of 83 events centered around $\sqrt{s} = 2.989$ GeV, and after correction for the branching fraction $BR(\phi\rightarrow K^+K^-) = 0.495$, the resulting production cross-section is $\sigma(\bar{p}p\rightarrow\phi\phi) = (25.0 \pm 8.3)$ nb. Besides the direct coupling of $\bar{p}p$ to $\phi\phi$, the formation and decay of an intermediate η_c might also contribute to the measured cross-section. In an earlier bubble-chamber experiment on $\bar{p}p\rightarrow K^+K^-K^+K^-$ between 1.6 and 2.2 GeV/c incident momentum only six events were identified [26]. These also included $K^+K^-\phi$ and $\phi\phi$ intermediate states. The production cross-section $\sigma(\bar{p}p\rightarrow K^+K^-K^+K^-) = (3.8 \pm 1.7)$ μb was calculated from the data.

Prominent information on $\phi\phi$ hadronic production stems from measurements of $\pi^-p\rightarrow\phi\phi n$, which were performed in four generations of experiments using a 22 GeV/c beam and the MPS facility at BNL—AGS. The ϕ-mesons were identified through their decays into K^+K^-, and the analysis required the mass of the K^+K^- pairs close to the ϕ mass. In the first of these experiments [27], 102 events were obtained. The $\phi\phi$ invariant-mass spectrum showed an enhancement between threshold and 2.4 GeV. The data also revealed the ratio of cross-sections to be $\sigma(K^+K^-\phi)/\sigma(\phi\phi) \leq 10$, which was interpreted as a lack of the expected OZI suppression. In fact, when taking into account decay modes other than $\phi\rightarrow K^+K^-$, this ratio is probably more like 5. With the second experiment [28], the event statistics was increased to 1203, which allowed a partial wave analysis of $\phi\phi$ production to be performed. All permitted values of $J\leq4$ and $L\leq3$ were included, resulting in 52 waves. The $\phi\phi$ mass spectrum was found to be essentially reproduced by only two amplitudes, both having $J^{PC}=2^{++}$, from which the following resonance parameters were extracted: $M_1 = (2.16 \pm 0.05)$ GeV with $\Gamma_1 = (0.31 \pm 0.07)$ GeV for the S-wave, and $M_2 = (2.32 \pm 0.04)$ GeV with $\Gamma_2 = (0.22 \pm 0.07)$ GeV for the D-wave. The sensitivity of this analysis was then increased by the third experiment [29], which led to a total of 3652 events. As a consequence, a strongly dominant S-wave and two D-waves, again all with $J^{PC}=2^{++}$, were found to comprise all of the observed cross-section. The resonance parameters of the S-wave were $M_1 = (2.05 \pm 0.07)$ GeV with $\Gamma_1 = (0.20 \pm 0.11)$ GeV; those for the two D-waves were $M_2 = (2.30 \pm 0.06)$ GeV with $\Gamma_2 = (0.20 \pm 0.06)$ GeV, and $M_3 = (2.35 \pm 0.03)$ GeV with $\Gamma_3 = (0.27 \pm 0.11)$ GeV. In contrast to this the $K^+K^-\phi$ background was seen to be mostly structureless. Finally, the fourth experiment [30] doubled the statistics, which now included a total of 6658 events (see figure 3). The refined partial wave analysis included all 114 waves with $J\leq6$ and $L\leq4$. Again three 2^{++} resonances contained all of the observed $\phi\phi$ strength. The dominant S-wave accounted for 45 % of the data; its resonance parameters were determined to be $M_1 = (2011 \pm 69)$ MeV with $\Gamma_1 = (202 \pm 65)$ MeV. The two D-waves comprised 20 % and 35 % of the data; their parameters were $M_2 = (2297 \pm 28)$ MeV with $\Gamma_2 = (149 \pm 41)$ MeV, and $M_3 = (2339 \pm 55)$ MeV with $\Gamma_3 = (319 \pm 75)$ MeV, respectively. The production of $K^+K^-\phi$ was evaluated simultaneously. Two thirds of its strength were found to be structureless and incoherent, whereas one third exhibited $J^{PC}=1^{--}$, and a few percent $J^{PC}=2^{++}$.

The view of $\pi^-p\rightarrow\phi\phi n$ as an OZI-suppressed process and the unexpectedly large signal seen in $(\phi\phi)$ vs $(K^+K^-\phi)$ led the BNL experimenters to strong conclusions. The OZI suppression was seen to be broken down due to the annihilation of $(q\bar{q})$ pairs into resonating gluons. The three broad

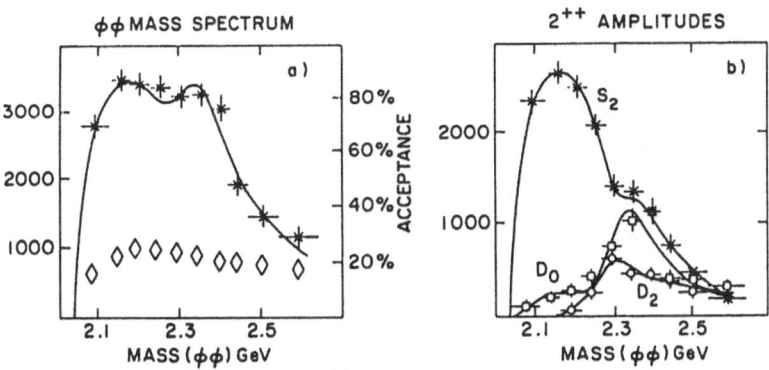

Figure 3. Results [30] from $\pi^-p \to \phi\phi n$. a) Acceptance-corrected $\phi\phi$ mass spectrum, and b) intensity for the three $J^{PC} = 2^{++}$ waves.

resonances that were unravelled in the amplitude analysis were thus interpreted as tensor glueballs. The authors labelled the states g_T, however, in ref. [3] they are listed as f_2. The strong and uncompromising claim for the discovery of at least one glueball has caused a great deal of controversy, for which ref. [31] gives illustrative examples. It has also inspired much theoretical work on $\phi\phi$ hadronic production as discussed previously. It appears unsatisfactory that none of the three resonances is clearly visible in the $\phi\phi$ mass spectrum, but they only showed up, as broad and overlapping as they are, resulting from a complicated amplitude analysis. Another problem, which is actually more challenging, is that the states have not been seen in the decay $J/\psi \to \gamma\phi\phi$, a channel that is certainly quark-disconnected and thus supposed to be gluon-enriched, so that the glueball interpretation of the BNL data becomes less likely. However, there do exist some experimental results consistent with the observations made at BNL. After the first experiment there, the same reaction $\pi^-p \to \phi\phi n$ was studied at CERN−SPS using the OMEGA spectrometer in a 16 GeV/c beam [32]. The $\phi\phi$ mass distribution, based on 153 event candidates, exhibited a threshold enhancement falling off at 2.5 GeV. From a spin−parity analysis the background-reduced event sample was found to be consistent with two interfering $J^P = 2^+$ resonances.

Experiment WA67 at CERN−SPS measured, also using the OMEGA spectrometer, the inclusive production $\pi^-Be \to \phi\phi + X$ at 85 GeV/c beam momentum [33]. The ratio of $(K^+K^-\phi$ or $K^+K^-K^+K^-)$ vs $(\phi\phi)$ production turned out to be of the order of one. From the $\phi\phi$ spectrum, comprising 13088 events up to 3.2 GeV, two broad resonances were extracted: $M_1 = (2231 \pm 10)$ MeV with $\Gamma_1 = (133 \pm 50)$ MeV as a mixture of S- and D-waves, and $M_2 = (2392 \pm 10)$ MeV with $\Gamma_2 = (198 \pm 50)$ MeV as a pure D-wave (see figure 4). Moreover, from the analysis of angular correlations in $\phi\phi$ decays it was found that the mass spectrum up to 2.5 GeV was consistent with a dominance of $J^P = 2^+$, and a signal observed in the higher mass range was attributed to $\pi^-Be \to \eta_c + X \to \phi\phi + X$.

Using a 400 GeV/c proton beam from the FNAL main ring, experiment E623 looked at the inclusive channel $pN \to \phi\phi + X$, where each ϕ decays into a K^+K^- pair [34]. The events were found to be concentrated at low invariant masses of the $\phi\phi$ system. This region was investigated further. The spectra of the above reaction were compared with others obtained in the same experiment [35]: $pN \to (K^+K^-\phi) + X$, $pN \to (\phi\phi)K^+K^- + X$, $pN \to (K^+K^-\phi)K^+K^- + X$, and $pN \to (\pi^+\pi^-\phi)K^+K^- + X$. A global fit to the data, comprising nearly 400 events above background, revealed a narrow enhancement with mass $M = (2141 \pm 16)$ MeV and width $\Gamma = (49 \pm 28)$ MeV (see figure 5). Also determined were the ratios of branching fractions for this state, the results being $BR(M \to \phi\phi)/BR(M \to K^+K^-\phi) = 0.39 \pm 0.24$ and $BR(M \to K^+K^-\phi)/BR(M \to \pi^+\pi^-\phi) = 0.49 \pm 0.16$. In view of the latter number, the observed state was found to be inconsistent with being of $(q\bar{q})$ nature. It also appeared to be not identifiable with any of the broad $\phi\phi$ resonances discussed above, nor with narrow structures seen in $\bar{K}K$ channels that are discussed further below.

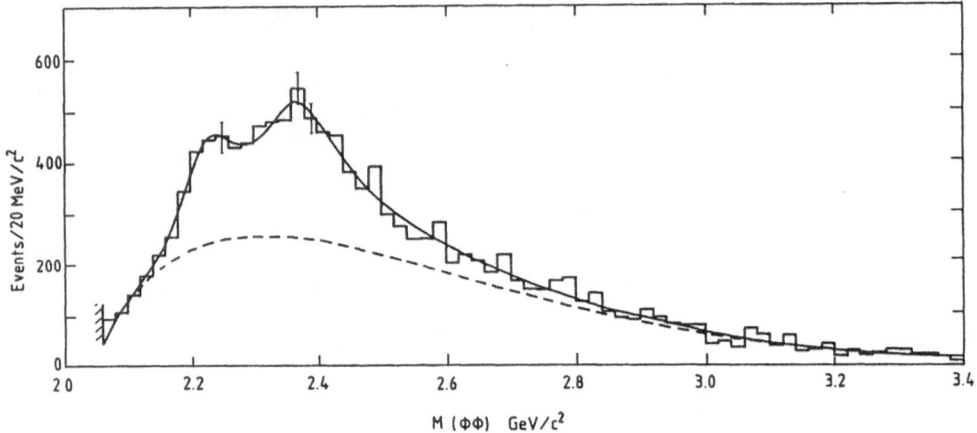

Figure 4.   Results [33] from $\pi^- \text{Be} \rightarrow \phi\phi + \text{X}$. Shown is the $\phi\phi$ mass spectrum with a fit to two Breit−Wigner curves and background.

The reaction $K^- p \rightarrow \phi\phi Y^0$, with Y^0 being a Λ or a Σ^0 hyperon, was measured for the first time at CERN using the OMEGA spectrometer in a 18.5 GeV/c beam [36]. The about 100 events were found to be concentrated in the low-mass region of the $\phi\phi$ spectrum. More recently, the $K^- p \rightarrow \phi\phi\Lambda$ reaction was also measured at SLAC using the LASS spectrometer in a 11 GeV/c beam [37]. Near 2.2 GeV a structure was visible in the $\phi\phi$ mass spectrum. An angular analysis then showed that the distributions were essentially consistent with a pure S-wave or a combination of waves giving $J^P = 2^+$, but the authors found the data to be inconsistent with a dominant $J^P = 0^-$ assignment.

The central production of neutral states was extensively studied with experiment WA76 using the OMEGA spectrometer at CERN−SPS. Exclusively produced states X^0 were investigated in

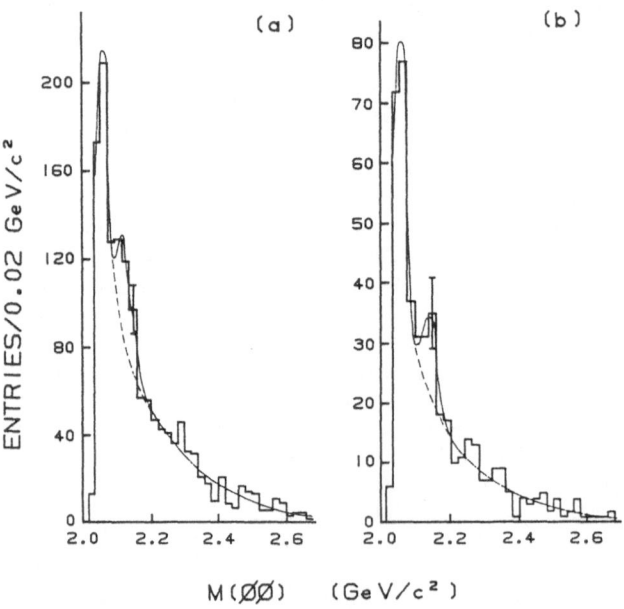

Figure 5. $\phi\phi$ mass spectra [35] with resonance and background fits. Shown are data from a) the reaction $pN \rightarrow (\phi\phi) + X$ and b) the reaction $pN \rightarrow (\phi\phi)K^+ K^- + X$.

$\pi^+p\rightarrow\pi^+(X^0)p$ at 85 GeV/c and also in $pp\rightarrow p(X^0)p$ at 85 and 300 GeV/c. Of particular interest in the context here is the production of those states that were identified as $\phi\phi$, $K^+K^-\phi$ or $K^+K^-K^+K^-$. The central system (X^0) is characterized by small values of the Feynman variable, $x_F = 2p_L/\sqrt{s}$. It is presumed to be produced by a double exchange process, which, at high centre-of-mass energies, is dominated by double Pomeron exchange. As the Pomeron is believed to be a multi-gluon state, Pomeron−Pomeron scattering may be a source of states other than $(q\bar{q})$. The events from the π^+ and pp experiments performed at 85 GeV/c showed similar features, so they were considered together [38]. By plotting K^+K^- mass spectra the production of one or two ϕ-mesons could be demonstrated. Although the statistics of $\phi\phi$ events was very low, there was an apparent accumulation of events near threshold. After correction for decay modes other than the detected $\phi\rightarrow K^+K^-$, the ratios of cross-sections were determined to be $\sigma(K^+K^-\phi)/\sigma(\phi\phi) = 1.5 \pm 0.6$ and $\sigma(K^+K^-K^+K^-)/\sigma(\phi\phi) = 0.3 \pm 0.2$. From the pp data taken at 300 GeV/c the ratio $\sigma(K^+K^-\phi)/\sigma(\phi\phi) = 1.0 \pm 0.3$ was extracted [39]. Combining the 85 GeV/c and the 300 GeV/c data into one $\phi\phi$ mass distribution, a shape similar to the one obtained from the π^-p experiments at BNL was found (see figure 6). The angular distribution of the $K^+K^-K^+K^-$ system was used to determine the spin and parity of the intermediate $\phi\phi$ state. As a result, $J^P = 2^+$ was favoured over $J^P = 0^-$, although other waves could not be ruled out.

All of the data obtained in hadronic production of $\phi\phi$ pairs up to about 2.5 GeV invariant mass appear to indicate the dominance of $J^P = 2^+$ or, at least, are not inconsistent with such an assignment. This has been challenged by some of the results from radiative decays $J/\psi\rightarrow\gamma\phi\phi$. The

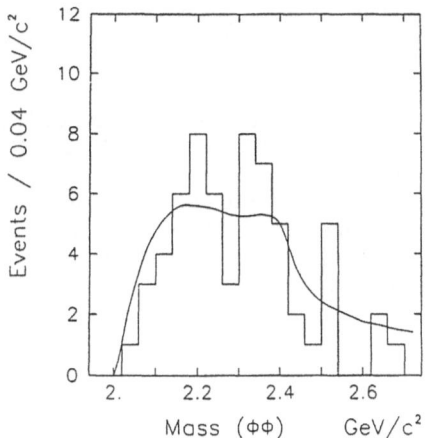

Figure 6.   Results [39] from $\pi^+p\rightarrow\pi^+(\phi\phi)p$ and $pp\rightarrow p(\phi\phi)p$. Shown is the combined $\phi\phi$ mass spectrum. The curve indicates the shape of the spectrum from ref. [30] obtained in $\pi^-p\rightarrow\phi\phi n$.

experiments DM2 at Orsay−DCI and MARK III at SLAC−SPEAR both studied various hadronic and radiative decay modes following the resonance formation $e^+e^-\rightarrow J/\psi$. The data sample collected by DM2 consisted of 8.6×10^6 events. For the study of $J/\psi\rightarrow\gamma\phi\phi$, the decay mode $\phi\phi\rightarrow K^+K^-K^+K^-$ was evaluated [40]. A spin−parity analysis of the $\phi\phi$ mass spectrum was performed in order to search for intermediate states (see figure 7a). The region up to 2.5 GeV was found to agree with the hypothesis of an S-wave giving $J^P = 2^+$. However, the probabilities for other waves were also significant in some cases, and the possible mixing between different waves was not taken into account. Near 2.2 GeV an enhancement of events was seen in the mass spectrum. Here a P-wave and $J^P = 0^-$ was the preferred assignment, and, moreover, even parity appeared to be strongly disfavoured. The mass region above 2.9 GeV revealed presence of the sequential decay $J/\psi\rightarrow\gamma\eta_c\rightarrow\psi\phi\phi$. A spin−parity analysis confirmed the assignment $J^P = 0^-$ for the η_c, whereas adjacent events between 2.68 and 2.9 GeV did not fit with the $J^P = 0^-$ hypothesis. More recently, the DM2 group also performed an analysis of the decay mode $\phi\phi\rightarrow K^+K^-K_SK_L$ [41]. This confirmed the presence of an enhancement in the $\phi\phi$ invariant-mass spectrum near threshold (see figure 7b). Taking the sum of the efficiency-corrected mass spectra of both decay modes, the extracted resonance parameters were M = (2238 ± 7) MeV and Γ = (80 ± 30) MeV. The new analysis was consistent with the pseudoscalar assignment for this state as concluded from final states with four charged kaons.

The J/ψ sample evaluated by MARK III consisted of 4.9×10^6 produced events [42, 43]. In the decays J/$\psi \rightarrow \gamma\phi\phi$, both the $\phi\phi$ final states $K^+K^-K^+K^-$ and $K^+K^-K_SK_L$ (with $K_S \rightarrow \pi^+\pi^-$) were studied. The efficiency-corrected mass spectra of $\phi\phi$ for each of the two modes exhibited the same two structures, one being the η_c and the other one located near 2.2 GeV (see figure 8). The parameters extracted for the low-mass state [42], when averaged over the two decay modes, were M = (2222 ± 27) MeV and Γ = (150 ± 190) MeV. A spin−parity analysis was performed, in which the angle between the ϕ decay planes in the $\phi\phi$ rest frame and the polar angle of the decay kaon in its ϕ rest frame were evaluated. The resulting distributions showed the characteristics of a pseudoscalar, $J^P = 0^-$, for this state and also for the η_c. No indication for the presence of a $J^P = 2^+$ state in the $\phi\phi$ spectrum was found. In fact, and somewhat contrasting the DM2 result, the entire region below 2.4 GeV appeared to be dominantly $J^P = 0^-$.

Pseudoscalar structures near threshold produced in radiative J/ψ decays were seen, besides $\phi\phi$, also in other vector−vector meson channels [42−45]: $\rho\rho$, $\omega\omega$, and $\overline{K}^*K^*$. In view of the quark content of ω- and ϕ-mesons and the $\omega-\phi$ mixing, the channel J/$\psi \rightarrow \gamma\phi\phi$ is naturally complemented by studies of J/$\psi \rightarrow \gamma\omega\phi$ and J/$\psi \rightarrow \gamma\omega\omega$. Suffering from very low statistics, however, no significant structures were observed in the $\omega\phi$ invariant-mass spectrum [43]. Radiative J/ψ decays to $\omega\omega$ appear to be an order of magnitude stronger. A prominent threshold enhancement near 1.8 GeV, again featuring $J^P = 0^-$, was seen in the $\omega\omega$ spectrum, but no structures were observed at higher masses [46].

The GAMS collaboration studied the reaction $\pi^-p \rightarrow \omega\omega n$ using a 38 GeV/c beam [47]. The ω-mesons were identified through their $\gamma\pi^0$ decay mode. Near 2 GeV invariant mass of the $\omega\omega$ system, two states were identified by an analysis that intentionally emphasizes different spin states (see figure 9). One of the states was measured to have the parameters M = (1924 ± 14) MeV with Γ = (91 ± 50) MeV, and the assignment $J^{PC} = 2^{++}$ was determined from the data. The other state occurred at M = (2060 ± 20) MeV with Γ = (170 ± 60) MeV, and it was identified with the 4^{++} candidate $f_4(2050)$ listed in the data tables of ref. [3].

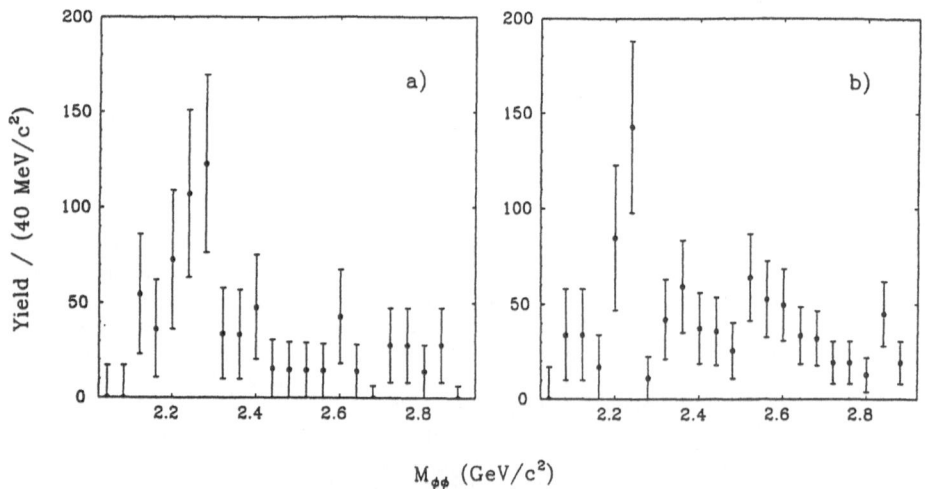

Figure 7. Results [41] from J/$\psi \rightarrow \gamma\phi\phi$ of the DM2 experiment. Shown are the efficiency-corrected $\phi\phi$ mass spectra, a) obtained from the $K^+K^-K_SK_L$ decay mode, and b) from the $K^+K^-K^+K^-$ mode.

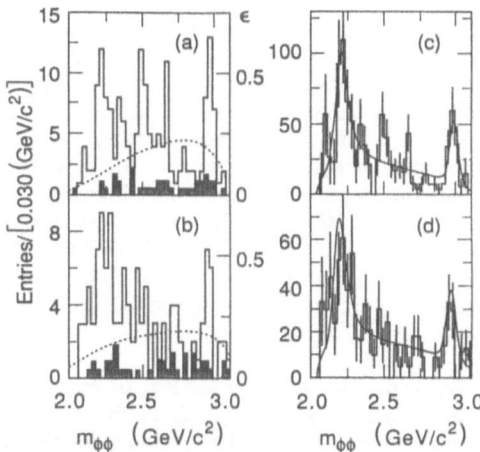

Figure 8.   Results [42] from J/$\psi \to \gamma\phi\phi$ of the MARK III experiment. Shown are the $\phi\phi$ mass spectra, a) obtained from the $K^+K^-K^+K^-$ decay mode, and b) from the $K^+K^-K_SK_L$ decay mode. The shaded area is the estimated background; the dashed curve shows the acceptance. The effiency-corrected spectra for the two final states with fits to Breit−Wigner resonances are shown in c) and d), respectively.

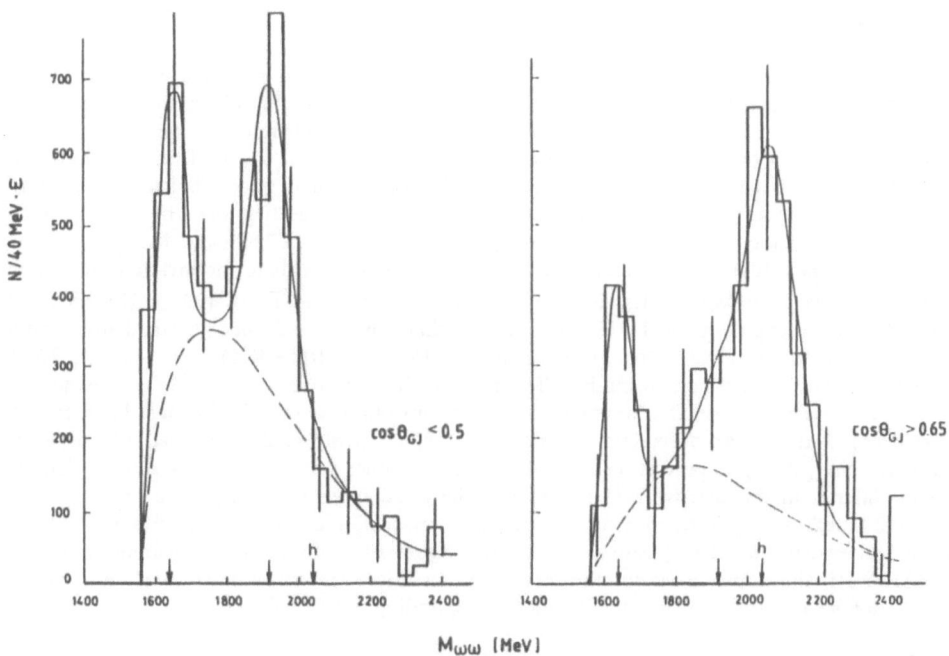

Figure 9.   Results [47] from $\pi^-p \to \omega\omega n$. Efficiency-corrected $\omega\omega$ mass spectrum, measured in two different polar angular regions, and fits with three resonances plus background.

Previous data on $\bar{K}^* K^*$ states

Having dealt mostly with $\phi\phi$ states in the previous section, we now turn to $\bar{K}^* K^*$ as another very interesting channel that involves a pair of vector mesons. The associated production of neutral K^* and $\bar{K}^*$ mesons can be identified through the decays $K^+\pi^-$ and $K^-\pi^+$, respectively. It was observed in central hadronic collisions, namely in $\pi^+ p \rightarrow \pi^+(\bar{K}^* K^*)p$ measured at 85 GeV/c, and also in $pp \rightarrow p(\bar{K}^* K^*)p$ at 85 and 300 GeV/c [48, 49]. Experiment WA76, which studied these processes, reported a production cross-section ratio $\sigma(\bar{K}^* K^*)/\sigma(\phi\phi) \approx 10$ from the 85 GeV/c data. In all spectra obtained, the $\bar{K}^* K^*$ events were found to be concentrated in a relatively narrow non-resonant enhancement near threshold. A similar threshold behaviour was also observed in the LASS experiment investigating the reaction $K^- p \rightarrow \bar{K}^* K^* \Lambda$ at 11 GeV/c [37]. Here, both neutral $(K^-\pi^+ K^+\pi^-)$ and charged $(K_S\pi^+ K_S\pi^-)$ modes were measured. The angular distributions of $\bar{K}^* K^*$ were found to be consistent with $L=0$ production, giving rise to $J^P = 2^+$, and inconsistent with the assumption of $J^P = 0^-$ dominance.

Studying various vector$-$vector meson states, the MARK III group also looked at the radiative decays $J/\psi \rightarrow \gamma \bar{K}^* K^*$ in the charged mode [44, 45]. A fit to the $\bar{K}^* K^*$ invariant-mass spectrum revealed some interesting structures below 2.5 GeV, their parameters being $M_1 = (1930 \pm 9)$ MeV with $\Gamma_1 = (75 \pm 21)$ MeV, $M_2 = (2067 \pm 12)$ MeV with $\Gamma_2 = (113 \pm 30)$ MeV, and $M_3 = (2335 \pm 31)$ MeV with $\Gamma_3 = (180 \pm 64)$ MeV. The $\bar{K}^* K^*$ spectrum appeared to be dominated by $J^P = 0^-$, as it was the case in radiative decays to $\phi\phi$. It is interesting to note that enhancements occurring in channels like $\bar{K}^* K^*$ and $\phi\phi$ close to their respective thresholds seem to be a universal feature of vector$-$vector meson states, as observed through several different production mechanisms. The nature of these enhancements, however, has yet to be understood. Possible explanations include radial excitations of pseudoscalar mesons. Another interesting feature is that radiative J/ψ decays seem to favour the assignment $J^P = 0^-$, whereas hadronic production channels are consistent with $J^P = 2^+$.

Previous data on $\bar{K}K$ states

The past years have seen a wealth of spectroscopic information provided by studies both of $\bar{K}^0 K^0$ and $K^+ K^-$ final states produced from various initial states. In this section we summarize experimental results that have been obtained in $\bar{K}K$ and related channels above 2 GeV invariant mass. In the context of the JETSET experiment, the channel $\bar{p}p \rightarrow K_S K_S$ is of prime interest. One reason for this is a narrow resonance originally called $\zeta(2230)$ and now tabulated in ref. [3] as $X(2220)$ because of its yet unconfirmed identity. The $\zeta(2230)$ was first seen by the MARK III experiment [50] in the radiative decays $J/\psi \rightarrow \gamma \bar{K}K$, where both the $K^+ K^-$ and $K_S K_S$ modes were looked at (see figure 10a). The resonance parameters extracted from fits to the invariant-mass spectra are, averaged over the two modes, $M = (2231 \pm 13)$ MeV and $\Gamma = (22 \pm 23)$ MeV. The significance of the signal in the $K^+ K^-$ and $K_S K_S$ channels was 4.5 and 3.6 standard deviations, respectively, and the product branching fractions $BR(J/\psi \rightarrow \gamma \zeta) \cdot BR(\zeta \rightarrow \bar{K}K)$ were $(4.2 \pm 1.7) \times 10^{-5}$ and $(3.1 \pm 1.6) \times 10^{-5}$ for the two modes. The ratio of these fractions, 1.3 ± 0.9, is consistent with the value 2 expected for an isoscalar state. The quantum numbers of $\zeta(2230)$ must lie in the series $J^{PC} = (\text{even})^{++}$ due to its observation in $K_S K_S$. A spin$-$parity analysis yielded $J \geq 2$ as a lower limit [43]. No $\zeta(2230)$ signal was seen by MARK III in other two-meson decay modes. However, the great interest in this narrow state stems from the possibility that it may be a non-$(q\bar{q})$ object. The DM2 experiment [51], by investigating the same production and decay channels as MARK III, did not find evidence for the $\zeta(2230)$ resonance (see figure 10b). Based on the parameters extracted by MARK III, the upper limits for the product branching fractions as determined by DM2 were 2.3×10^{-5} for $K^+ K^-$ and 1.6×10^{-5} for $K_S K_S$ (corresponding to 95 % C.L.).

Three things are interesting to note here, which appear to make the observations by the two experiments not inconsistent. Firstly, the MARK III values and the DM2 upper limits for the product branching fractions are consistent within one standard deviation. Secondly, a spike is visible in the $K_S K_S$ mass spectrum of DM2, which is basically one bin below the $\zeta(2230)$ mass value from MARK III. Thirdly, the $K_S K_S$ spectra obtained by the two experiments look very much alike over the entire mass region (see figure 10 b and d), and the channel contains very little background due to its restriction to $J^{PC} = (even)^{++}$. In particular, both $K_S K_S$ spectra exhibit an excess of events between 2 and 2.4 GeV, which is not explained by any known source of background. A fit performed by DM2 using a Breit–Wigner curve in this region revealed the parameters M = (2197 ± 17) MeV and Γ = (201 ± 51) MeV, the product branching fraction being as large as 1.5×10^{-4}. Coherent fits by MARK III to $\xi(2230)$ and the broader 2.2 GeV structure were also reported [52]. The resonance parameters extracted for the latter, M = (2184 ± 64) MeV and Γ = (413 ± 211) MeV, are consistent with the DM2 data.

An earlier observation in the $K^+ K^-$ channel of a structure near 2.2 GeV with a width of the order of 200 MeV stems from a measurement [53] of $\pi^- p \to K^+ K^- n$ at 10 GeV/c using the OMEGA spectrometer at CERN–PS. The complementary reaction $\pi^- p \to K_S K_S n$ was measured in a 22 GeV/c beam at BNL–AGS, but no structure was observed in $K_S K_S$ above 2 GeV [54]. This latter null result has been contrasted by studies of the same reaction, $\pi^- p \to K_S K_S n$, measured at 40 GeV/c with the MSS facility at Serpukhov [55]. A little structure could be seen in the $K_S K_S$ mass

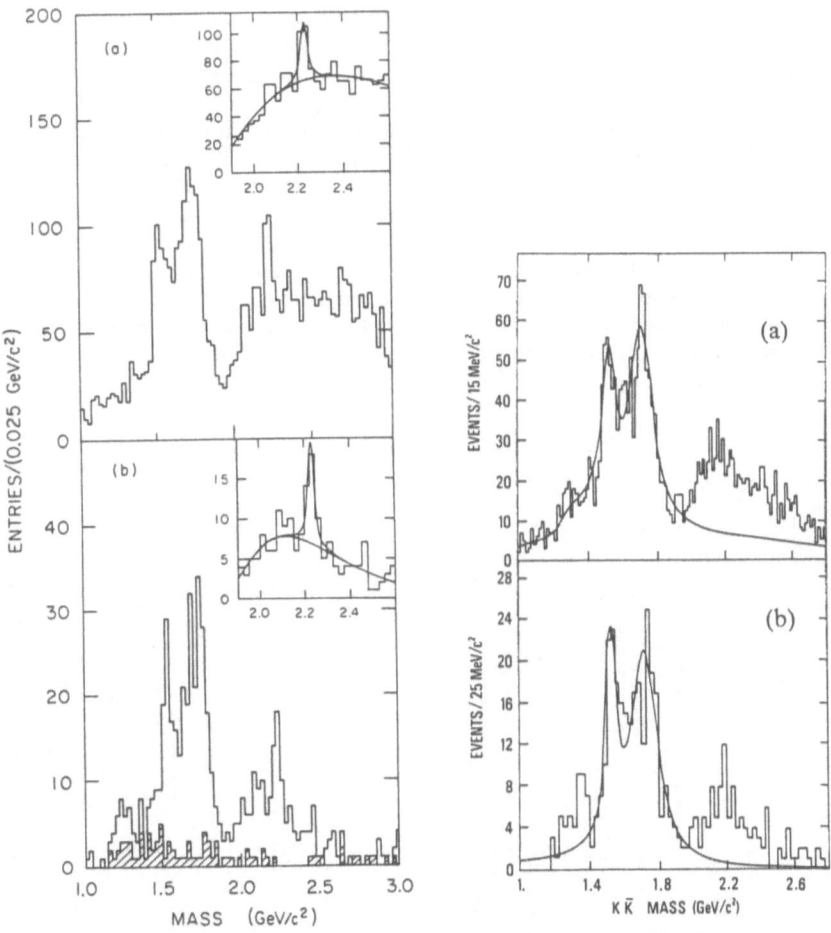

Figure 10. Results from $J/\psi \to \gamma \bar{K}K$ of the MARK III and DM2 experiments. Left: $\bar{K}K$ invariant mass spectra obtained by MARK III [50], where a) the $K^+ K^-$ and b) the $K_S K_S$ modes are shown. The 4π background is shown cross-hatched. Fits to the $\xi(2230)$ region are displayed in the insets. Right: $\bar{K}K$ invariant mass spectra obtained by DM2 [51], where a) the $K^+ K^-$ and b) the $K_S K_S$ modes are shown with fits to the lower-mass region.

185

spectrum near 2.2 GeV. Based on an amplitude analysis, the authors identified this to be a D-wave resonance having $J^{PC} = 2^{++}$ with parameters M = (2230 ± 20) MeV and Γ = (80 ± 30) MeV, and, moreover, they identified it with the ζ(2230) resonance of MARK III.

Using the LASS spectrometer at SLAC, the reactions $K^-p \rightarrow K_S K_S \Lambda$ and $K^-p \rightarrow K^+K^-\Lambda$ were studied at 11 GeV/c incident momentum, the main motivation being to investigate the spectrum of (s$\bar{s}$) strangeonium states. The final data samples from these two reactions contained 441 and 12294 events, respectively. The analysis of the $K_S K_S$ angular distribution [56] in terms of spherical harmonic moments in the t-channel helicity frame indicated the presence of a structure near 2.2 GeV with J≥2, so that $J^{PC} = 2^{++}$ or 4^{++} became most likely (see figure 11a). For the K^+K^- final state a similar analysis was performed [57], more detailed in terms of helicity amplitudes, which confirmed this structure (see figure 11b). It appeared to be most pronounced in the L=4 wave and, even more so, when plotting the sum of interference terms between L=4 and L≤3. The peak strikingly visible in this latter plot is, however, not easily to be traced back to the invariant-mass spectrum from which it essentially originates. A resonance fit to this G-wave amplitude yielded the parameters M = (2209 ± 16) MeV and Γ = (60 ± 82) MeV. For this state the 4^{++} assignment was the most likely one. The conclusion was that it be mostly of (s$\bar{s}$) nature and a member of the ^{3}F ground-state nonet. Furthermore, it seemed possible that the state observed here and the ζ(2230) state seen in $J/\psi \rightarrow \gamma \bar{K}K$ are one and the same. Both $J^{PC} = 2^{++}$ or 4^{++} would be possible in such a case, since from the Constituent Quark Model one expects both ^{3}F (s$\bar{s}$) states in this mass region.

It was also attempted to detect the direct formation of ζ(2230) in $\bar{p}p$ interactions. The reaction $\bar{p}p \rightarrow K^+K^-$ was studied at BNL−AGS [58] and by experiment PS170 at LEAR [59]. Evaluating the ratio of cross-sections for (K^+K^-) vs $(\pi^+\pi^-)$, no evidence for ζ(2230) formation was found. Upper limits for the product branching fraction BR($\bar{p}p \rightarrow \zeta$)·BR($\zeta \rightarrow K^+K^-$) were typically around 2×10^{-4}, when Γ ≈ 30 MeV and J=2 were assumed for ζ(2230). The reaction $\bar{p}p \rightarrow K_S K_S$, investigated by experiment PS185 at LEAR, contains a non-resonant background smaller by one order of magnitude compared with K^+K^-, hence this channel is expected to be more sensitive to a $\bar{K}K$ resonance. Still, no evidence for ζ(2230) formation has been found in the excitation function or the angular distributions [60]. Assuming, for instance, Γ ≈ 30 MeV and J=2 for the state, an upper limit BR($\bar{p}p \rightarrow \zeta$)·BR($\zeta \rightarrow K_S K_S$) ≈ 2×10^{-5} was calculated from the data.

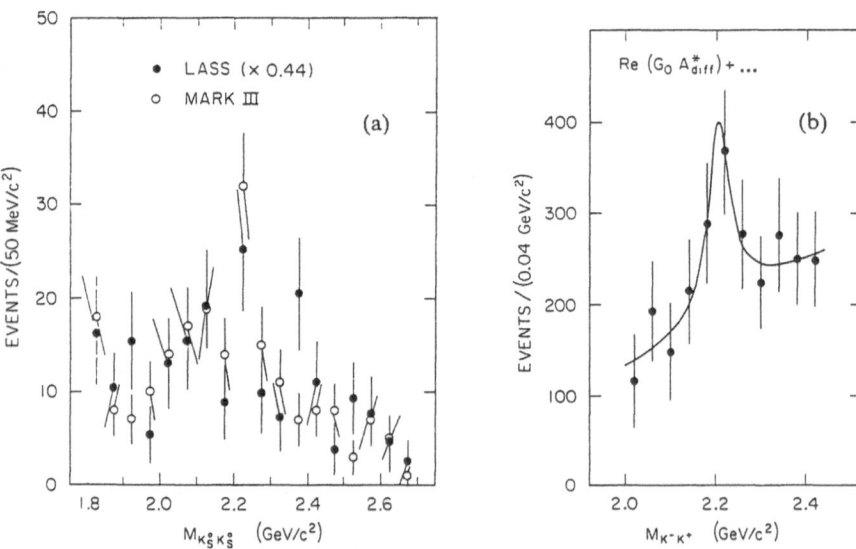

Figure 11.   Results [57] from $K^-p \rightarrow \bar{K}K\Lambda$. a) Acceptance-corrected mass spectrum for the $K_S K_S$ mode and comparison with $J/\psi \rightarrow K_S K_S$ data from ref. [50]. b) Mass dependence of interference terms between particular partial waves in the K^+K^- mode.

In addition to $\bar{K}K$ states discussed in the previous section, channels containing other pairs of pseudoscalar mesons, such as $\eta\eta'$ or $\pi\pi$, can provide information that allows the flavour-dependent decay of intermediate states to be probed. The production of neutral hadrons in $\pi^-p\rightarrow X^0n$ and their subsequent decays into multi-photon final states was studied by joint CERN−IHEP experiments using the GAMS electromagnetic calorimeters. Data at 38 GeV/c beam momentum were taken at Serpukhov, those with a 100 GeV/c beam at CERN−SPS. A structure centered at (2220 ± 10) MeV decaying into $\eta\eta'$, seen as a sub-sample of 4γ events, was observed in both data sets [61]. The statistics being very low, the width of this state appeared to be comparable to the instrumental resolution of about 100 MeV at this energy. From anisotropic angular distributions the lower limit $J\geq2$ was concluded. In view of the strange-quark content of η and η' it is tempting to identify this structure with the state $\xi(2230)$ seen in $J/\psi\rightarrow\gamma\bar{K}K$ and/or with the 2.2 GeV structure seen in $K^-p\rightarrow\bar{K}K\Lambda$.

From an early experiment using a missing-mass spectrometer in order to study the inclusive reaction $\pi^-p\rightarrow p + X^-$, the observation of three resonances originally called S, T and U was reported in the mass region between 1.9 and 2.4 GeV [62]. Also in $\bar{p}p$ interactions enhancements were seen in the energy dependence of the total cross-section, as well as in that of the elastic, the inelastic, and the annihilation cross-sections. One experiment, by investigating these structures in high-statistics studies, extracted the following resonance parameters for the T- and U-meson regions from the data [63]: $M_T = (2193 \pm 1)$ MeV with $\Gamma_T = (98 \pm 7)$ MeV, and $M_U = (2359 \pm 1)$ MeV with $\Gamma_U = (165 \pm 13)$ MeV. Another experiment studied specifically the annihilation channels $\bar{p}p\rightarrow\pi^0\pi^0$ and $\bar{p}p\rightarrow\pi^0\eta$ in these energy regions [64]. The latter process did not exhibit any energy-dependent structure in the differential cross-section. The $\pi^0\pi^0$ channel, however, was analyzed in terms of partial-wave amplitudes. Here, the results suggested the presence of a 2^{++} resonance with mass M = 2.15 GeV and width Γ = 0.25 GeV. From a comparison of $\pi^0\pi^0$ and $\pi^+\pi^-$, the authors interpreted this resonance as the isospin-0 contribution to the T-meson region. Data obtained in measurements of $\pi^+p\rightarrow\pi^+\pi^0\pi^0p$ were analyzed in terms of $\pi^+\pi^-\rightarrow\pi^0\pi^0$ scattering amplitudes. This again gave some evidence for a resonance in the energy region just above 2 GeV [65], the parameters extracted from the fits being M = (2015 ± 28) MeV and Γ = (186 ± 81) MeV.

A high-mass structure decaying into $\pi^+\pi^-$ was seen by MARK III in studies of $J/\psi\rightarrow\gamma\pi^+\pi^-$ (see figure 12a). The parameters of this resonance with yet ambiguous interpretation [66] were determined as M = (2086 ± 15) MeV and Γ = (210 ± 63) MeV. However, the higher-statistics

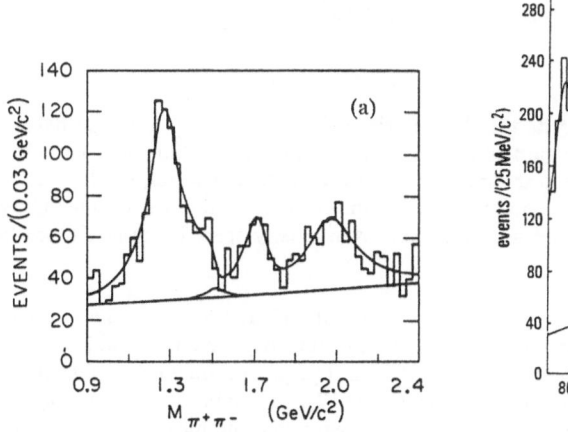
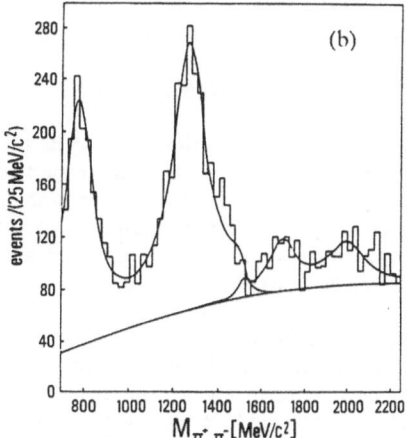

Figure 12.   Results from $J/\psi\rightarrow\gamma\pi^+\pi^-$ of the MARK III and DM2 experiments. Shown are the $\pi^+\pi^-$ invariant mass spectra, a) with a four-resonance fit obtained by the MARK III experiment [66], and b) with a five- resonance fit obtained by the DM2 experiment [67].

data from MARK III reported in ref. [43] led to the values M = (2089 ± 18) MeV and Γ = (127 ± 106) MeV. The DM2 measurements of $J/\psi \rightarrow \gamma\pi^+\pi^-$ also revealed a relatively broad structure in the $\pi^+\pi^-$ invariant-mass spectrum [67]. Its resonance parameters were somewhat dependent on the number of Breit−Wigner curves introduced to fit the entire mass spectrum (see figure 12b). The fits gave average values M = (2024 ± 27) MeV and Γ = (290 ± 50) MeV. The authors suggested to identify this state with the 4^{++} candidate tabulated as $f_4(2050)$ in ref. [3].

Outlook: Enriched physics with an improved machine

In the preceeding chapters we have demonstrated the richness of the physics to be pursued with the JETSET experiment, in particular by measurements of $\bar{p}p$ annihilations into $\phi\phi$, $\omega\phi$, $\omega\omega$ and $\pi^0\phi\phi$, and into $\bar{K}K$, $\bar{K}K^*$ and $\bar{K}^*K^*$. Although these studies take full advantage of the high luminosity and the good final-state mass resolution obtainable in a jet-target experiment, there are some limitations due to the machine parameters to be encountered. The main deficiencies of the LEAR accelerator in its actual configuration are the large size of the vacuum chamber and the upper limit of the beam momentum. Improvements in these respects, however, may be impossible without severe negative consequences in other areas, such as the acceptance of the machine or the availability of low-momentum beams. The advantage of beam momenta higher than the present LEAR limit, say 4 GeV/c, is obvious. An increase of the invariant-mass range up to $\sqrt{s} \approx 3$ GeV would open the window to a largely unexplored regime in the spectroscopy of conventional and non-conventional hadrons. Formation studies such as $\bar{p}p \rightarrow \phi\phi$ and, even more so, production experiments like $\bar{p}p \rightarrow \pi^0\phi\phi$ would greatly profit from such an 'energy upgrade'.

There are several strong arguments in favour of a small-sized and correspondingly thin-walled beam pipe. Firstly, the innermost tracking device ('vertex detector') can be positioned close to the interaction region, which leads to a better precision in the reconstruction of production vertices. Secondly, triggers for 'long-lived' neutral particles such as K_S or Λ can be established more efficiently. Thirdly, the detection of charged particle tracks can be extended down to small angles. This last point becomes more and more important as one approaches the threshold of the reaction under study, since in that case all particles are emitted within a narrow forward cone centered around the beam axis. Moreover, in many cases the early decays of low-momentum particles make both triggering and tracking difficult.

In the following two sections we briefly discuss two experiments, both of which would take full advantage of a small-sized and thin-walled vacuum chamber. Both cases would need very high luminosities as available in internal jet-target experiments. For instrumental reasons, however, they are both beyond the scope of the phase-1 physics programme of JETSET. They are being discussed here as interesting possibilities − among others − for later stages.

Threshold studies and $\bar{K}K$ molecules

The dynamics of a reaction close to its threshold is simplified in that only a few partial waves are supposed to contribute. But this region is always a particularly interesting one, because the excitation function may reveal structures near threshold due to quasi-bound states, or the excitation function of another reaction may exhibit a structure due to the opening of the first channel ('cusp'). At and around the $K^+K^-K^+K^-$ threshold we encounter seven closely spaced thresholds (see table 1), which call for a detailed investigation.

The f_0 and a_0 particles have masses only a few MeV below the K^+K^- invariant mass. As these 0^{++} states are almost degenerate and seem to differ only by isospin, they have been subject to many speculations as to their origin. The theoretical investigation of $(qq\bar{q}\bar{q})$ spectra showed that the only bound states of this kind are those having a $\bar{K}K$-like structure [17]. It has been suggested that the f_0 and a_0 are such $\bar{K}K$ 'molecules' or 'dimesons', one argument being that the measured and calculated widths agree rather well but a $(q\bar{q})$ state would be much broader [68]. Both f_0 and a_0 seem to be strongly coupled to the $\bar{K}K$ channels. If they are of non-$(q\bar{q})$ nature, one may eventually expect enhanced branching fractions for $\phi \rightarrow \gamma f_0$ or $\phi \rightarrow \gamma a_0$. A continuous scan in $\sqrt{s}$ for all channels and across all thresholds listed in table 1 would certainly help to unravel many mysteries and open

Table 1. Thresholds in the vicinity of $K^+K^-K^+K^-$

Final state	$\sqrt{s}$ (MeV)	p (MeV/c)
$f_0 f_0$	1952	559.09
$K^+K^- f_0$	1963.29	603.82
$a_0 a_0$	1966	614.24
$K^+K^- a_0$	1970.29	630.53
$K^+K^-K^+K^-$	1974.58	646.56
$K^+K^- \phi$	2006.70	760.25
$\phi\phi$	2038.82	866.03

questions related to multi-kaon final states − even though it may be difficult for any experiment to differentiate between the four-quark and the $\overline{K}K$-molecule interpretation of the f_0 and a_0 states.

High-statistics $\bar{p}p \to \overline{\Lambda}\Lambda$ for a CP violation search

Up to here all discussion in this report has focussed on strong interactions and the underlying theory of QCD. But also aspects of electroweak interactions can be addressed, which constitute the other building block of the Standard Model. A fundamental question in this context is the physical origin of CP non-conservation, since no system, other than the $K^0 - \overline{K}^0$ system, has exhibited CP violation so far. An experimental programme that can be devoted to this [69−71] is the investigation of decay asymmetries in hyperon−antihyperon systems such as $\overline{\Lambda}\Lambda$ or $\overline{\Xi}^+\Xi^-$. At LEAR energies the reaction of choice is $\bar{p}p \to \overline{\Lambda}\Lambda$. The produced hyperons emerge with large polarizations [72] of the order of $|P| \approx 0.5$. Assuming the hadronic production process to be charge-conjugation invariant, the Λ and $\overline{\Lambda}$ polarizations are equal. Parity conservation requires that the polarization vectors are transverse to the production plane. The $\bar{p}p$ initial state, and thus the $\overline{\Lambda}\Lambda$ final state too, have a definite CP property, so that final-state interactions cannot generate a misleading signal [73]. Owing to baryon-number conservation, there is no $\Lambda - \overline{\Lambda}$ mixing, and therefore any observed signal constitutes a measure of $\Delta S = 1$ CP violation.

The non-leptonic weak decay $\Lambda \to p\pi^-$ ($\overline{\Lambda} \to \bar{p}\pi^+$) is parity-violating. For a sample of Λ hyperons with polarization P, the angular distribution of the decay protons in the Λ rest frame is given by $W(\theta_p) = (4\pi)^{-1} [1 + (\alpha P)_\Lambda \cos\theta_p]$, where θ_p is measured between the normal to the production plane and the proton momentum vector, and $\alpha = 0.642 \pm 0.013$ is the $\Lambda \to p\pi^-$ decay asymmetry parameter. An interesting observable is $A = (\alpha + \bar{\alpha})/(\alpha - \bar{\alpha})$, which would signal CP violation if measured to be non-zero. The absolute size of this signal, as predicted in the framework of the Standard Model, is small. Typical values [74, 75] are of the order of a few 10^{-4}, and they are somewhat dependent on yet poorly known quantities such as hadronic matrix elements, the t-quark mass, and the 'direct' CP violation parameter $|\varepsilon'/\varepsilon|$.

Experimentally the ratio A is relatively easy to determine. The best result obtained so far stems from experiment PS185 at LEAR, where data taken at 1.546 GeV/c (4063 events) and 1.695 GeV/c (11427 events) were used to evaluate the ratio A. The statistics of $\bar{p}p \to \overline{\Lambda}\Lambda \to \bar{p}\pi^+ p\pi^-$ events combined from the two beam momenta [72] gave the average value $\langle A \rangle = \langle (\alpha + \bar{\alpha})/(\alpha - \bar{\alpha}) \rangle = -0.024 \pm 0.057$. On the basis of statistical arguments one can estimate how many events must be accumulated and analyzed in order to reach the level of sensitivity at which a CP violation effect can conceivably be expected. For a number of N events, the statistical uncertainty on the ratio A is $\sigma_A = (\alpha P)^{-1}\sqrt{[3/(2N)]}$. Assuming $|P| = 0.3$ as the average of the angular dependent polarization, the uncertainty $\sigma_A \approx 10^{-4}$ corresponds to $N > 4\times10^9$ events. It is interesting that the result from PS185 as quoted above is consistent with this estimate and an average polarization $|P| = 0.27$.

The question arises how such a huge number of events can be accumulated within a reasonable running time [69−71].
- The optimum beam momentum in a jet-target experiment on $\bar{p}p \to \overline{\Lambda}\Lambda$ is 1.65 GeV/c, which is just below the $\overline{\Lambda}\Sigma^0$ threshold. The production cross-section is $\sigma(\overline{\Lambda}\Lambda) \approx 80 \ \mu b$.

189

- The 'overall efficiency' of the experiment is about 10 %. This number is a little lower than what has been learned from PS185 in 'real life'. It includes the double-branching fraction $BR(\Lambda \to p\pi^-) \cdot BR(\overline{\Lambda} \to \overline{p}\pi^+) = 0.41$, the on-line trigger efficiency of 0.6, the data-acquisition live-time of 0.7, and the off-line reconstruction efficiency for four-track ($2V^0$) events of 0.6.
- With at least 10^{11} antiprotons stored in the machine, a revolution frequency of 4 MHz, and a jet-target density of 10^{14} atoms/cm², the peak luminosity is 4×10^{31} cm^{-2}·sec^{-1}.

From these numbers one calculates an acquisition rate for 'good' events of 320 sec^{-1} at peak luminosity. The accumulation of 4×10^9 events would thus require an effective data-taking time of 1.25×10^7 sec. This is not far from the 'canonical' value of 10^7 sec, which is often assumed as the equivalent to being 'on-beam' for one full year. Preliminary studies indicate that the detector elements suited for the triggering and for the recording of charged particle tracks do not require much R&D effort, the main difficulties of the experiment being in the areas of fast data acquisition and pre-processing. Computing power, however, continues to grow rapidly.

Acknowledgements

The author warmly thanks R. Landua, J.−M. Richard and R. Klapisch for having organized this enjoyable school with its stimulating atmosphere. The JETSET Collaboration owes much to the engineers and technicians at the home institutes; without their dedicated efforts the completion of the apparatus would not have been accomplished in time. The Collaboration also thanks the LEAR team for their excellent cooperation during the installation of the experiment. The work reported here is supported in part by CERN, the German Bundesministerium für Forschung und Technologie, the Italian Istituto Nazionale di Fisica Nucleare, the Norwegian Research Council, the Swedish Natural Science Research Council, and the United States National Science Foundation.

References

1.   R.A. Eisenstein et al. (JETSET Collaboration), Proc. BNL Workshop on Glueballs, Hybrids and Exotic Hadrons, Upton (New York), 1988, ed. S.−U. Chung (AIP Conference Proceedings No. 185, American Institute of Physics, New York, 1989) p. 636.

2.   D.W. Hertzog et al. (JETSET Collaboration), Proc. 3rd Conference on the Intersections between Particle and Nuclear Physics, Rockport (Maine), 1988, ed. G.M. Bunce (AIP Conference Proceedings No. 176, American Institute of Physics, New York, 1988) p. 334.

3.   Particle Data Group, Phys. Lett. 204B (1988) 1.

4.   S. Baird et al., Proc. 4th Workshop on Physics with Low-Energy Antiprotons, Villars-sur-Ollon, 1987, eds. C. Amsler et al. (Harwood Academic Publishers, Chur, 1988) p. 91.

5.   P. Lefèvre, Proc. International School of Physics with Low-Energy Antiprotons on Antiproton − Nucleon and Antiproton − Nucleus Interactions, Erice, 1988, eds. F. Bradamante et al. (Plenum Press, New York, 1990).

6.   F.E. Close, Rep. Prog. Phys. 51 (1988) 833.

7.   M.S. Chanowitz, Proc. CCAST (World Laboratory) Symposium/Workshop on Charm Physics, Beijing, 1987, eds. M. Ye and T. Huang (Gordon and Breach Science Publishers, Montreux, 1988) p. 161.

8.   T.H. Burnett and S.R. Sharpe, preprint DOE$-$ER$-$40423$-$04$-$P90 (1990), submitted to Annual Reviews of Nuclear and Particle Science.

9.   T. Barnes, F.E. Close and F. de Viron, Nucl. Phys. B224 (1983) 241.

10.   M. Chanowitz and S.R. Sharpe, Nucl. Phys. B222 (1983) 211.

11.   N. Isgur and J. Paton, Phys. Rev. D31 (1985) 2910.

12.   G. Schierholz, Proc. BNL Workshop on Glueballs, Hybrids and Exotic Hadrons, Upton (New York), 1988, ed. S.$-$U. Chung (AIP Conference Proceedings No. 185, American Institute of Physics, New York, 1989) p. 281.

13.   D. Robson, Nucl. Phys. B130 (1977) 328.

14.   M.S. Chanowitz and S.R. Sharpe, Phys. Lett. 132B (1983) 413.

15.   N. Isgur, R. Kokoski and J. Paton, Phys. Rev. Lett. 54 (1985) 869.

16.   R.L. Jaffe, Phys. Rev. Lett. 38 (1977)195 and Phys. Rev. Lett. 38 (1977) 617.

17.   J. Weinstein and N. Isgur, Phys. Rev. D27 (1983) 588.

18.   B.L. Ioffe, these Proceedings.

19.   S. Okubo, Phys. Lett. 5 (1963) 165. G. Zweig, CERN report 8419/TH$-$412 (1964). I. Iizuka, Prog. Theor. Phys. Suppl. 37$-$38 (1966) 21.

20.   J.F. Donoghue, Proc. Europhysics Conference on High-Energy Physics, Bari, 1985.

21.   H.J. Lipkin, Nucl. Phys. B244 (1984) 147 and Nucl. Phys. B291 (1987) 720.

22.   C.B. Dover and P.M. Fishbane, Phys. Rev. Lett. 62 (1989) 2917.

23.   W. Roberts and G. Karl, Phys. Rev. D39 (1989) 1877.

24.   J. Ellis, E. Gabathuler and M. Karliner, Phys. Lett. 217B (1989) 173.

25.   C. Baglin et al. (R704 Collaboration), Phys. Lett. 231B (1989) 557.

26.   J. Davidson et al., Phys. Rev. D9 (1974) 77.

27.   A. Etkin et al., Phys. Rev. Lett. 40 (1978) 422 and Phys. Rev. Lett. 41 (1978) 784.

28.   A. Etkin et al., Phys. Rev. Lett. 49 (1982) 1620.

29.   A. Etkin et al., Phys. Lett. 165B (1985) 217.

30.   A. Etkin et al., Phys. Lett. 201B (1988) 568.

31.   See Proc. International Conference on Hadron Spectroscopy, College Park (Maryland), 1985, ed. S. Oneda (AIP Conference Proceedings No. 132, American Institute of Physics, New York, 1985): S.J. Lindenbaum and R.S. Longacre, p. 51; J.F. Donoghue, p. 460.

32.   T.A. Armstrong et al., Nucl. Phys. B196 (1982) 176.

33.   P.S.L. Booth, Nucl. Phys. B273 (1986) 677 and Nucl. Phys. B273 (1986) 689.

34. T.F. Davenport, Phys. Rev. D33 (1986) 2519.

35. D.R. Green et al., Phys. Rev. Lett. 56 (1986) 1639.

36. T. Armstrong et al., Phys. Lett. 121B (1983) 83.

37. D. Aston et al., preprint SLAC$-$PUB$-$5150 (1989), presented at 3rd International Conference on Hadron Spectroscopy, Ajaccio, 1989.

38. T.A. Armstrong et al., Phys. Lett. 166B (1986) 245.

39. A. Kirk, Thesis, University of Birmingham (1989). T.A. Armstrong et al. (WA76 Collaboration), Phys. Lett. 221B (1989) 221.

40. D. Bisello et al. (DM2 Collaboration), Phys. Lett. 179B (1986) 289 and Phys. Lett. 179B (1986) 294.

41. D. Bisello et al. (DM2 Collaboration), preprint LAL 90$-$12 (1990), submitted to Physics Letters B.

42. Z. Bai et al. (MARK III Collaboration), preprint SLAC$-$PUB$-$5159 (1990), submitted to Physical Review Letters.

43. G. Eigen, Proc. International School of Physics with Low-Energy Antiprotons on Spectroscopy of Light and Heavy Quarks, Erice, 1987, eds. U. Gastaldi et al. (Plenum Press, New York, 1989) p. 183.

44. G. Eigen, Proc. 24th International Conference on High-Energy Physics, Munich, 1988, eds. R. Kotthaus and J.H. Kühn (Springer-Verlag, Berlin and Heidelberg, 1989) p. 590.

45. U. Mallik (MARK III Collaboration), Proc. BNL Workshop on Glueballs, Hybrids and Exotic Hadrons, Upton (New York), 1988, ed. S.$-$U. Chung (AIP Conference Proceedings No. 185, American Institute of Physics, New York, 1989) p. 325.

46. R.M. Baltrusaitis et al. (MARK III Collaboration), Phys. Rev. Lett. 55 (1985) 1723.

47. D. Alde et al., preprint CERN$-$EP/90$-$39 (1990), submitted to Physics Letters B.

48. T.A. Armstrong et al. (WA76 Collaboration), Z. Phys. C34 (1987) 33.

49. T.A. Armstrong et al. (WA76 Collaboration), preprint CERN$-$EP/89$-$108 (1989), submitted to Zeitschrift für Physik C.

50. R.M. Baltrusaitis et al. (MARK III Collaboration), Phys. Rev. Lett. 56 (1986) 107.

51. J.E. Augustin et al. (DM2 Collaboration), Phys. Rev. Lett. 60 (1988) 2238.

52. T.A. Bolton, Thesis, MIT (1988), referenced in: D.G. Hitlin, Proc. BNL Workshop on Glueballs, Hybrids and Exotic Hadrons, Upton (New York), 1988, ed. S.$-$U. Chung (AIP Conference Proceedings No. 185, American Institute of Physics, New York, 1989) p. 88.

53. C. Evangelista et al., Nucl. Phys. B154 (1979) 381.

54. R.S. Longacre et al., Phys. Lett. 177B (1986) 223.

55. B.V. Bolonkin et al., Sov. J. Nucl. Phys. 46 (1987) 451.

56. D. Aston et al., Nucl. Phys. B301 (1988) 525.

57. D. Aston et al., Phys. Lett. 215B (1988) 199.

58. J. Sculli et al., Phys. Rev. Lett. 58 (1987) 1715.

59. G. Bardin et al., Phys. Lett. 195B (1987) 292.

60. H. Schledermann, Dissertation, University of Freiburg (1989). H. Schmitt et al., Proc. 9th European Symposium on Antiproton−Proton Interactions and Fundamental Symmetries, Mainz, 1988, eds. K. Kleinknecht and E. Klempt, Nucl. Phys. B (Proc. Suppl.) 8 (1989) 162.

61. D. Alde et al., Phys. Lett. 177B (1986) 120.

62. G. Chikovani et al., Phys. Lett. 22 (1966) 233.

63. J. Alspector et al., Phys. Rev. Lett. 30 (1973) 511.

64. R.S. Dulude et al., Phys. Lett. 79B (1978) 329 and Phys. Lett. 79B (1978) 335.

65. N.M. Cason et al., Phys. Rev. Lett. 48 (1982) 1316.

66. R.M. Baltrusaitis et al. (MARK III Collaboration), Phys. Rev. D35 (1987) 2077.

67. J.E. Augustin et al. (DM2 Collaboration), Z. Phys. C36 (1987) 369.

68. J. Weinstein and N. Isgur, Univ. of Tennessee preprint UTPT−89−03 (1989).

69. N.H. Hamann, Proc. 3rd Conference on the Intersections between Particle and Nuclear Physics, Rockport (Maine), 1988, ed. G.M. Bunce, (AIP Conference Proceedings No. 176, American Institute of Physics, New York, 1989) p. 417.

70. N.H. Hamann, Proc. Workshop on CP Violation at KAON Factory, TRIUMF, Vancouver, 1988, ed. J.N. Ng, report TRI−89−3 (1989), p. 29.

71. D.W. Hertzog, Proc. Symposium/Workshop on Spin and Symmetries, TRIUMF, Vancouver, 1989, eds W.E. Ramsay and W.T.H. van Oers, report TRI−89−5 (1989), p. 89.

72. W. Dutty, Dissertation, Univ. of Freiburg (1988). G. Sehl, Dissertation, Jül−Spez−535 (1989).

73. J.F. Donoghue, B.R. Holstein and G. Valencia, Phys. Lett. 178B (1986) 319.

74. J.F. Donoghue, X.G. He and S. Pakvasa, Phys. Rev. D34 (1986) 833.

75. J.F. Donoghue, Proc. 3rd Conference on the Intersections between Particle and Nuclear Physics, Rockport (Maine), 1988, ed. G.M. Bunce, (AIP Conference Proceedings No. 176, American Institute of Physics, New York, 1989) p. 341.

PS199: $\bar{p}p \to \bar{n}n$ SPIN PHYSICS AT LEAR

The PS199 Collaboration:

R.Birsa, F.Bradamante, S.Dalla Torre-Colautti, M.Giorgi, M.Lamanna, A.Martin,
A.Penzo, P.Schiavon, F.Tessarotto
INFN Trieste and University of Trieste, Trieste, Italy

M.P.Macciotta, A.Masoni, G.Puddu, S.Serci
INFN Cagliari and University of Cagliari, Cagliari, Italy

T.Niinikoski, A.Rijllart
CERN, Geneva, Switzerland

A.Ahmidouch, E.Heer, R.Hess, C.Lechanoine-Le Luc, C.Mascarini, D.Rapin
DPNC, University of Geneva, Geneva, Switzerland

J.Arvieux, R.Bertini, H.Catz, J.C.Faivre, R.A.Kunne, F.Perrot-Kunne
CEN DPhN and LNS, Saclay, Gif-sur-Yvette, France

M.Agnello, F.Iazzi, B.Minetti
INFN Turin and Turin Polytechnic, Turin, Italy

T.Bressani, E.Chiavassa, N.De Marco, A.Musso, A.Piccotti
INFN Turin and University of Turin, Turin, Italy

(presented by Massimo Lamanna)

ABSTRACT

The experiment PS199, studying the spin structure of the charge-exchange reaction $\bar{p}p \to \bar{n}n$ at LEAR (CERN) is described and preliminary results are discussed.

INTRODUCTION

The extensive research program proposed by the collaboration PS199 at LEAR has started: the aim is to investigate the spin structure of the $\bar{p}p \to \bar{n}n$ charge-exchange reaction. The asymmetry A_{on} of this reaction will be measured at 11 momenta in the range from 500 MeV/c to 1500 MeV/c of the incoming antiproton momentum using a polarized target. Double scattering measurements using the neutron counters as active targets for both outcoming neutron and antineutron will also be performed at selected momenta in restricted angular regions to measure D and D_t. In this momentum range no spin-observable has ever been measured.

Medium-Energy Antiprotons and the Quark–Gluon Structure of Hadrons
Edited by R. Landua *et al.*, Plenum Press, New York, 1991

EXPERIMENTAL SETUP

The experimental apparatus is sketched in Fig.1. The neutron and the antineutron produced in the charge-exchange reaction are detected with three neutron counters NC1-3 and two antineutron counters ANC1-2.   The beam is provided by the LEAR storage ring at CERN;   the typical intensity is 10^6 $\bar{p}$/sec characterized by high monochromaticity and low emittance.

The beam is defined by two scintillator counters. The first one, called B_0, is a 3×5 cm² counter, 0.5 cm thick, placed before the cryostat defining the incoming beam. The second one is a disk of 1 cm diameter, 0.5 cm thickness, viewed by to photomultipliers called B_1 and B_2 sitting inside the cryostat in front of the target (19 cm from the centre of the target).

The target is a pentanol ($C_5H_{11}OH$) target operated in frozen-spin mode (Ref.1). Pentanol beads are contained in a 12 cm long, 1.8 cm diameter can and are cooled down to 80 m°K in a He^3-He^4 dilution refrigerator cryostat inserted between the poles of a magnet. This is a dipole magnet providing a magnetic field orthogonal to the horizontal plane up to 2.5 T and highly homogeneous (better than 10^{-5}) in the target region. The polarization of the target is performed at 2.5 T. The maximum polarization value achieved is 96%, but the typical working value was 80%. As soon as the desired value of the polarization is reached, the microwave source is switched

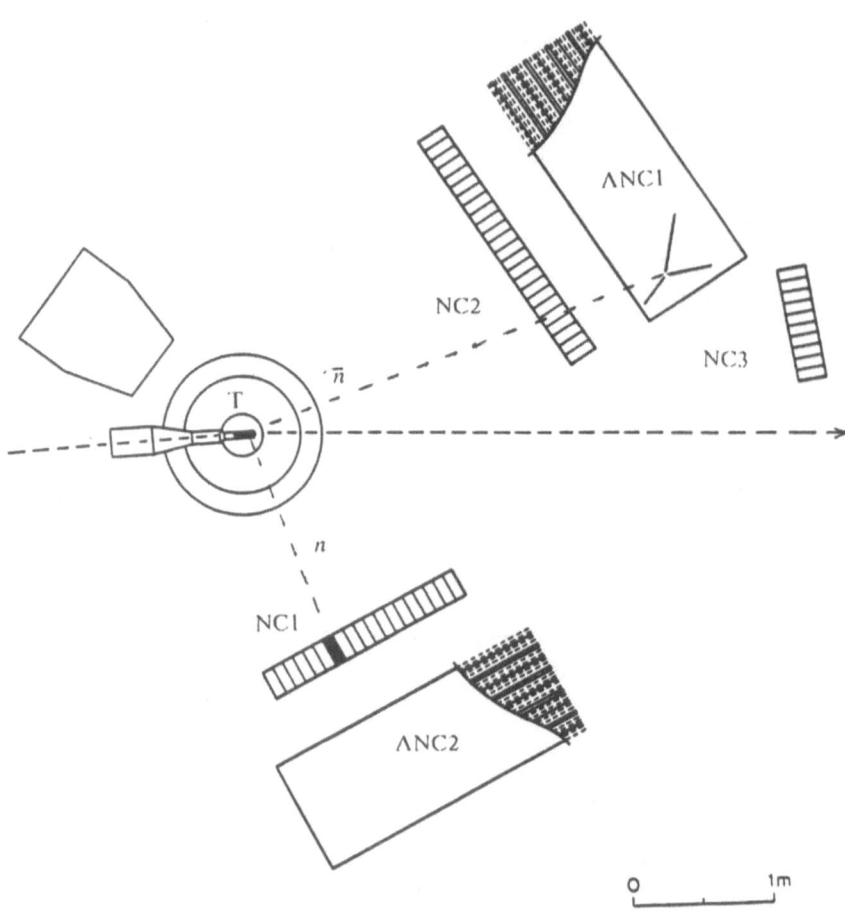

Figure 1.   PS199 experimental set-up. ANC: antineutron detectors. NC: neutron counters. T: target region and veto system.

196

off and the target cooled down to the minimum value: at this point the spins are "frozen" (i.e., if the temperature is low enough, the relaxation time can be of the order of hours or even of days with a relatively small holding field. In the present setup the relaxation time of our target is of the order of 12 hours with a field of 3.785 kG). The possibility to operate the polarized target with a low magnetic field is an essential feature of the experiment allowing us to focus the beam in the long target even at the lowest foreseen momentum (500 MeV/c).

To estimate the background due to the non-polarized nucleons of the nuclei in the target (essentially the carbon content of pentanol and the liquid helium used for cooling) a dummy target filled with FEP (fluorinated-ethyline-polypropylene) is used. For calibration and normalization purposes a liquid hydrogen target (12 cm long, 3.2 cm diameter) is also employed.

To define the charge-exchange reaction trigger we need to reject the annihilations (charged and neutral mesons in the final state) and the elastic channel. The reactions with at least one charged particle are discarded by surrounding the interaction region with a box of three scintillation counters. The γ's from π^o decay are rejected with two lead-scintillator sandwiches placed outside the geometrical acceptance of the apparatus, above and below the interaction region against the magnet poles. The efficiency of the pole-faced vetos on neutral annihilations has been calculated and is of the order of 95%: the limited acceptance of the pole-face veto is compensated from the high multiplicity of the annihilation into neutrals. The scintillator slabs vetoing the charged particles and the pole-faced sandwiches form the veto box (VETO).

Two identical antineutrons detectors (ANC1 and ANC2) were built (Ref.2). Each of them is a sandwich of 20 planes of limited streamer tubes (LST), 5 hodoscopes (consisting of 6 scintillation counter with one PM on each side) and 4 iron slabs. The sensitive surface of the LST planes is 1660×2000 mm^2, the scintillator slabs dimensions are $330 \times 1600 \times 10$ mm^3. Each LST plane is equipped with strips controlled with a digital read-out system permitting to read the hit patterns on two projections at the same time: 160 strips (10.5 mm pitch) for the coordinate orthogonal to the wires of the chambers (side view) and 192 strips (10 mm pitch) for the top view to a total of 14080 channels. The hodoscopes signal are used essentially for triggering purposes and TOF measurements. The LST planes and the hodoscopes are arranged in substructures called modules: they are sandwiches of four LST planes with one hodoscope in between; the depth of a module is 15.5 cm. Five modules interleaved by four 3 cm thick iron slabs form a detector (Fig.2).

The neutron counters (NC1,NC2 and NC3) are made up of a total of 53 vertical scintillator bars. Each scintillator is 80 mm wide, 200 mm thick and the length ranges from 400 to 1300 mm, and is equipped by a 5" photomultiplier from each side (up and down) (Ref.3). A hit is defined as an up-down coincidence in both neutron counters' bars and antineutron detectors' slabs to cut down the noise rate. A software threshold is also applied in the offline analysis to guarantee the stability in time of the efficiency in neutron detection. The vertical coordinate (i.e. along the scintillator) is given by the difference between the time measured with the top and the bottom photomultipliers of each bar.

DATA ACQUISITION

The trigger used to collect the sample for the measurement of A_{on} at 700 MeV/c, 800 MeV/c and 900 MeV/c in the first physics run of the collaboration (September 1989) was given by the coincidence[1]:

$$B_0 \cdot B_1 \cdot B_2 \cdot \overline{VETO} \cdot [\, NC1 \cdot ANC1 + NC2 \cdot ANC2 + NC1 \cdot NC3 \,]$$

---

[1] *VETO* is the OR condition on the set of veto detectors

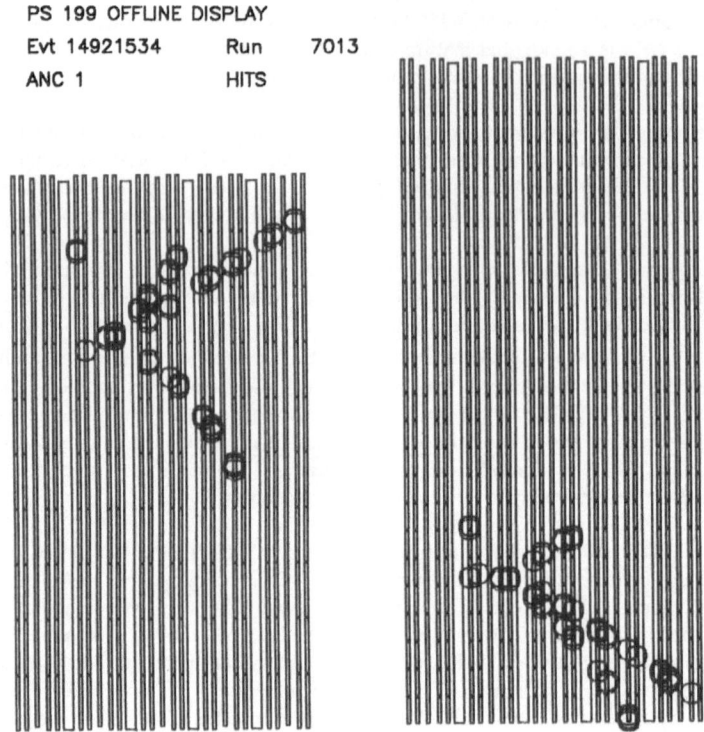

Figure 2.   PS199 antineutron annihilation . Left: side view Right: top view

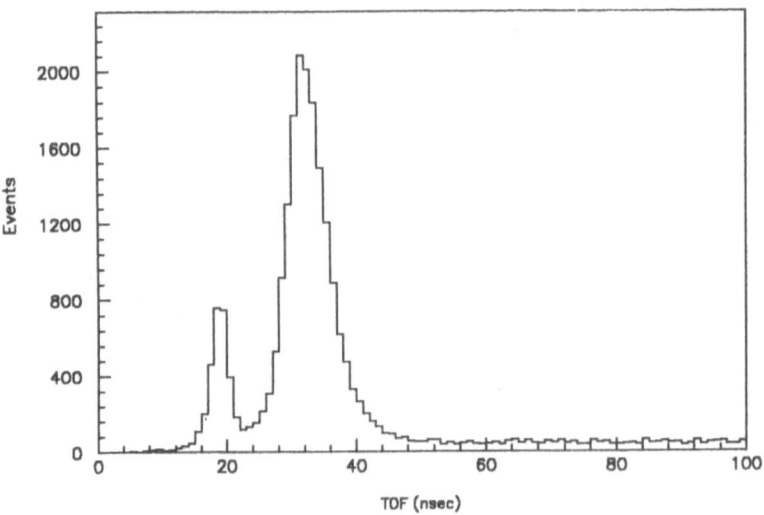

Figure 3.   PS199 antineutron and gamma peek in TOF spectra

Each NC is the OR of the up-down coincidences of the neutron counter bars. ANC is the majority condition given by the request to have at least two scintillator counters firing in the antineutron detector. Its typical rate was 200 Hz of triggers when running with a beam giving a flux of 0.5 M antiprotons per second in the target ($B_0 \cdot B_1 \cdot B_2$) The amount of information is compacted online with two J11-based CAMAC processors (CES Starburst) reducing the number of words (down to ~ 400) to be stored on the tape by the μVax II and allowing a better handling of the high interrupt rate reducing the number of data transfer from CAMAC to the μVax. The major monitoring tasks are performed by the μVax itself and by a Vaxstation 2000. The laser calibration of the scintillation counters are controlled by a PC producing calibration files. The polarization value is continuosly transfered to the μVax and stored onto data tapes.

EVENT RECONSTRUCTION AND RESULTS

The data reduction process starts with the analysis of the ANC coordinates and the selections of antineutron annihilation candidates. The ANC information is analyzed by looking for tracks coming out from the absorbers, separately on the top and the side view of each detector. Set of tracks pointing to the same region are ordered in classes of patterns (topologies). Up to now, we have considered the $\bar{n}$ candidates only in the sample of events with at least a 3 tracks topology (a *star*) on one view unambiguosly associated with a topology in the other view with at least 2 track. (In Fig.2 is showed an event with a 4 tracks star on each side).

Different requirements to define $\bar{n}$ candidates have been investigated and their performances evaluated. With the rather selective requirement we chose, the efficiency of the ANC to detect antineutrons turns out to be of the order 20%. The contamination of neutrons (i.e. neutrons detected as antineutrons) is estimated to be less than 10^{-4}, by counting the number of events with one $\bar{n}$ candidate in ANC1 and one $\bar{n}$ candidate in ANC2 with the correct kinematical correlation. Gammas from non-vetoed annihilations could contaminate the sample of $\bar{n}$ stars: their interaction in the antineutron detectors can produce charged tracks satisfying the requirements for $\bar{n}$, but they can be discarded with an appropriate time of flight cut (Fig.3).

In the set of events with an antineutron candidate, the information of the neutron counters is analyzed. Only events with a single cluster of at most 2 bars hit with the correct timing in the neutron counter in coincidence with the antineutron detector in which the antineutron pattern was recognized (ANC1 with NC1 and ANC2 with NC2), are used to select charge-exchange candidates. The events selected by the $NC3 \cdot NC1$ trigger are for the detection of charge-exchange events in the very forward region (in the range between 4° and 10°), and the analysis of this sample (characterized by the use of NC3 to detect antineutrons) is not yet finalized.

Tracking back the antineutron annihilation point and the neutron interaction point to the centre of the target, the directions of the particles of the final state are estimated. Using the mean beam direction it is possible to test the hypothesis that the target proton was free: the Fermi motion of the unpolarized bound protons in the nuclei of the target region spoils the coplanarity[2], the angular correlation and the time of flight of the particles in the final state. The event selection of the final sample is made essentially using these three quantities.

For the time being, the data collected at 700 MeV/c have been analyzed. Fig.4 gives the angular correlation signal ($\Delta\theta = \theta_{\bar{n}}^{meas} - \theta_{\bar{n}}^{calc}$) after applying all cuts of the data with polarized target and dummy target; the coplanarity cut is $|copl| < 3.5°$.

---

[2] (the coplanarity being defined as

$$copl = \frac{r_b^{\rightarrow} \cdot r_n^{\rightarrow} \times r_{\bar{n}}^{\rightarrow}}{|r_n^{\rightarrow} \times r_{\bar{n}}^{\rightarrow}|}$$

where $r_b^{\rightarrow}$ is the beam versor and $r_n^{\rightarrow}$ and $r_{\bar{n}}^{\rightarrow}$ are the neutron and antineutron directions).

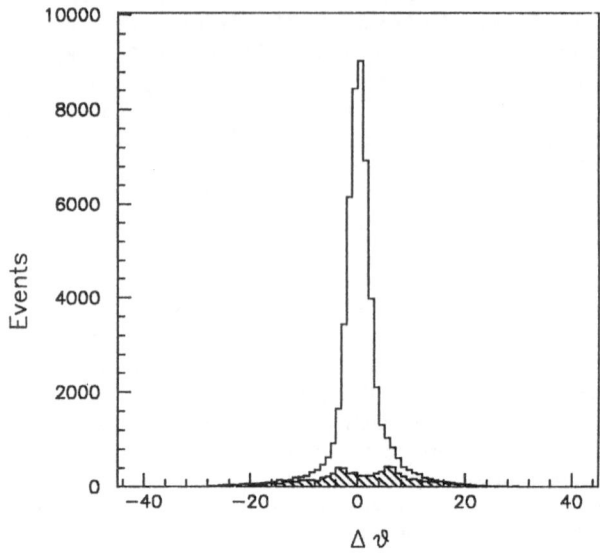

Figure 4. PS199 Angular correlation. Polarized Target (hollow area) Dummy Target (hatched area)

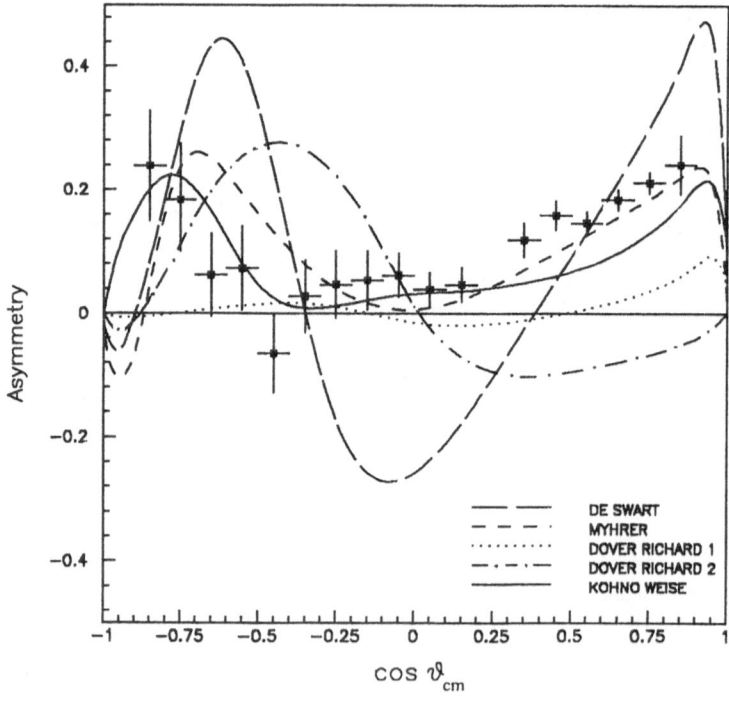

Figure 5. PS199 Asymmetry of charge-exchange at 656 MeV/c. Experimental results and theoretical predictions. The quoted value is the momentum of the incoming beam in the middle of the target, the spread is 66 MeV/c.

200

The background over signal plus background ratio r has been evaluated at all angles and its mean value over the whole angular range is 6.1 %. The final set of events to evaluate the asymmetry has been determined by applying a cut in the angular correlation ($\Delta\theta < 5°$). The stability of the results on the definition of good events has been verified.

The asymmetry A_{on} has been evaluated using

$$A_{on} = \frac{1}{(1-r)P_T} \frac{N^+B^- - N^-B^+}{N^+B^- + N^-B^+}$$

where $N^\pm$ and $B^\pm$ are the number of events and relative useful beam for the two polarization; P_T is the mean value of the target polarization (80% for both orientation).

The experimental points measured by the collaboration PS199 are given in Fig.5 with some theoretical predictions. The references to the original papers are listed in the bibliography (Ref.4-7). The polarization exceeds the 20% level in both forward and backward hemisphere and the models seem to be able, as best, to reproduce only the qualitative feature of the data; the difficulties for the theoretical models state the power of the tests on spin-dependent observables to understand the basic ingredients and mechanisms of the antibarion-barion interaction in the medium energy domain; the situation in the differential cross-section measurements in all the $N\overline{N}$ physics is basically different: as a rule, the agreement with experimental data is fair and the spread among theoretical prediction small.

With the same set-up used for the asymmetry measurements and a liquid hydrogen target, some differential cross-sections are measured using, up to now, some calibration data on the liquid hydrogen target. The calibration data have been collected in September; at present these sets of data are being analyzed to obtain differential cross-sections.

CONCLUSION

The results presented here are just a first glimpse of the information we hope to collect on the spin structure of the charge-exchange reaction. The scan of the A_{on} and the measurements of D and D_t will be performed in 1990 and the collaboration is ready and willing to start new challenging measurements in this exciting field of physics.

References

1.   T.O.Niinikoski and F.Udo, Nucl.Instr. and Methods **134** (1976)219

2.   R.Birsa et al., "The Limited Streamer Tubes system in Experiment PS199 at Lear", to be submitted to Nucl.Instr. and Methods

3.   G.Barbier et al., "The Neutron Counters system in Experiment PS199 at Lear", to be submitted to Nucl.Instr. and Methods

4.   C. Dover and J.M. Richard, Phys.Rev. **C21** (1980) 1466

5.   O.D. Dalkarov and F. Myhrer, Nuovo Cimento **40A** (1977) 152

6.   R.G.e Timmermans, Th.A. Eijeken and J.J. De Swart, Preliminary results obtained with Nijmegen P-Matrix Model.

7.   M. Kohno and W.Weise. Nucl.Phys. **A454** (1986) 429

THE OBELIX SPIRAL PROJECTION CHAMBER

G. Bendiscioli,[4] P. Boccaccio,[2] A. Coc,[1] E. David,[1]
V. Filippini,[4] U. Gastaldi,[1,2] M. Lombardi,[2] C. Marciano,[4]
G. Maron,[2] M. Morando,[3] G. Pasquali,[3] L. Peruzzo,[3]
R.A. Ricci,[2] A. Rotondi,[4] Ch. Sabev,[1] P. Salvini,[4]
G. Sartori,[3] L. Vannucci,[4] G. Vedovato,[4] and A. Zenoni[1,4]

1. CERN, Geneva, Switzerland
2. INFN, Lab. Nazionali di Legnaro, Italy
3. INFN, Sez. di Padova, Italy
4. INFN, Sez. di Pavia, Italy

(Presented by V. Filippini)

INTRODUCTION

An X-ray detector for p$\bar{\text{p}}$ atomic transitions is necessary for performing a program of differential measurements proposed to observe broad states
of glueballs, hybrids, and quasi-nuclear baryonium, if they are produced
in p$\bar{\text{p}}$ annihilation at rest at appreciable rates and with marked dependence
on the quantum numbers of the initial state.[1]

The X-ray detector is required to feature the following properties:

- High angular acceptance
- Low energy threshold
- Low mass
- High granularity
- High spatial resolution in three dimensions
- Fair energy resolution
- Good X-ray identification

These properties, which are also an asset for a good annihilation
vertex detector, are necessary not only to identify and measure efficiently protonium X-rays, but also to make the detection efficiency for a
given exclusive final state of annihilation independent of the presence of
X-rays in the event. This last feature is very import in order to
factorize efficiencies and acceptances in the differential measurements.[1]

A program of meson spectroscopy differential measurements with two
different distributions of angular momentum of the initial state of protonium is part of the scientific program of the OBELIX experiments,[2] which
includes also studies of interactions of antinucleons with nucleons and
nuclei.

Our groups are commissioning the central detector of OBELIX, which
is briefly presented in this contribution.

Medium-Energy Antiprotons and the Quark–Gluon Structure of Hadrons   203
Edited by R. Landua *et al.*, Plenum Press, New York, 1991

DETECTOR CHARACTERISTICS

The detector--called Spiral Projection Chamber (SPC)[3] or X-ray Drift Chamber (XDC)[4] depending whether its use for charged particles or for X-rays is emphasized--surrounds a cylindrical H_2 gas target at STP, and is positioned along the axis of the Open Axial Field Magnet[5] in the center of the OBELIX experiment. The SPC is similar to the Time Projection Chamber (TPC) invented by Nygren and coworkers,[6] but has a different geometry and accordingly different advantages.

The design[7] of the OBELIX SPC was based on the operation experience from the ASTERIX SPC[6] and aimed at major improvements in the hardware capabilities, calibration procedures, and event reconstruction software.

The detector will be used as an X-ray drift chamber to identify and measure the energy of soft X-rays (0.5-15 keV) emitted in the atomic cascade of $p\bar{p}$ atoms formed by stopped antiprotons in the H_2 target surrounded by the SPC. The detection of the last X-ray transition of protonium permits us to constrain or fix the angular momentum of the initial state of a $p\bar{p}$ annihilation reaction at rest. It is therefore foreseen to have a possibility of triggering the general data acquisition system on preselected X-ray patterns. The SPC will be used to count the multiplicity of charged particles emitted in $p\bar{p}$ and $\bar{p}A$ annihilations, to measure direction and ionization dE/dx of prongs, to reconstruct accurately in three dimensions the annihilation vertex, and also to veto events with antiprotons scattered too far away from the $\bar{p}$ beam axis and entering the SPC active volume.

The detector is a 90-cell cylindrical projection chamber (L = 60 cm, $\emptyset_i$ = 6 cm, $\emptyset_e$ = 29 cm) with radial drift field. The counter gas is separated from the H_2 target gas by a thin mylar tube ($\emptyset$ = 6 cm, ~6 µm thickness) to allow good transmission for soft X-ray and low-momentum prongs. The mylar tube also acts as the internal cathode surface for the drift chamber. Charge division is applied to the 90 wires, and 100 MHz flash analog-to-digital converters (FADC) are used to sample and measure the shape of the pulses generated by primary ionization deposited in the active volume of the chamber by X-rays or by charged particles. 90 cathode strips are wound on the inside of the external cathode surface ($\emptyset_e$ = 29 cm) of the chamber so that each strip crosses 85 sensing wires. The strips are equipped on one side with the same front-end and read-out electronics as the anodes. The information of the strips extends the dynamic range of amplitude analysis of the anode wire pulses (8 bit linear, 10 bit nonlinear) and improves the z resolution by the center of gravity technique.[9]

Primary ionization clusters are localized in three dimensions: $\emptyset$ by wire number, r by drift time (delay of the leading edge of the associated pulse), and z by charge division first and by anode-strip center of gravity crossing next. The adsorption point of an X-ray (and the generation point of a large cluster produced by the passage of an ionizing particle) is localized inside one of the about 10^7 pixels into which the active volume of the detector is sliced electronically. The form of the equidrift lines in the drift cells gives the pixels a "gondoletta" shape with radial extension 500 µm (determined by the drift velocity and the electronic sampling time), a z extension of 3mm, and an angular extension of 4°.

The axial symmetry of the detector and the equality of all the drift cells make it possible to realize simple triggers based on the hit multiplicity at all radii in the active volume (X-ray: isolated hit in space; prong multiplicity: number of hits at the same drift time; neutral particle decay vertex--e.g., $K^0 \rightarrow \pi^+\pi^-$: variation by two of the hit multiplicity as a function of the drift time). Commercial electronics exploiting the analog signals of the wires is employed for producing triggers on X-ray, K^0 or Λ, and prong multiplicities.

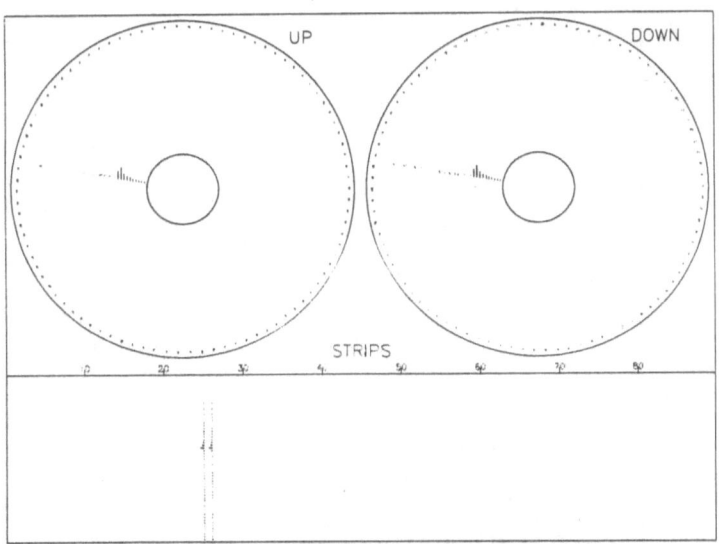

Fig. 1. Display of one event with a 5.5 keV Mn^{64} X-ray in the SPC.

INITIAL PERFORMANCES

Initial measurements outside the OAFM magnet with tagged X-ray sources and cosmics indicate that the design performances can be reached. In particular, a resolution better than σ_z = 2 mm has been measured for a 5.5

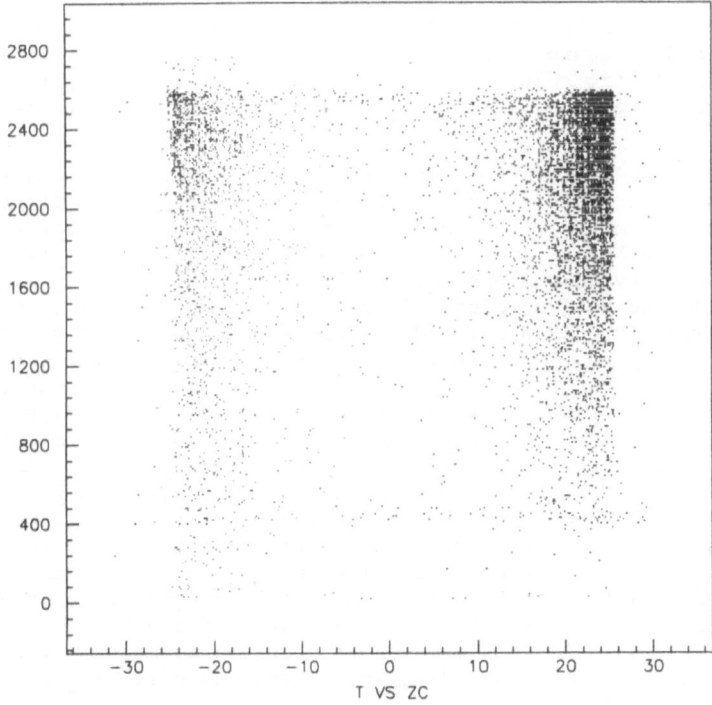

T VS ZC

Fig. 2. (r,z) scatter plot of absorption points of X-rays emitted by the two Mn^{64} sources positioned on the SPC axis.

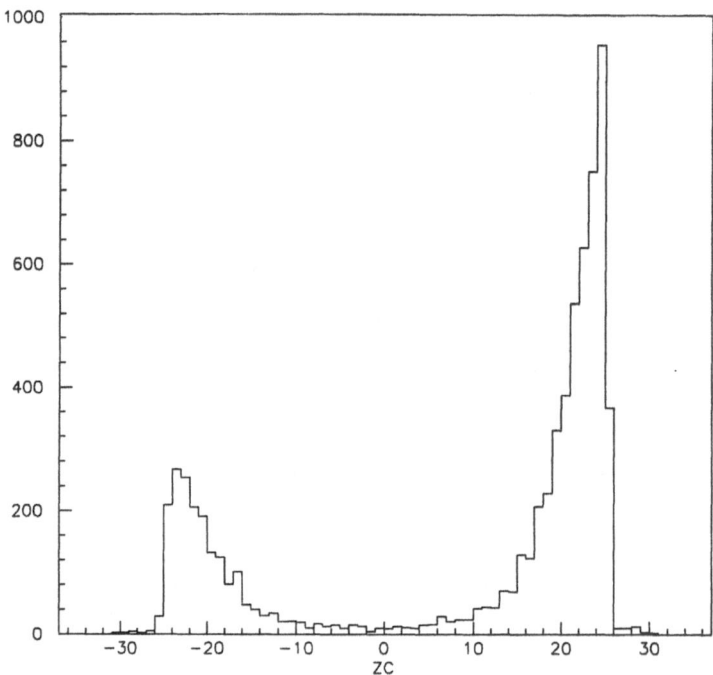

Fig. 3. z distribution of all the points of the scatter plot of Figure 2.

keV X-ray by use of crossing of the center of gravity of the strips' signal and the hit anode wire.

Figure 1 shows the event display of one event containing a 5.5 keV X-ray from a Mn[64] source deposited on the axis of the SPC onto the flat surface of a plastic scintillator, which provides the trigger and the timing of the data acquisition via the 0.8 MeV gamma ray emitted in coincidence.

For test and calibration purposes we had two such sources and scintillators positioned upstream and downstream of the SPC on its axis at distances of 248 mm from the target center. The support of the sources screened all the X-rays emitted in the half hemispheres not facing the target center. By this means the distribution of the absorption points of X-rays inside the active volume of the SPC was expected to feature a sharp cutoff at values z = ±248 mm, and in correspondence of the separation mylar tube (r = 3 cm). Figure 2 shows a scatter plot for a run without magnetic field of the absorption points of X-rays (z versus drift time-- corresponding in first approximation to a z versus radius plot) that features the expected sharp cutoffs. Figure 3 shows the z distribution of the points of Figure 2, illustrating the two sharp cutoffs in z that reflect the localization of the two sources and the good z resolution obtained with the anode crossing the strips' center of gravity for all drift times. Figure 4 shows the distribution of the absorption points in the radial direction with a z cut selecting X-rays emitted in directions nearly orthogonal to the SPC axis. This distribution features an exponential slope that gives the mean free paths of 5.5 keV X-rays in the SPC gas (Ar-C_2H_6). This curve makes it possible to measure the minimum and maximum drift time, and indicates a signal-to-noise ratio better than 200 near to the mylar membrane.

The continuum line in Figure 5 shows the amplitude distribution of X-rays from events where only one cell of the SPC had fired. This selec-

206

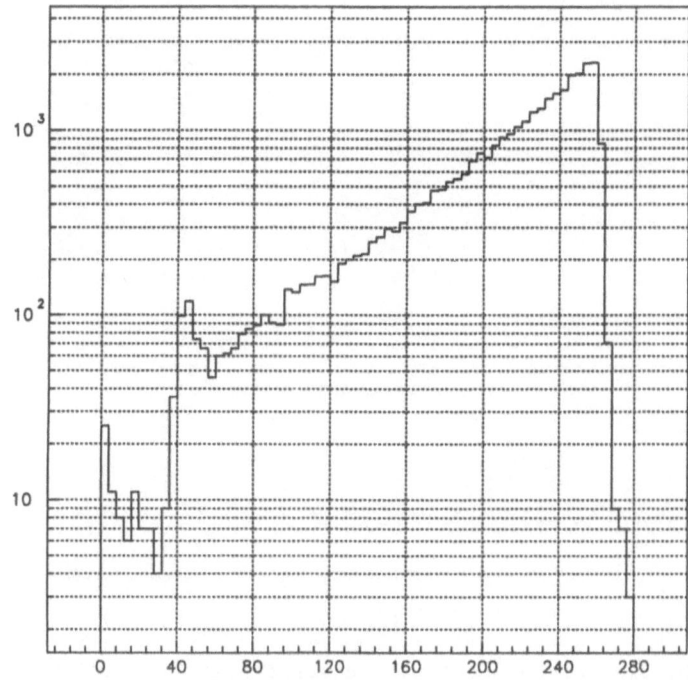

Fig. 4. Radial distribution of the absorption points of X-rays emitted by one source in directions nearly orthogonal to the SPC axis.

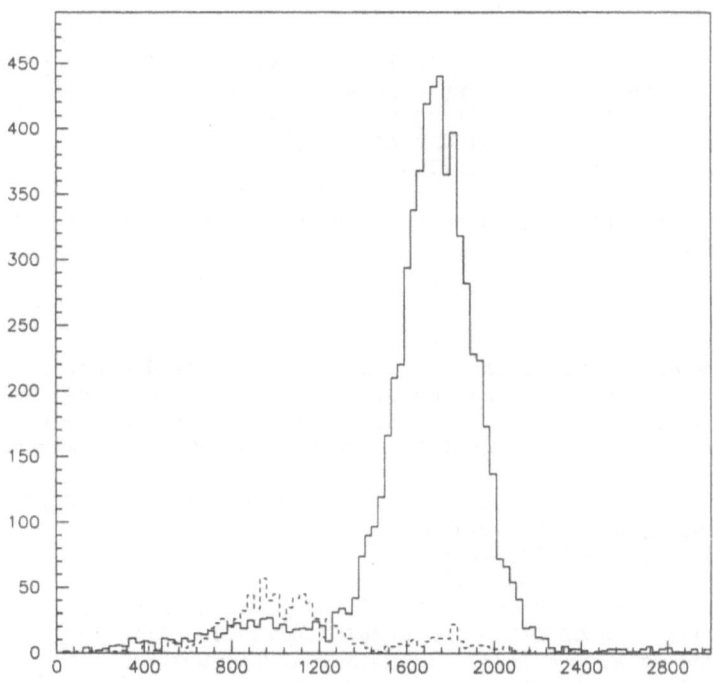

Fig. 5. Mn^{64} X-ray amplitude spectra.

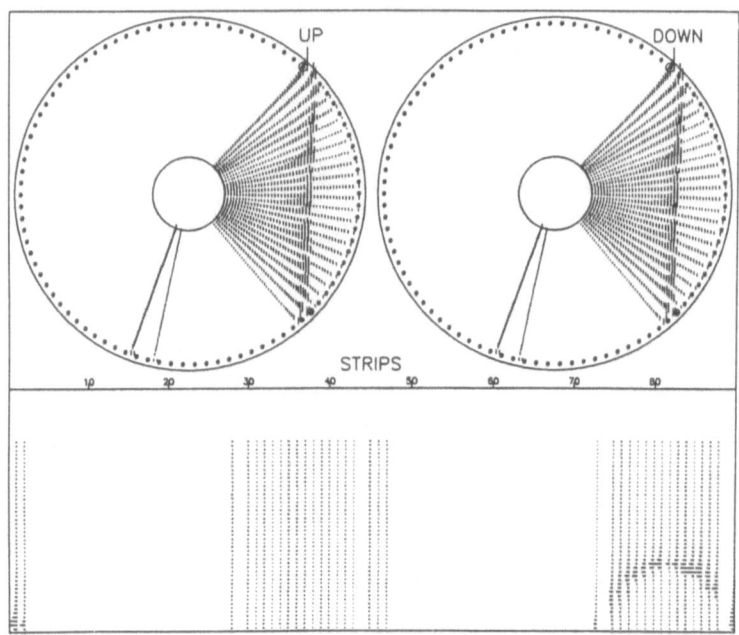

Fig. 6. Display of an event with a cosmic ray crossing the SPC and an external scintillator used for triggering and timing.

tion rejects nearly completely the escape peak at 2.5 keV. The dashed line in Figure 5 shows the amplitude of the events in which two and only two nonadjacent cells of the SPC had fired. This selection picks up events with a 2.5 keV energy deposition in one cell, and a 3 keV Ar fluorescence X-ray absorbed in another.

Figure 6 shows the event display of one event with a cosmic ray crossing the detector. The upper part of the figure is self-explanatory, taking into account that amplitudes integrated over 50 ns sampling time are displayed in the vertical direction. The lower part of the figure displays the signals of the strips.

The detector has been exposed to a 105 MeV/c antiproton beam at LEAR in November 1989.

REFERENCES

1. U. Gastaldi, in Spectroscopy of Light and Heavy Quarks, U. Gastaldi, R. Klapisch, and F. Close, eds., Plenum, New York (1989) 311.
    U. Gastaldi, Why LEAR is competitive in light meson spectroscopy, these proceedings and LNL-INFN (REP)031/90(1990) and references therein.
2. R. Armenteros et al., OBELIX Collab., Proposal CERN/PSCC/86-4.
3. U. Gastaldi, NIM 188(1981)459.
4. U. Gastalid, NIM 157(1978)441.
5. M. Gordon et al., NIM 196(1982)303.
6. D. Fancer et al, NIM 161(1979)383 and references therein.
7. U. Gastaldi, in OBELIX Notes CERN/OX/01/86(1986)140,398.
8. S. Ahamad et al., ASTERIX Coll., NIM A286(1990)76.
    U. Gastaldi et al., The X-ray drift chamber of the ASTERIX experiment at LEAR, to be submitted to NIM.
9. G. Charpak et al., NIM 167(1979)455.

THE FILTER TARGET TEST EXPERIMENT

K. Zapfe

Max-Planck-Institut für Kernphysik
P.O. Box 10 39 80
D-6900 Heidelberg
Rep. Germany

ABSTRACT

For measurements of the spin dependence in $\bar{p}p$ interaction it has been proposed to polarize antiprotons by inserting a gas target of polarized hydrogen into the LEAR ring. The polarized target acts as an antiproton spin filter, provided the $\bar{p}p$ total cross section is spin dependent. The filter method will be tested with protons instead of antiprotons in the Heidelberg Test Storage Ring TSR. The status of the components for this test experiment is reported.

1. INTRODUCTION

In 1985 a collaboration of the MPI together with seven other german and american institutes has proposed the FILter Target EXperiment FILTEX (Döbbeling et al., 1985) to measure the spin dependence in $\bar{p}p$ interaction at low momenta at the LEAR-ring. The proposal was given the status of a conditional approval with the condition, that the feasibility of the method to polarize stored antiprotons be demonstrated for protons in another ring.

The motivation for studying $\bar{N}N$ scattering was based on the idea, that if one have a meson exchange model for NN interactions, a clear prediction could be made for the corresponding $\bar{N}N$ reactions by performing a G-parity transformation for the individual meson contributions to the NN potential. In the existing theoretical models the experimental data for the long- and medium-range part of the $\bar{N}N$ potential are conformable described. In the short-range part ($r \leq 1$ fm) the hadronic interaction causes additional contributions such as the charge exchange reactions (e.g. $\bar{p}p \rightarrow \bar{n}n$) and the annihilation (e.g. $\bar{p}p \rightarrow \pi^+\pi^-$), which break the symmetry between the NN and the $\bar{N}N$ potential. In this range the models show discrepancies, which can be solved by measuring new observables only.

Up to now nearly all $\bar{N}N$ experiments have been unpolarized (spin averaged). These measurements are inadequate to constrain the theoretical models in sufficient detail. As a complete spin dependent description of the $\bar{N}N$ interaction is required, both a polarized antiproton beam and a polarized proton target are needed.

Medium-Energy Antiprotons and the Quark–Gluon Structure of Hadrons
Edited by R. Landua *et al.*, Plenum Press, New York, 1991

209

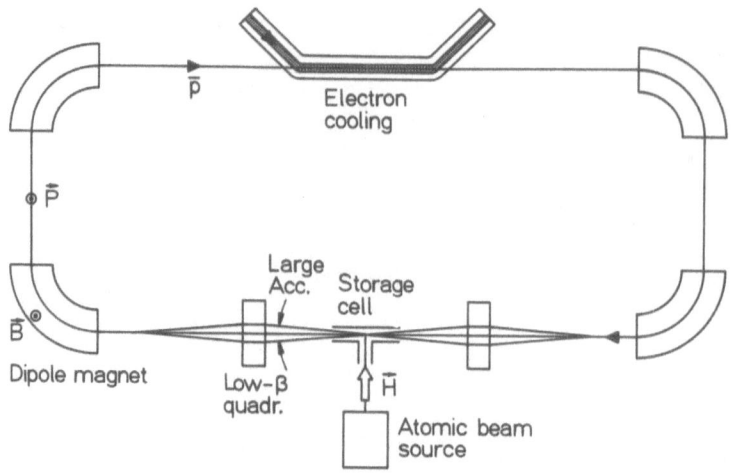

Fig. 1. Scheme of a storage ring with an internal target.

With such a polarized antiproton beam it would be possible to study the spin dependence of the elastic and charge exchange channels as well as the spin dependence of two-body annihilations.

To get a polarized antiproton beam we will install an internal polarized hydrogen gas target consisting of a thin walled storage cell fed by an atomic beam source in a storage ring (Haeberli, 1985). This is shown schematically in figure 1.

In a first step the analysing power $P(\theta)$ will be measured from scattering an unpolarized beam scattered from a polarized target. The differential cross section is given for this case by:

$$\frac{d\sigma}{d\Omega} = \frac{d\sigma_0}{d\Omega}(1 + P(\theta) \cdot P_{Tn}) \qquad (1.1)$$

where $d\sigma_0/d\Omega$ denotes the spin independent cross section and P_{Tn} the known target polarization perpendicular to the scattering plane.

If the $\bar{p}p$ interaction is spin dependent, the circulating beam can be polarized by the Filter Method as described in the following chapter. With this polarized beam we will measure in a second step several spin correlation coefficients the differential cross section, which is now given by:

$$\frac{d\sigma}{d\Omega} = \frac{d\sigma_0}{d\Omega}[1 + P(\theta)(P_{Bn} + P_{Tn}) + A_{00nn}(\theta)P_{Bn}P_{Tn} + A_{00ss}(\theta)P_{Bs}P_{Ts}$$

$$+ A_{00sk}(\theta)(P_{Bs}P_{Tk} + P_{Bk}P_{Ts}) + A_{00kk}(\theta)P_B P_{Tk}], \qquad (1.2)$$

where the unit vectors $\hat{n}, \hat{k}$ and $\hat{s}$ are chosen according to the Madison convention (1971):

$$\hat{n} = \frac{\vec{k}_i \times \vec{k}_f}{\left|\vec{k}_i \times \vec{k}_f\right|} \qquad \hat{k} = \frac{\vec{k}_i}{|\vec{k}_i|} \qquad \hat{s} = \hat{n} \times \hat{k}. \qquad (1.3)$$

2. THE FILTER METHOD

The total hadronic cross section for the scattering of polarized antiprotons from polarized protons is given by (Bilenky and Ryndin, 1963):

$$\sigma_{tot} = \sigma_{0tot} + \sigma_{1t}(P_{Bn} \cdot P_{Tn}) + \sigma_{2t}(P_{Bk} \cdot P_{Tk}), \qquad (2.1)$$

where σ_{0tot} is the spin independent part and σ_{1t} and σ_{2t} are the spin dependent parts of the cross section. The third term in this expansion differs from zero for longitudinal beam and target polarization only. In the following we consider a vertical target polarisation of $P_T = +1$, where this term does not contribute. For an unpolarized antiproton beam with $m_j = \pm\frac{1}{2}$ the cross sections for the two spin components (beam and target spin parallel or antiparallel) simplifies to:

$$\sigma_{tot\pm} = \sigma_0 + \sigma_1 \cdot (\pm 1) = \sigma_0 \left(1 \pm \frac{\sigma_1}{\sigma_0}\right) \quad \text{for} \quad m_j = \pm\frac{1}{2}. \qquad (2.2)$$

If the total cross section of the $\bar{p}p$ interaction is spin dependent, the difference between the scattering of the two spin components permits to us the polarized target as a spin filter. The expected time dependence of the polarization P and the intensity I are given by (Brückner et al., 1985):

$$P(t) = tanh\frac{t}{\tau_1} = tanh\left(\frac{\sigma_1}{\sigma_0}\frac{t}{\tau_0}\right) \qquad (2.3)$$

$$I(t) = I_0 exp(-t/\tau_0) cosh\left(\frac{\sigma_1}{\sigma_0}\frac{t}{\tau_0}\right). \qquad (2.4)$$

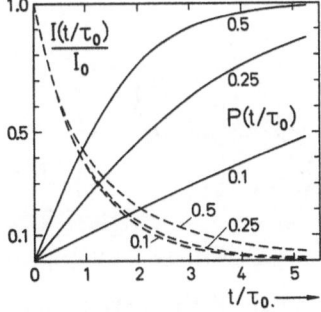

Fig. 2. Antiproton intensity and polarization with σ_1/σ_0 as parameter (Brückner, 1985).

In terms of the target thickness n (atoms/cm^2) and the particle revolution frequency f, the beam lifetime against strong interaction τ_0 and the polarization build up time are given as:

$$\tau_0 = \frac{1}{\sigma_0 n f} \quad and \quad \tau_1 = \frac{1}{\sigma_1 n f},\qquad(2.5)$$

respectively. P(t) and I(t) are plotted in figure 2 as function of t/τ_0 for different values of the unknown spin dependence σ_1/σ_0. If other mechanisms for particel losses different from strong interaction like large anlge Coulomb scattering come into play, the intensity falls off more quickly whereas the build-up of polarization is not affected.

In order to estimate the required target thickness of a polarized gas target, we assume that the beam lifetime τ_1, which gives the order of the time to polarize the beam, should be 10 hours. For an antiproton momentum of p = 235 MeV/c (E = 30 MeV) the total hadronic cross section is σ_0 = 300 mbarn. With this values and a revolution frequency of f = 1 MHz equation 2.5 leads to an average target thickness of

$$n = 1 \cdot 10^{14} \text{atoms/cm}^2.$$

With polarized free atomic beams the best area densities achieved up to now are around 10^{12}atoms/cm^2. To increase the thickness by about two orders of magnitude, we will inject the atomic beam into a thin-walled, T-shaped storage cell (Haeberli, 1985) as indicated in figure 3.

3. THE TEST EXPERIMENT

To test the filter method, we will install a polarized target in the Heidelberg Heavy Ion Test Storage Ring TSR to polarize protons instead of antiprotons. Due to the Pauli Principle a very strong spin dependence exists at low energies, where the test experiment will be performed with 30 MeV protons in autumn 1990.

The TSR storage ring (Arnold et al., 1986) consists of four straight sections serving primarily as injection, electron cooling and experimental area and section for acceleration or bunching with an rf cavity. The circumference of the ring is 55.4 m. The maximum magnetic rigidity of the dipole magnets is Bρ = 1.5 Tm, which corresponds to a kinetic energy of 30 MeV/A assuming a typical charge to mass ration of q/A = 1/2.

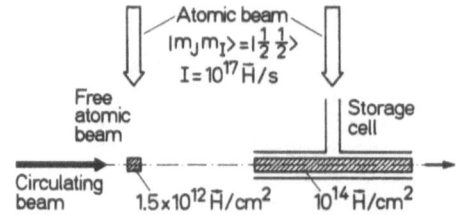

Fig. 3. Principle of the storage cell target.

To minimize the scattering losses in the target region and to get a beam of small diameter low-β quadrupoles will be used. To compensate the heating of the beam by the target, continuous electron cooling of the beam is required as shown in figure 1.

Several test runs of the TSR with protons have been performed. In the normal mode the observed lifetime for stored protons of 21 MeV with electron cooling was more than 36hours. The particle losses are primarily determined by single scattering with the residual gas and by neutralisation in the electron cooler. In the low-β mode the lifetime will drop by a factor of four due to the reduced acceptance of the ring. In test runs in this mode a fixed tube of the same diameter as the FILTEX storage cell was filled with 1 mA by multiturn injection and cooler stacking. In the FILTEX test experiment the cell will be opened up for injection to enable HF-stacking.

4. THE ATOMIC BEAM SOURCE

To get a target thickness of 10^{14}atoms/cm^2, it is necessary to inject $1 \cdot 10^{17}$atoms/s in one hyperfine state into the acceptance of the storage cell. As this is about a factor of 3-7 higher than one has achieved so far (Mathews, 1979; Singy et al., 1989), a new high intensity atomic beam source for high gas throughput of Q = 5 mbar l/s was constructed and fabricated in Munich and is now installed in Heidelberg.

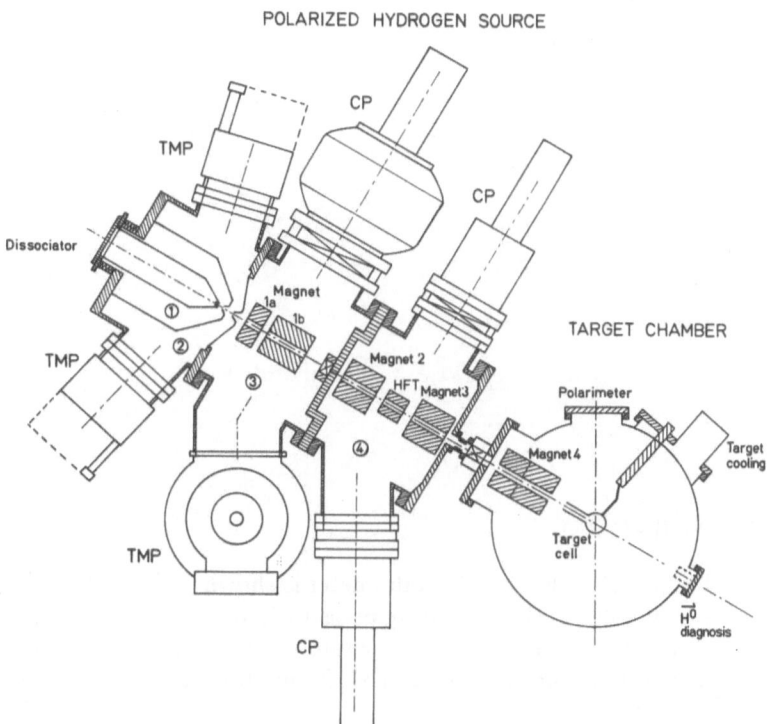

Fig. 4. FILTEX atomic beam source and target chamber (TMP = turbo molecular pump, CP = cryo pump).

This atomic beam source (ABS) together with the FILTEX target chamber and the target cell is shown in figure 4. The ABS (Graw et al., 1988; Korsch, 1990) consists of four vacuum chambers with high pumping speed of 19.000 l/s. In a pyrex glass tube with a cooled aluminum nozzle the hydrogen molecules are dissociated into atomic hydrogen. The atoms coming out from the dissociator are formed to a beam by two skimmers. A magnet system consisting of four sextupole-magnets together with a high frequency transition behind the second sextupole polarize the beam and guide it into the storage cell. This magnet system was designed based on measured beam properties, e.g. the velocity distribution (Schick, 1990), and will be installed in May 1990.

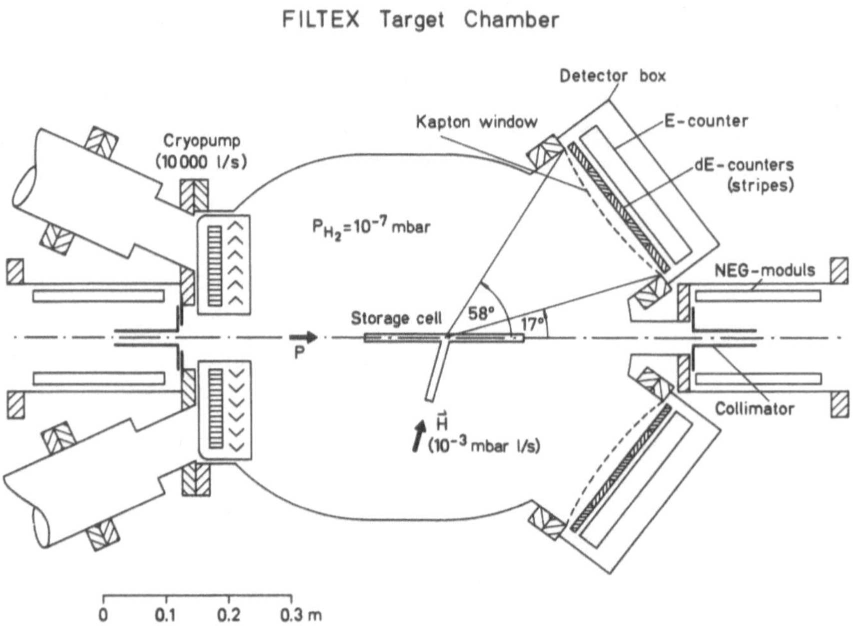

FILTEX Target Chamber

Fig. 5. FILTEX target chamber.

5. THE TARGET CHAMBER

The spherical target chamber of 60 cm diameter is shown in figure 5. It is pumped by a cryo pump with 10.000 l/s. The resulting pressure in operation will be 10^{-7} mbar. To reduce the pressure to the normal ring vacuum of 10^{-11} mbar, three differential pumping stages of Non-Evaporating-Getter (NEG) modules and collimators will be installed on each side of the target.

The circulating protons pass through the target cell in the middle of the chamber. The scattered protons enter through circular, 125 μm thick kapton windows into the detection system consisting of two detectors in the vertical ($\phi = 90^0, 270^0$) and two in the horizontal plane ($\phi = 0^0, 180^0$). Each detector consits of five parallel, 1 mm thick scintillator stripes for θ measurement and a 10 mm thick scintillator as stopping detector. The angle acceptance is $17^0 < \theta < 58^0$ at $\phi = 0^0, 180^0$ and $29^0 < \theta < 62^0$ at $\phi = 90^0, 270^0$.

6. THE TARGET CELL

The T-shaped target cell, consisting of 0.5 mm thick aluminum, is shown in figure 6. It is 250 mm long and has a diameter of 11 mm. The feed tube for the polarized hydrogen gas is 100 mm long with a diameter of 10 mm. The small whole opposite to the feed tube is needed for diagnostics. The atoms diffuse slowly out of the cell and perform about 500 wall collisions. The attainable depolarisation is minimized by coating the cell with Teflon.

If an atomic flux of intensity I (atoms/s) is injected into a storage cell of conductance C and lenght l, the area density n is given by (Haeberli, 1985):

$$n = \frac{I}{C} \cdot l \qquad (6.1)$$

To get a high density, the conductance must be minimized. Therefore a small beam size at the target is of greatest importance. On the other hand, the acceptance of the ring should not be limited by the cell. As the cell diameter of 11 mm is matched to the beam envelope, the cell will be opened to an inner diameter of 40 mm for the filling of the storage ring. The cell will be cooled down to 100 K to further reduce the conductance.

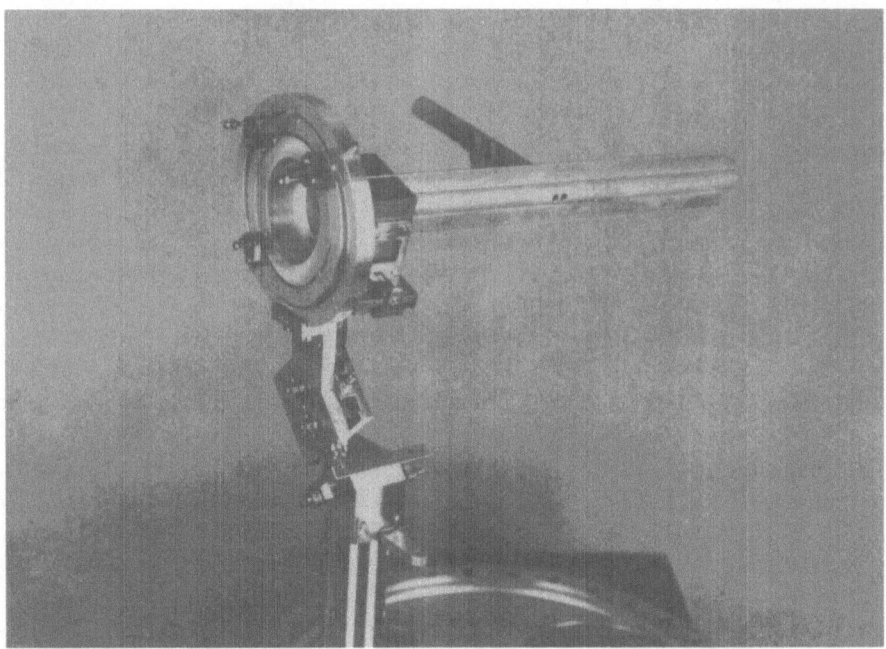

Fig. 6. Prototype storage cell with clam shell mechanism.

A weak guide field of 0.5 Tm defines the direction of the target polarization. By alternating the field direction, the polarization is switched between $+P_{Tn}$ and $-P_{Tn}$.

The target polarization will be measured by means of a Balmer polarimeter (Graw et al., 1988; Luck, 1990). The target atoms are excited by electron bombardement and the target polarization is deduced from the degree of circular polarization of the H_α- Balmer light.

7. CONCLUSIONS

The measurements of the performance of the atomic beam source are very promising. Together with the optimized magnet system we expect an average target density of 10^{14} atoms/cm^2 in the storage cell. As the TSR delivers at least 1 mA stored protons into the storage cell, a luminosity of several 10^{29} /cm^{2}s will be reached. This will enable us to test the filter method and to measure the polarization build up for the stored protons within a short time.

REFERENCES

Arnold, W., Baumann, P., Blum, M., Brix, P., Friedrich, A., Grieser, M., Habs, D., Jaeschke, E., Krämer, D., Krüchten, B.V., Martin, C., Matl, K., Povh, B., Repnow, R., Rudnik, U., Schmidt-Rohr, U., Sailer, H., Schuch, R., Steck, M., Steffens, E. and Wiedner, C.A., 1986, in: *Proc. 13th Int. Conf. on High Energy Accelerators*, Novosibirsk

Bilenky, S.M., Ryndin, R.M., 1963, *Phys. Lett.*, 7:217

Brückner, W., Döbbeling, H., Dworschak, K., Kneis, H., Nomachi, N., Paul, S., Povh, B., Ransome, R., Shibata, T.-A., Steffens, E., Treichel, M. and Walcher, Th., 1985, Proposal for measurement of spin dependence of p̄p interaction at low momenta, in: *Proc. III LEAR Workshop*, Tignes, U. Gastaldi, R. Klapisch, J.M. Richard, J. Tran Thanh Van, eds., Editions Frontieres, Gif sur Yvette: 245

Döbbeling, H., Dworschak, K., Jänsch, H., Jaeschke, E., Krämer, D., Nomachi, M., Paul, S., Povh, B., Repnow, R., Shibata, T.-A., Steffens, E., Pinsky, L.S., Mayes, B.W., Mutchler, G.S., Kruk, J., Poth, H., Wolf, A., Haeberli, W., Wise, T., Fick, D., Korsch, W., Walcher, Th., Graw, G., Schiemenz, P., and Ransome, R., 1985, *Proposal CERN/PSSC/85-80*, and 1986, Addendum

Graw, G., 1980, Schiemenz, P., Korsch, W., Luck, W., Fick, D., Giroux, J., Jänsch, H., Povh, B., Steffens, E., Haeberli, W., 1988, Intense polarized hydrogen source for the FILTEX target, in: *Proc. IV LEAR Workshop*, Villars 1987, C. Amsler, G. Backenstoss, R. Klapisch, C. Leluc, D. Simon and L. Tauscher, eds., Harwood, Chur: 221

Habs, D., Baumann, W., Berger, J., Blatt, P., Faulstich, A., Krause, P., Kilgus, G., Neumann, R., Petrich, W., Stokstad, R., Schwalm, D., Szmola, E., Welti, K., Wolf, A., Zwickler, S., Jaeschke, E., Kraemer, D., Bisoffi, G., Blum, M., Friedrich, A., Geyer, C., Grieser, M., Heyng, H.W., Holzer, B., Ihde, R., Jung, M., Matl, K., Ott, W., Povh, B., Repnow, R., Steck, M., Steffens, E., Dutta, D., Kühl, T., Marx, D., Schröder, S., Gerhard, M., Grieser, R., Huber, G., Klein, R., Krieg, M., Schmidt, N., Schuch, R., Babb, J.F., Spruch, L., Arnold, W. and Noda, A., 1989, *Nucl. Instr. and Meth.* B43:390

Haeberli, W., 1985, Free and stored atomic beams as internal polarized target, in: *Proc. Workshop on Nucl. Phys. with Stored, Cooled Beams*, Indiana 1984, P. Schwandt and H.O. Meyer, eds., AIP Conf. Proceed. No. 128, New York: 251

Korsch, W., 1990, Ph. D. Thesis, Universität Marburg

Luck, W., 1990, Ph. D. Thesis, Universität Marburg

Madison Convention in polarization phenomena in nuclear reactions, 1971, H.H. Barschall and W. Haeberli, eds., Madison, University of Wisconsin Press: XXV-XXIX

Mathews, H.G., 1979, Ph. D. Thesis, Universität Bonn

Singy, D., Schmelzbach, P.A., Grüebler and Zhang, W.Z., 1989, *Nucl. Instr. Meth. A* 278:349

Schick, M., 1990, Diplomathesis, Universität Heidelberg

THE COMPLETE EXPERIMENT IN N̄N SCATTERING

R.A. Kunne

DPNC, University of Geneva, Geneva, Switzerland
(Present address: LNS, Saclay, France)

ABSTRACT

An overview is given of the existing N̄N → N̄N data. Their interpretation with potential models is shown to be difficult. The possibility to do the "complete" N̄N experiment is investigated.

WHAT DO WE KNOW OF N̄N SCATTERING?

The territory of N̄N → N̄N scattering is badly explored, as abundant anti-proton beams are only available since 1982 and as polarized beams do not exist. Only differential cross sections and asymmetries have been measured. Nothing what so ever, is known of other spin observables.

Of the six possible N̄N → N̄N reactions, the p̄p elastic scattering reaction is the easiest to measure (figure 1). Measurements of the asymmetry between 500 and 1600 MeV/c incoming momentum [1] [2] and of ρ (the ratio of the real to the imaginary part of the scattering amplitude) between 0 and 1100 MeV/c [3] are added to the (mainly) pre-LEAR differential cross section measurements (for instance [4] and [5]).

The next best investigated reaction is the charge exchange reaction p̄p → n̄n. Here the experimental problem is the detection of two neutral particles in the final state: a problem of efficiency. At the time of this school LEAR has produced one asymmetry measurement at 650 MeV/c [6] and a range of differential cross sections [4].

The elastic reaction p̄n → p̄n can not be measured directly as there are no free neutrons targets. It has to be inferred from p̄A → p̄A where A is a nucleus, like deuterium, ⁴He:.... The calculation of the p̄n amplitude from the p̄-nucleus reaction involves the application of a model (like Glauber theory) and can only be trusted for forward angles. Only indirect measurements of total reaction cross sections exist [7]. n̄p elastic scattering may be measured (at low energies at least)

Medium-Energy Antiprotons and the Quark–Gluon Structure of Hadrons
Edited by R. Landua *et al.*, Plenum Press, New York, 1991

219

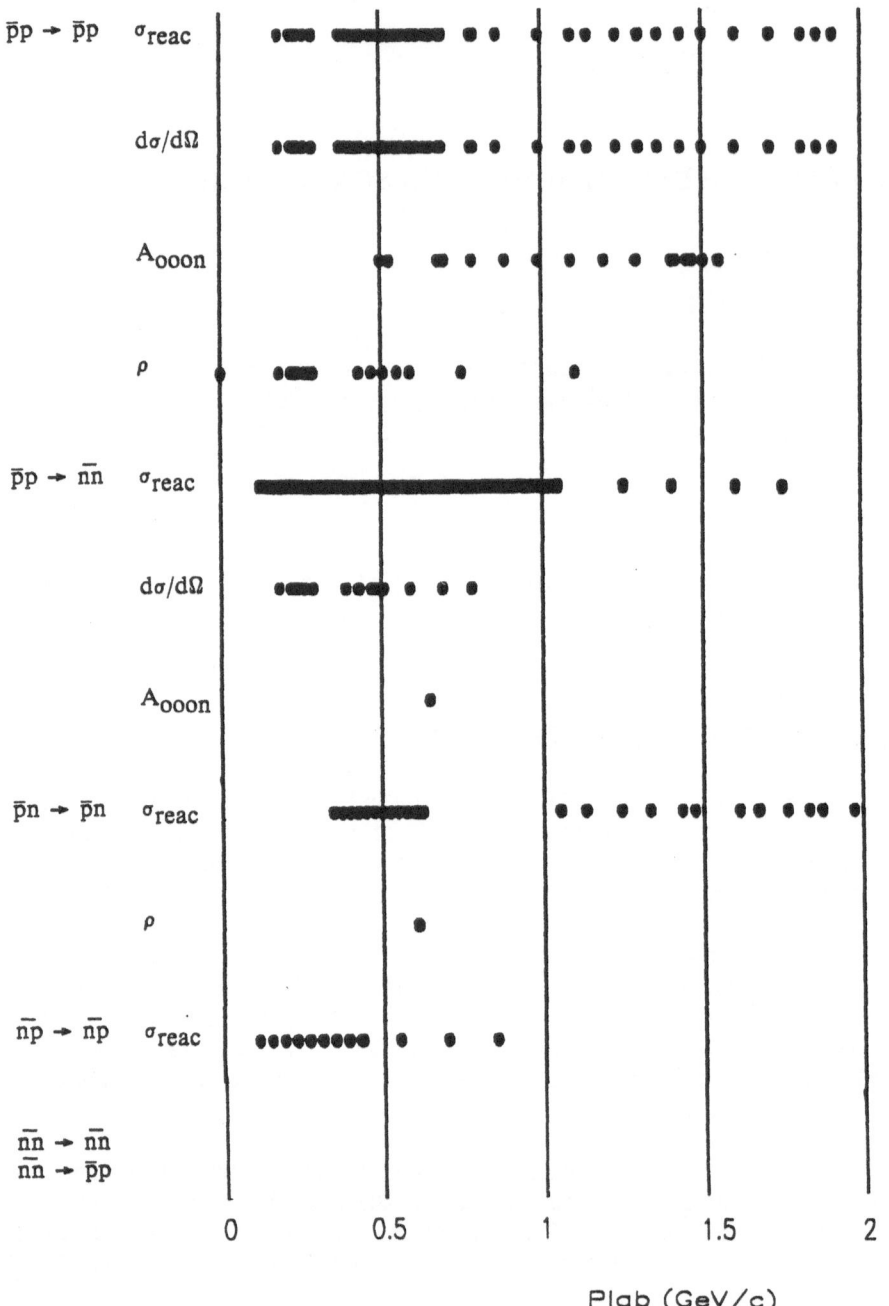

Fig. 1. Summary of existing $\bar{N}N \to \bar{N}N$ data. Pre-LEAR data are included only if LEAR didn't provide any better data.

by making an antineutron beam using the $\bar{p}p$ charge exchange reaction, while tagging on the neutron. This technique has been used [8], but the low cross sections prevent the measurement of more than the total reaction cross sections.

The remaining two reactions, $\bar{n}n \rightarrow \bar{n}n$ and $\bar{n}n \rightarrow \bar{p}p$, are theoretician's dreams and nightmares for the experimentalist. The problems of producing an antineutron beam and not having a free neutron target are combined. Hence there are no data available, nor likely forthcoming.

A CRITIQUE OF POTENTIAL MODELS

The data collected in 40 years of investigating of the NN-force with proton and neutron beams, are reasonably well explained by potential models. Several models are on the market, of which the Paris [9], the Nijmegen [10] and the Bonn [11] models are the best known. Potential models attempt to describe the NN-force in terms of meson exchanges.

The exchange of the lightest meson is well understood. Not so clear is the situation in the case of the exchange of more massive mesons, like the ρ, ω, δ,.... and the exchange of more than one meson. For those exchanges the coupling constants can not be measured directly, but have to be determined as free parameters. Often a hypothetical meson, the σ, with the right quantum numbers and adjustable coupling and mass is introduced to describe the two-pion exchange. Other groups attempt a rigorous calculation.

For the short range part of the potentials, in the region where quark and gluon degrees of freedom start to play an role, more theoretical ambiguities arise. Cut-offs of the longer range potentials and a parametrization of the short range potential provide a solution. Despite these problems the existing NN potentials can be adapted to describe all NN data in a satisfactory way.

Attempts to calculate $\bar{N}N$ potentials based on NN potentials were already started early [12]. The meson exchange part of a NN potential can be transformed into a $\bar{N}N$ potential. The coupling strength between meson and antibaryon has either the same or opposite sign as the meson-baryon coupling, depending on the G-parity of the exchanged meson. The resulting $\bar{N}N$ potential is more attractive than the NN potential. The short range part of the $\bar{N}N$ potential can not be obtained from NN in the same way. However, it was assumed that this part will be washed out by annihilation effects. Therefore simple parametrizations were deemed sufficient.

Indeed the $\bar{N}N$ potentials models were, at the time of their conception, quite good in explaining the existing total and differential cross section data [13] [14] [15] [16]. This changed when the LEAR data $\bar{p}p$ elastic scattering asymmetries came in [1]. The predictions of all models disagreed largely with the data (see figure 2). However, most potentials may easily be adapted to postdict the data, thanks to the parameters in the phenomenological short range potentials. Figure 3 is an example of the dependency of the asymmetry on the form of the short range part given by Myhrer [17]. This figure shows clearly that any asymmetry can be fitted by carefully adjusting the form of the short range potential. This means that the models do not have any predictive power and do not give an

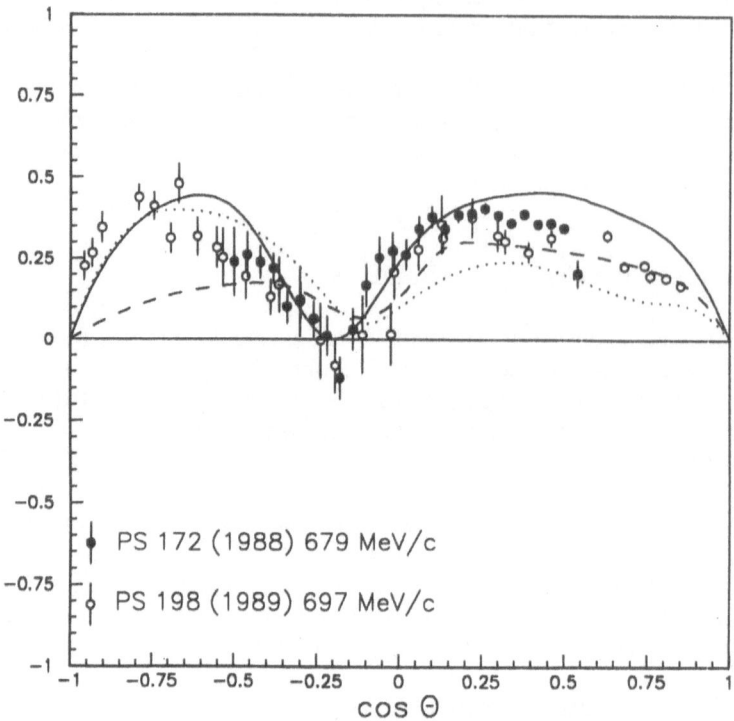

Fig. 2. Comparison of the asymmetry data at 679 MeV/c [1] and at 697 MeV/c [2] with predictions from the Paris [13] (solid line), Dover – Richard II [14] (dashed line) and Bonn models [16] (dotted line).

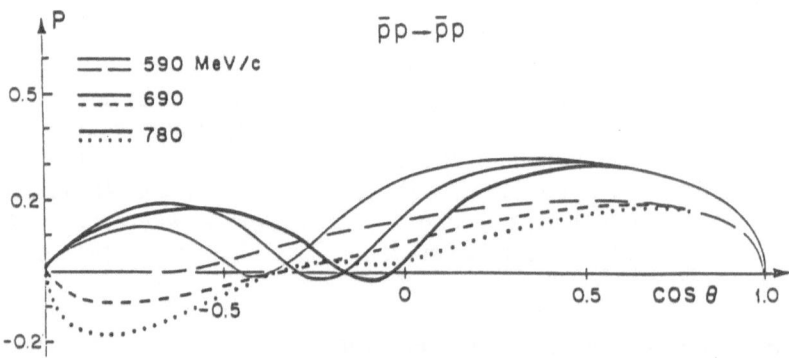

Fig. 3. The elastic p̄p asymmetry calculated at three momenta with the Nijmegen model and an imaginary potential which strength is adjusted to fit the differential cross sections. The solid curves have a linear cut-off of the potential below r = 0.63 fm, the dashed lines have a constant cut-off.

explanation of the physics behind the data. The general agreement is growing, that the only feasible way out is to look for specific spin observables and reactions, which might say something about the long range meson exchange parts of the potentials, without interference of the short range part. A theoretical calculation of the short range potential starting from QCD basics seems out of the question.

THE SCATTERING MATRIX

To bring the experimental knowledge of $\bar{N}N$ scattering to the level of that of NN scattering, measurements of more spin observables are obviously needed. The question of how many experiments are needed may be replied using the description of scattering in terms of amplitudes.

The scattering of a reaction $\bar{N}N \to \bar{N}N$ can be described with the M-matrix formalism. The scattering matrix M is a 4x4 matrix in the combined spin space of the two particles. In its most general form — describing the scattering of two massive, spin 1/2 particles — the scattering matrix involves 16 independent amplitudes that are functions of the centre of mass momenta $\vec{k}_i$ and $\vec{k}_f$ of the incoming and scattered particle (or equivalently any two kinematic variables, like the Mandelstam variables s and t, or the energy and the scattering angle θ_{cm}). For reactions that have certain symmetries less amplitudes are needed. For instance for the reaction $\bar{p}p \to \bar{p}p$, only five complex amplitudes a,b,c,d,e are sufficient:

$$M(\vec{k}_i, \vec{k}_f, \vec{\sigma}_1, \vec{\sigma}_2) = \quad 1/2 \ [(a+b) + (a-b) (\vec{\sigma}_1 \, \vec{n}) (\vec{\sigma}_2 \, \vec{n})$$

$$+ (c+d) (\vec{\sigma}_1 \, \vec{m}) (\vec{\sigma}_2 \, \vec{m}) + (c-d) (\vec{\sigma}_1 \, \vec{l}) (\vec{\sigma}_2 \, \vec{l})$$

$$+ e (\vec{\sigma}_1 + \vec{\sigma}_2) \vec{n} \,]$$

where $\vec{\sigma}_1$ and $\vec{\sigma}_2$ are the Pauli spin vectors operating on nucleon 1 and 2; $\vec{n}$, $\vec{m}$ and $\vec{l}$ are the unit vectors in the centre of mass frame in the directions defined by $\vec{k}_i \times \vec{k}_f$, $\vec{k}_i - \vec{k}_f$ and $\vec{k}_i + \vec{k}_f$. Many other parametrizations that are linear combinations of a,b,c,d,e are quoted in the literature [18]. Further amplitudes, corresponding to different combinations of the spin vectors $\vec{\sigma}_1$ and $\vec{\sigma}_2$ with the direction vectors $\vec{l}$, $\vec{m}$ and $\vec{n}$, are zero for $\bar{p}p$ elastic scattering due to rotational invariance, time reversal invariance, parity conservation and invariance under charge conjugation of the reaction. This means that five moduli and the five corresponding phases fully determine the scattering.

On the experimental side 4^4 different spin observables may be measured. These correspond to four spin measurements (spin direction $\vec{l}$, $\vec{m}$, $\vec{n}$ or no measurement) for each of the four particles. An spin observable is denoted as X_{srbt}, where the indices indicate the spin state of the four particles (scattered, recoil, beam and target) and X gives the type of reaction. Spin observables are expressed in terms of the amplitudes of the scattering matrix, e.g.:

Cross section: $\quad d\sigma/d\Omega = I_{0000} \quad = 1/2\,(|a|^2 + |b|^2 + |c|^2 + |d|^2 + |e|^2)$

Asymmetry: $\quad d\sigma/d\Omega \cdot A_{000n} \quad = \mathrm{Re}\,(a^{*}e)$

Depolarization: $\quad d\sigma/d\Omega \cdot D_{0n0n} \quad = 1/2\,(|a|^2 + |b|^2 - |c|^2 - |d|^2 + |e|^2)$

An experiment that measures enough spin observables to reconstruct the scattering matrix, is said to be "complete". What is the minimum number of spin observables that have to be measured? At first glance, the answer seems to be nine, as changing the five amplitudes by a common phase, does not change the spin observables. Actually the mathematical answer is five [19]. Five experiments give the five moduli of the amplitudes and it can be shown that unitarity of the reaction leads to five integral equations between the phases of the amplitudes. However, the integrals involved in these equations have to be taken over the whole range of scattering angles. This means that the spin observables have to be measured over the whole angular range between 0° and 180°, which in experimental practice is impossible.

For pp and np elastic scattering the reconstruction of the amplitudes has been done at several energies. It turns out that 12 to 15 spin observables have to be measured in order to solve the ambiguities introduced by incomplete angular ranges and measuring errors.

THE COMPLETE EXPERIMENT IN p̄p ELASTIC

To do the complete experiment in the elastic p̄p channel, four ingredients are needed: a polarized proton target, a polarized antiproton beam and analyzers of the polarization of the outgoing proton and antiproton.

Polarized Proton Targets

The polarization of protons in a fixed target is a standard technique for the last twenty years. Protons, bound in an organic molecule like pentanol, are polarized using microwave radiation. The polarization (90%, obtainable in 2-3 hours) is "frozen" at temperatures of 50 mK in a field of 0.3-0.5 T. The relaxation time is of the order of 10 days. The proton spins can be given any orientation, by slow rotation of the holding field. The background of scattering on non-polarized protons bound in the other nuclei of the molecule, may easily be subtracted with a dummy target, which does not contain hydrogen.

Polarized Antiproton Beams

Polarized antiproton beams do not exist as yet. Apart from the obvious, but dirty way of polarizing antiprotons by scattering on a polarized target, there are currently two ideas to make a polarized beam.

The first one is the idea of the Spin Splitter (figure 4), based on the possibility of separating particles with opposite spins in the field of a quadrupole

224

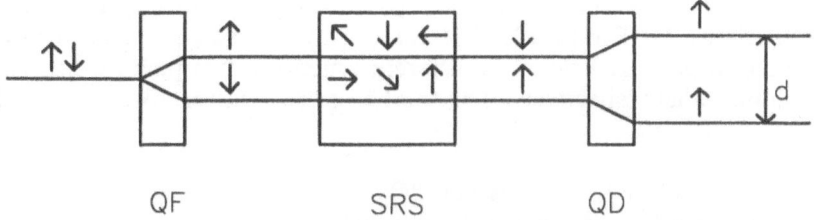

Fig. 4. Schematic view of Spin Splitter. QF and QD denote focussing and defo – cussing quadrupoles respectively. SRS is a spin rotating solenoid.

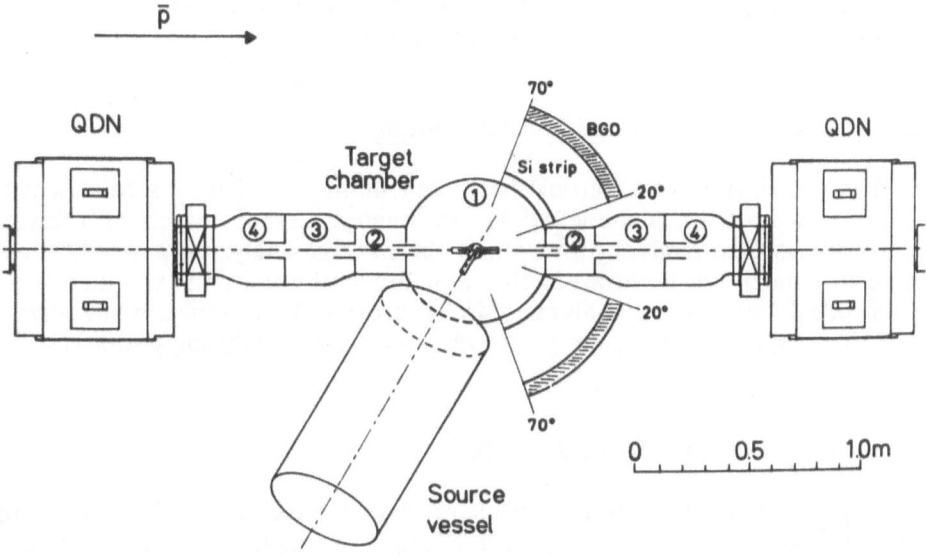

Fig. 5. Schematic view of Spin Filter.

(Stern-Gerlach effect). The setup [20] consists of two quadrupoles of opposite polarity to separate the spins without changing the optics of LEAR, and a solenoid placed between them. This solenoid rotates the spins of the antiprotons 180°. In LEAR a separation of 2.5 mm/hour can be obtained (assuming p = 200 MeV/c and a quadrupole strength of $\Delta B/\Delta x = 20$ T/m). Although the proposal was rejected by the PSCC, tests with protons are foreseen at the Indiana University Cooler Ring.

The second idea to make a polarized antiproton beam in Lear is called the Spin Filter (figure 5). The idea [21] is based on the prediction that absorption of antiprotons in a polarized target is different when the two particle spins are parallel and antiparallel respectively. The total cross section for absorption is denoted as

$$\sigma^+ = \sigma^0 + \sigma^1 \quad \text{spins parallel}$$
$$\sigma^+ = \sigma^0 - \sigma^1 \quad \text{spins antiparallel}$$

σ^1 / σ^0 is predicted to be 6% [22]. The polarized is a gas jet target currently being developed at the Test Storage Ring in Heidelberg. With a density of 10^{14} atoms/cm^2 and twelve hours of filtering the polarization will be 10%. The price to pay is a loss of 90% of the initial beam. An attractive feature of the proposed experimental test at LEAR is the possibility of measuring spin correlation parameters using BGO detectors placed around the target cell.

Polarization Analyzers of Proton and Antiproton

The polarization of a particle beam can be measured by a second scattering and a measurement of the left-right asymmetry. Carbon targets of a few cm thickness are used to analyze protons and the average analyzing power [23] as a function of momentum is shown in figure 6. For antiprotons the situation is less good: carbon is not a good analyzer [24]. Hydrogen, for instance in the form of CH scintillators, can be used, but gives only about 10% analyzing power [1].

A PROPOSAL FOR AN EXPERIMENT

Experiment PS199 currently measures asymmetries in the charge exchange channel. In two stages the setup (figure 7) may be adapted to measure spin observables in the elastic channel. In the first stage the parameters $D = D_{onon}$ and $D_t = K_{noon}$ will be measured. In this case only a polarized target is needed. The proton scatters to the right in the polarized target. Scintillator counter array NC3 functions as life target and the scattering angle is measured in NC1. The antiproton scatters to the left where NC2 functions as life target. The annihilation star in ANC2 proofs that the particle really was an antiproton. The trigger is crucial in selecting only doubly scattered particles. The kinematical correlation between the hits in NC2 and NC3 selects the elastic events. An array of 64 scintillating fibers placed around the target narrows this selection. The second scattering of the proton is selected by looking at the correlations between NC1 and NC3. In a later stage, using a polarized beam with a solenoid spin rotator, many other spin observables may be investigated.

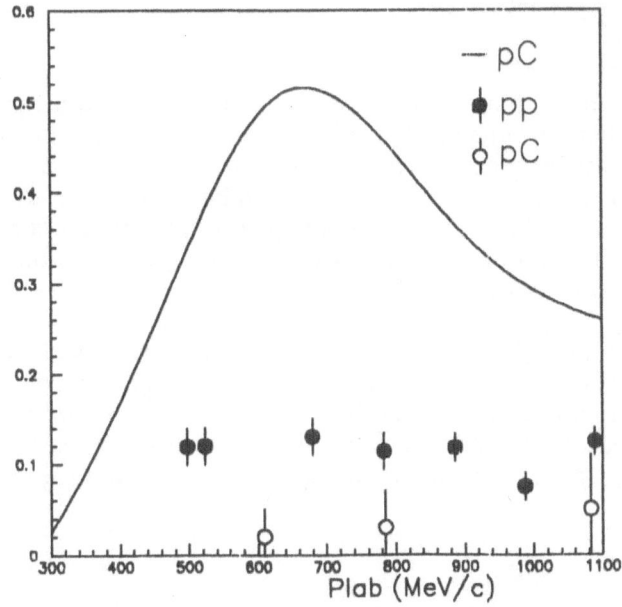

Fig. 6. Average analyzing power as a function of momentum for protons on carbon, antiprotons on carbon and antiprotons on hydrogen.

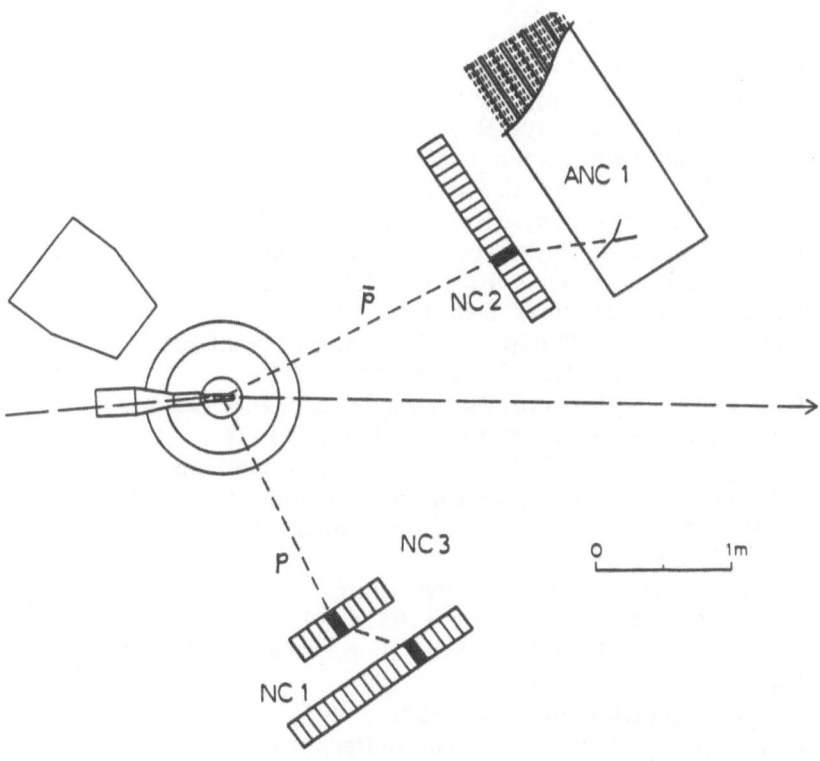

Fig. 7. Experimental setup of the PS199 experiment as used to measure spin observables. ANC1 functions as antibaryon detector. NC1, NC2, NC3 are arrays of scintillators used as life targets.

CONCLUSION

Potential models have learned us some basic antiproton physics like the range of annihilation and the need of tensor forces to describe the interaction. But they have too many free parameters (coming from the definition of the phenomenological short range part that describes annihilation and the quark and gluon degrees of freedom), to have real predictive power. This failure already shows up when the models are compared with asymmetry data, the spin observable that is the easiest to measure experimentally.

A thorough investigation of the $\bar{p}p$ elastic channel seems therefore in order. The complete experiment, i.e. one that measures enough spin observables to reconstruct the amplitudes, is feasible in a couple of years, when polarized antiproton beams become possible. In the mean time observables like $D = D_{onon}$ and $D_t = K_{noon}$ might be measured by an upgrade of experiment PS199 at LEAR.

REFERENCES

[1]   R.A.Kunne et al., Phys. Lett. 206B (1988) 557.
[2]   R.Bertini et al., Phys. Lett. B228 (1989) 531.
[3]   T.P.Gorringe et al., Phys. Lett. 162B (1985) 71.
        L.Linssen et al., Nucl. Phys. A469 (1987) 726.
        P.Schiavon et al., Nucl. Phys. A505 (1989) 595.
[4]   W.Brückner et al., Phys. Lett. 166B (1986) 113.
[5]   T.Kageyama et al., Phys. Rev. D35 (1987) 2655.
[6]   M.Lamanna, Contribution to this School.
[7]   G.Bendiscolli et al., Nucl. Phys. A469 (1987) 669.
[8]   T.Armstrong et al., Phys. Rev. D36 (1987) 659.
[9]   M.Lacombe et al., Phys. Rev. C23 (1981) 2405.
[10]   M.M.Nagels et al., Phys. Rev. D12 (1975) 744.
[11]   R.Machleidt et al., Phys. Rep. 149 (1987) 1.
[12]   R.A.Bryan, R.J.N.Phillips, Nucl. Phys. B5 (1968) 201.
[13]   J.Côté et al., Phys. Rev. Lett. 48 (1982) 1319.
[14]   C.B.Dover, J.M.Richard, Phys. Rev. C25 (1982) 1952.
[15]   P.H.Timmers et al., Phys. Rev. D29 (1984) 1928.
[16]   T.Hippchen et al., IXth European Symposium on Proton − Antiproton Interactions, Mainz, F.R.Germany, September 5 − 10, 1988.
[17]   F.Myhrer, invited talk, Few Body Problems in Physics, Vancouver, Canada, July 2 − 8, 1989.
[18]   J.Bystricky et al., Journ. de Phys. (Paris) 39 (1978) 1.
[19]   L.Puzikov et al., Nucl. Phys. 3 (1957) 436.
[20]   A.Penzo et al., IV LEAR Workshop, Villars − sur − Ollon, Switzerland, September 6 − 13, 1987.
[21]   K.Zapfe, Contribution to this School.
[22]   C.B.Dover, In: Low Energy Antimatter, p. 1.
[23]   TRIUMF Kinematics Handbook, p. VIII − 1.
[24]   A.Martin et al., Nucl. Phys. A487 (1988) 563.

WHY LEAR IS COMPETITIVE IN LIGHT MESON SPECTROSCOPY

Ugo Gastaldi

INFN, Laboratori Nazionali di Legnaro
Via Romea 4, I-35020 Legnaro (PD)

INTRODUCTION

Spectroscopy of light mesons has played an important role in the
development of particle physics and was traditionally a high-energy physics
activity. With the migration of particle physicists to higher and higher
energies naturally the question arises whether fundamental information can
still be provided by light meson spectroscopy or if continuation of activity in this sector represents only a work of completion of systematics and
of polishing. An answer to this question emerges considering some of the
open problems in the foundations of particle physics and the lines of
experimental and theoretical research necessary to attack these problems.

FUNDAMENTAL ISSUES IN PARTICLE PHYSICS AND LIGHT MESON SPECTROSCOPY

First, consider particles. The top quark and antiquark remain to be
discovered and their observation requires high-energy experiments at c.m.
energies in excess of twice the top quark mass. The mechanism of mass
generation remains to be understood. Work on the neutrino and antineutrino
mass is carried out at both low and high energies. The observation of the
Higgs particles, which are considered responsible for the generation of the
masses of quarks, leptons, and massive gauge bosons, requires experiments
at higher and higher energies. Experimentation at both low and high
energies is necessary for providing clues concerning the origin of CP
violation and the mechanism of mixing between quarks of different families.

Next, consider interactions. While the graviton remains to be observed,
the other gauge bosons that mediate the electromagnetic, the weak, and the
strong forces between fundamental particles have experimental evidence.
The photon has been well known for a long time, and in recent years beautiful experimental evidence has been provided of the existence of the
Z^0, W^+, W^- gauge bosons of the weak interaction through the observation of
their leptonic and hadronic decays and of gluons (gauge bosons of the
strong interaction between quarks) by the observation of the associated
jets in deep inelastic interactions. However the fundamental property of
the gauge bosons responsible for weak and strong interactions, i.e., their
direct interaction with other gauge bosons that mediate the same force,
has not yet been proved experimentally. For W^+, W^-, and Z^0 the experimental proof will require studying W^+W^- pair production at LEP operated at
c.m. energies in excess of twice the W mass and demonstrating delicate
interference patterns between W^+W^- production via decay of a virtual photon

Medium-Energy Antiprotons and the Quark–Gluon Structure of Hadrons
Edited by R. Landua *et al.*, Plenum Press, New York, 1991

229

formed in an e^+e^- annihilation and W^+W^- production via the decay of a virtual Z^0 formed in an e^+e^- annihilation and coupling directly to a W pair. This point is of central importance for the foundation of the electroweak theory, and it is indeed one of the main motivations for bringing LEP to and running it at the highest design energies. For gluons there are indirect indications of gluon-to-gluon direct coupling from angular distributions of jets produced in high-energy $p\bar{p}$ interactions that can be interpreted as gluon jets due to hard gluon-gluon scattering. However a convincing proof of direct gluon-gluon coupling is still missing, and it could be obtained at low energies because of the zero gluon rest mass. The revival of interest in light meson spectroscopy in recent years is mainly linked to the existence of the possibility of proving the nonabelian nature of QCD, which gives rise to direct coupling between gluons. This feature of direct coupling is the main qualitative difference between QED and QCD, as a photon cannot interact directly with another photon because of the absence of a term quadratic in the fields in the QED lagrangian, while the contact term in the QCD lagrangian implies vertices with direct couplings of three and four gluons.

GLUEBALLS AND HYBRIDS AS PROOF OF THE DIRECT COUPLING BETWEEN GLUONS

Because of the possibility of direct coupling of one gluon with another (three-gluon vertex), a gluon can play the role of a constituent particle in a hadronic structure. Consequences of the basic feature of QCD of being nonabelian in the gauge fields are the possibilities of existence of glueballs (made of two or three constituent gluons with zero total color charge) and of hybrids (made of a quark, an antiquark, and a constituent gluon with zero total color charge). The constituent gluons are bound to the other constituent gluons in glueballs and to the constituent quark and antiquark in hybrids by the exchange of gauge gluons, much the same way as constituent quarks in mesons and baryons are thought to be kept together by the exchange of gauge gluons. Because of the extreme difficulties in handling the QCD lagrangian, theory is unable at present to make definite and compelling predictions about masses, widths, and branching ratios of decay channels for glueballs and hybrids, as well as for conventional $q\bar{q}$ mesons, and is also not in the position to prove color confinement as a consequence of the basic QCD lagrangian. It is not easy therefore to identify any new hadronic structure as a glueball or a hybrid, because first the possibility that it is a conventional meson has to be discarded, second the possibility exists that an authentic glueball or hybrid mixes with a conventional meson with the same quantum numbers, and finally because there is also the possibility for the new structure to be an exotic meson without constituent gluons. However glueballs and hybrids can have discrete quantum number configurations not accessible to conventional $q\bar{q}$ mesons due to selection rules. A well established bosonic resonance with quantum numbers not accessible to $q\bar{q}$ states would be the best candidate for an exotic meson, and if it could also be ascertained that it is a glueball or a hybrid, it would be of tremendous conceptual importance, as it would provide the experimental proof of the nonabelian nature of QCD. Whether gluonium or hybrid states (especially those with exotic quantum numbers) exist and are directly observable in invariant mass plots depends on the details of the dynamics, which technically are not yet under control. In particular, whether some gluonium or hybrid states could be narrow, and therefore more easily observable, and whether their production rate in a given reaction could be sizable cannot be firmly predicted. However light meson spectroscopy has discovered a large number of states with quantum numbers accessible to $q\bar{q}$ systems, of which the greatest majority appear well understood in terms of the quark model, but a number of states are in excess to those that cane be accommodated in SU(3) multiplets. This fact hints at the existence of mesonic structures of new types.

GLUEBALLS, HYBRIDS, AND OTHER LIGHT BOSONIC STRUCTURES

Besides glueballs and hybrids, other light bosonic structures have been predicted to exist in addition to $q\bar{q}$ mesons. These structures are multiquark states with $qq\bar{q}\bar{q}$ constituents and baryonium states with $qqq\bar{q}\bar{q}\bar{q}$ constituents. These objects are compatible with the QCD lagrangian and would represent a new category of quark matter, but their observation is of less fundamental importance than that of an authentic glueball or hybrid state, as their existence would not prove direct gluon-gluon coupling due to the nonabelian structure of QCD. Again the values of mass, width, and decay channel branching ratios for four-quark and baryonium states cannot be calculated rigorously, but are evaluated on the basis of models. Four-quark and baryonium states are of high relevance in connection with glueballs and hybrids because of the possibilities of confusion and of mixing for states with exotic quantum numbers and because of the need to interpret the nature of observed states with nonexotic quantum numbers that are in excess of the available slots in the SU(3) nonets of classical $q\bar{q}$ mesons.

A STRATEGY FOR GLUEBALLS AND HYBRID OBSERVATION

Because of the fundamental importance of the observation of even only one authentic glueball or hybrid state, and because of the excitement caused by the observation of structures in excess of those that can fit into the $q\bar{q}$ SU(3) meson classification scheme, a lot of experimental work is underway at different facilities with different beams and experimental apparatuses in formation and production experiments. There is a consensus that glueball and hybrid finding and assessment requires complementary experiments that would permit us to identify with different formation or production mechanisms both well-known classical mesonic resonances and new states, and to measure several of their decay channels.

In the absence of definite predictions concerning special decay channels, efforts have been made, or are in progress, to extend detection, identification, and trigger capabilities to different decay channels containing also (or dominantly) neutral particles that decay into gammas and/or kaons. However in order to convince everybody of the existence of glueballs and hybrids it will be necessary to find states with exotic quantum numbers.

Formation experiments do not give access to states with exotic quantum numbers, while production experiments can. However there is no guarantee that there are narrow states with exotic quantum numbers, and if glueballs and hybrids with exotic quantum numbers are broad and are not noticeable for that reason, the search for them may be inconclusive.

Fortunately differential measurements, besides making possible high-precision measurements when the peaks in the spectra are narrow, offer a tool to unravel broad signals, if the formation or production branching ratio of some state that is observed in a given decay channel depends appreciably on the discrete quantum numbers of the intial state.

DIFFERENTIAL MEASUREMENTS IN FORMATION AND PRODUCTION

In order to perform differential measurements in meson spectroscopy, the detection conditions for the decay products of a resonance searched in a given decay channel must be frozen, and the formation or production conditions of the resonance have to be changed. In other words, polarization experiments of a new type appear necessary to give a conclusive answer to the question of the direct coupling between gluons. In these experiments

the selected final state is no longer the elastic scattering or charge exchange of traditional spin polarization experiments, but instead an exclusive final state that contains the decay particles of one of the decay
channels of a mesonic resonance, and the polarization that is changed is
not necessarily the spin polarization of classical polarization experiments,
but is the distribution of angular momentum or of isospin or of spin or of
parity of the initial state. Needless to say, this approach not only gives
a long-term perspective to the issue of giving an experimental answer to
the question of the existence of glueballs and hybrids, but will furnish
quantitative tools to study their spectroscopy. Moreover it will provide
as a byproduct a data bank invaluable for the study of the process of
hadronization by giving for narrow conventional light mesons the dependence
of their production rates on the discrete quantum numbers of the initial
state.

For the purpose of differential measurements in formation experiments
in colliding-beam facilities, the beam polarizations or the beam particles
should be changed, and two energy scans should be done through the same
energy windows under two different polarization conditions (e.g., switching
from e^+e^- to p^+p^- in storage rings, or changing the polarization of the
beams in e^+e^- and p^+p^- storage rings). For the same purpose, in formation
experiments with fixed targets, the target and beam polarization should be
changed (e.g., in $p\bar{p}$ collisions with a polarized hydrogen jet target of a
polarized $\bar{p}$ beam, if it will be possible to get stored polarized $\bar{p}$ beams)
and two energy scans should be run through the same energy window.

For the purpose of differential measurements in production experiments
--which are more important in the context of glueball and hybrid searches,
as they provide access to exotic quantum numbers--it is necessary to vary
the distribution of discrete quantum numbers of the initial state (like
spin or angular momentum or isospin or parity) and to reconstruct the decay
spectra into a given channel of a possible resonance for the two data
samples with different distributions of initial states, having selected from
the two data samples events with exclusive final states of the same type.
The variation of the distribution of discrete quantum numbers of the initial
state requires changing or selecting the beam or target particles, or the
target and/or beam polarization, or the distribution of angular momenta, or
the parity distribution.

Some of the differential measurements in the above-indicated sense are
already demonstrated as feasible in forward production with charge or
strangeness exchange, at e^+e^- machines in gamma-gamma and gamma-gammastar
production experiments and in $p\bar{p}$ experiments at rest. Certain types of differential measurements, like those with variation of the angular momentum of
the initial state in $p\bar{p}$ at rest, offer access to a limited energy window.
Others, like two-photon production, suffer from too low luminosity. Others
are not yet proven as feasible because of the necessity to overcome the
problems of obtaining polarized antiproton beams. Others, like those to be
performed by comparing $p\bar{p}$ and $\bar{p}n$ interactions, require specialized detector
developments for measuring accurately very-low-momentum spectator protons
from deuterium targets. There are then tools already available to attack
the search for low-mass glueballs and hybrids also of broad widths, and new
tools are being or could be developed to extend that search at higher
energies. In this perspective LEAR has a unique position both because of
present possibilities and prospective developments.

DIFFERENTIAL MEASUREMENTS IN LIGHT MESON SPECTROSCOPY AT LEAR

For a general view of the interest and problems in light meson spectroscopy the reader is referred to the proceedings of recent workshops and

232

schools.[1,2,3] An overview of detectors used worldwide in light meson
spectroscopy is given in ref. 4, while for an overview of the LEAR program,
which includes a review of light meson spectroscopy, see ref. 5 and the
references cited therein. A recent paper[6] covers differential measurements
in meson spectroscopy with low-energy antiprotons, which have been advocated
by the author on various occasions. In that paper, which contains an extended set of references, are discussed the general principles of differential measurements in production and their feasibility, difficulties, and
limits with low-energy antiprotons. Results achieved by the ASTERIX experiment in p$\bar{\text{p}}$ at rest by applying even partially the idea of differential
measurements are recalled or anticipated. The experimental program of
differential measurements proposed with the OBELIX detector and how it
motivated and affected the detector design are discussed in detail. Also,
new possibilities of differential measurements with polarized $\bar{\text{p}}$ beams are
suggested. Below we recall schematically some of the main points developed
in ref. 6, which is also the premise and the references pool for the outlook section.

p$\bar{\text{p}}$ at Rest: KGB(L)

Experiments with p$\bar{\text{p}}$ at rest have the highest possible event rate among
those with antiprotons, as practically all the antiprotons of a low-momentum
beam can be stopped in a hydrogen target, and all those that are stopped
annihilate there. In a gas target at NTP about 50% of all annihilations
occur in S wave. Annihilation for the events with an X-ray transition of
the L lines of protonium detected in coincidence occurs in P wave in typically 99% of the cases. Therefore in a gas target that can be surrounded
with an XDC[7] detector--which has a high detection efficiency for protonium
X-rays and a low X-ray background caused by annihilation products--two-event
samples can be collected with different distributions of angular momentum of
the initial states. This possibility, which has been proven and exploited
by the ASTERIX experiment, has been called Angular Momentum Polarization
(AMP[6]). The method of differential measurements, where the same decay
channels from the same exclusive final state are investigated and compared for two data sets with different angular momentum polarization,
has been called KGB(L).[6]* Application of this method on top of classical
methods for the analysis of final states has permitted us to establish the
existence of a new meson, the AX(1565),[8] which is not too broad and produces
a peak directly observable in the mass spectrum of the data with X-rays in
coincidence. Application of this method, even in an incomplete way, in the
case of the E(1420) meson, which again gives a peak directly observable in
the mass spectrum of its decay products, has given a new clue to establish
its quantum numbers[9] via the observation of the variation of the production
branching ratio by changing the AMP of the initial states. Measurements
with better statistics and quantitative application of KGB(L) should make
it possible, for instance, to observe in p$\bar{\text{p}}$ annihilations at rest in gas
the D(1285) meson, which has not been observed in annihilations in liquid,
and in general to observe and study broad structures not directly visible
in the mass spectra, but emerging when using one spectrum with a given AMP
as background for the other with a different AMP. A quantitative study of
the annihilation branching ratios from the various S-wave and P-wave initial
states is necessary to use KGB(L) for searching for broad states. There is
no guarantee of course of finding broad states; however, just the fact that
annihilation is studied with about 90% P-wave AMP has prompted a new state
in the analysis of the simplest final state that can be studied in production.[8] New states are very likely just around the corner, analyzing with
the KGB(L) approach yielding more complex final states. The study of the

---

*From the initials of colleagues who have been vital for the realization
 of the central detector of the ASTERIX experiment.

AMP dependence of the production branching ratios of all accessible mesons already observed in excess for the SU(3) nonets classification will give one more clue to establishing their quantum numbers and quark and gluon content. KGB(L) requires an adequate detector for the selection of the initial state and complete detectors for the identification and measurement of the final states. For the initial state the ideas for a XDC/SPC detector[7,10] have materialized in the ASTERIX XDS/SPC,[11,12] and a detector with improved performance is being commissioned by the OBELIX collaboration. Concerning the final state, two complementary general-purpose detectors are in preparation. One, Crystal Barrel, optimizes gamma detection with high calorimetric energy resolution and triggering on neutral particles decaying into gammas. The other, OBELIX, optimizes kaon detection and trigger capabilities, and gamma detection with high granularity and conversion point spatial resolution.

p$\bar{\text{p}}$ versus $\bar{\text{p}}$n Annihilations: KGB(I)

Measurement of the low-momentum spectator proton in antiproton deuterium interactions makes possible full kinematic reconstruction of $\bar{\text{p}}$n annihilations. Switching from p$\bar{\text{p}}$ to $\bar{\text{p}}$n data looking at the same decay channels in exclusive final states of the same type permits us to vary the isospin content of the initial state (I = 1 in $\bar{\text{p}}$n, I = 0 and 1 in p$\bar{\text{p}}$). Differential measurements are therefore possible, and the method has been called KGB(I).[6*] The event rates at rest can be as high as in KGB(L). The energy window can be enlarged appreciably by measurements in flight, but of course with lower interaction rates as a thin target is necessary to measure the spectator proton. Unlike KGB(L), in KGB(I) the final states of the two sets of measurements cannot be exactly identical because of the difference in total charge. Therefore only a part of the experimental factors discussed in ref. 6 factorize away in this kind of differential measurement. An SPC[10] detector can be employed to detect and measure recoil protons with very low momentum thresholds.

pp in Flight with a Polarized Stored Beam and Jet Target: KGB(S)

Polarized hydrogen jet targets have long been operated,[13] and could give viable luminosities. Schemes for polarizing a circulating antiproton beam have been proposed and are investigated experimentally with circulating proton beams at Indiana and at TARN. Under the hypothesis that one scheme or the other will give polarized antiproton beams in LEAR, we have considered in ref. 6 differential measurements in production where the distribution of spin singlet and triplet initial states is changed without altering the other experimental conditions. This method has been labeled KGB(S), and would have all the main advantages of KGB(L), plus the possibility of having a larger phase space by working at higher energy. The detector of the final state of course would suffer from the limitations imposed by the beam pipe and by the jet, and it would be hard to cover all final states with the same performances achievable in experiments at rest. Of course, in addition to differential production experiments that require only one fixed beam energy, differential formation experiments could be run mainly to study the spin dependence of the formation cross section of resonances. That would give us a powerful tool to establish their quantum numbers.

OUTLOOK

The importance of giving a conclusive proof of the existence of direct gluon-gluon interactions requires a global approach that encompasses the possibility that glueballs and hybrids are broad. This approach needs differential measurements where LEAR is very competitive, in the low-mass region, and can be rather competitive, for masses between protonium and

2.4 GeV/c^2. The low-mass region can be covered with p$\bar{\text{p}}$ and $\bar{\text{p}}$n annihilations at rest and the use of KGB(L) and KGB(I) differential measurements. These measurements cannot exploit the full antiproton production rate of 10^7 $\bar{\text{p}}$ sec^{-1} available at CERN because of rate limitations of the present detectors. At higher masses KGB(I) differential measurements can be employed with substantially high interaction rates in production experiments in flight since all the available antiproton intensity can be exploited as only a percentage of the beam interacts in a thin target. For antiproton economy, jet target operation with hydrogen and deuterium jets would be advantageous in order to exploit KGB(I) in flight, but the combined measurement of the slow spectator protons together with the annihilation products over a large solid angle is very difficult because of the constraints of the jet and of the beam pipe. The two experiments to be used mainly in production which are presently in preparation at LEAR, Crystal Barrel and OBELIX, can apply KGB(L) and KGB(I). OBELIX has been conceived and designed with this in mind as its main thrust for meson spectroscopy,[6] besides the improvements over ASTERIX in final state reconstruction, identification, measurement, and trigger possibilities.

Looking at the historical development of light meson spectroscopy (and in particular at the time that was needed to carry out part of the physics program for the LEAR experiments of the first generation that discovered something new and significant), it becomes clear that this chapter of physics will not be exhausted in a few years, and that the detectors should still be active on the floor for cross-confirmations and observations of other decay channels when data analysis will have given indications of exotics.

In order to perform KGB(L), Crystal Barrel will have to operate at lower beam momentum and be equipped with a XDC/SPC detector[7,10] for detecting protonium X-rays. In the long range, Crystal Barrel would not be able to profit from the huge event rate possible at rest because of the rate limitations imposed by the crystals; however, the experiment would be extremely valuable for differential measurements in flight, owing to its unique capabilities for final states with several neutrals and the high, but tolerable, interaction rates in a gas target.

OBELIX is instrumented with detectors developed at hadron machines, like the AFS JET chambers that can stand interaction rates of 10^6 sec^{-1}, and logistic provisions for a smooth upgrading of its detection, rate, and trigger capabilities are integrated in its design (Z chambers on the periphery of the AFS JET chambers, Cerenkov between AFS JET and TOF, radial segmentation of the SPC in order to reduce the rate limitations caused by the long radial drift path). Once structures have been identified with beam intensities below 10^5 sec^{-1}, more and more selective triggers on the final state could be employed to gather sizable statistics with both of the two initial state distributions necessary for differential measurements. A smooth upgrading of detection, rate, and trigger capabilities would permit us to investigate other rare decay channels--again with two different initial state distributions and the same detection and trigger biases on the final state--exploiting interaction rates up to 10^6 sec^{-1}.

Completion of work in the field of light meson spectroscopy requires not only peak search in new unexplored channels, but also patient and quantitative work in order to go beyond the excitement of the first observed exotics to bring things to the point where new states will be predicted and subsequently observed experimentally at the predicted place. On the experimental side the bottleneck in such a perspective will very likely be brainpower to carry out the data analysis thoroughly. For young physicists each annihilation or decay channel would provide a good argument for a thesis, and data taking could start at the very beginning of their PhD work.

There are discussions concerning the future of LEAR and fears that the facility could be disaffected or shipped elsewhere some years hence. I don't share these fears, for at least two reasons. The first is that the task in front of light meson spectroscopy is fundamental. The laboratory where the mediators of the weak force have been discovered in p$\bar{\text{p}}$ annihilations at the SPS collider and where evidence of their direct coupling will be searched for in the next few years at LEP is the same where, in 1963, before the invention of QCD, a resonance called E (for Europe) was discovered in p$\bar{\text{p}}$ annihilations at rest.[14] That resonance has not yet found a position in conventional multiplets, and it is a good candidate to be a glueball. It is a must to assess the nature of the coupling of the mediators of strong interactions between quarks, and the program of meson spectroscopy should not be interrupted but rather strengthened and should feature a strategy that is based on properties where LEAR offers unique advantages. The danger is otherwise not so much that the facility may be shut down, but rather that the program of spectroscopy is carried out incompletely, or worse, in the style of hurried hunting for only narrow peaks of which the baryonium saga has witnessed several examples.

The second reason is that LEAR offers a unique test facility for vertex detectors for LHC. One could fill the beam pipe that goes together with a vertex detector with hydrogen gas, and with a 100 MeV/c incoming beam all antiprotons would annihilate within a few centimeters of the detector center. Charged pions and kaons and gammas with momenta from typically 100 to 700 MeV/c are emitted in the average isotropically, thus reproducing--apart from the average energy--the conditions typical of central collisions but with the enormous practical advantage of operating at an extracted beam. The beam intensity could be increased continuously following the degree of reliability and of performance of the detector under test assessed at lower intensities. LEAR can accept from the accumulator some 10^{10} $\bar{\text{p}}$ per transfer, and then deliver a continuous spill that can last a few hours and can be shortened at will, so that even the extremely challenging situations of 10^9 interactions per second could be realistically tested in most of their aspects. Simple reactions like p$\bar{\text{p}}$ going to n^+n^- or K^+K^- or K^0K^0 which occur with branching ratios from 10^{-4} to a few 10^{-3} can be used to verify alignments, calibrations, and triggers. Fortunately there is space in the surroundings of the LEAR machine that could be devoted to these important tests, without interfering with an ongoing research program.

I hope that in a few years viable schemes to polarize the $\bar{\text{p}}$ beam circulating in LEAR will have been found. In that case KGB(S) differential measurements would be possible in production experiments at the highest momenta accessible with LEAR. This would open up one more window to discover mesons with exotic quantum numbers in a mass range larger than the one accessible at rest, and constitutes to my view a physics motivation more compelling than the ones that stimulate the groups that explore $\bar{\text{p}}$ beam polarization.

Finally, it seems appropriate to remember that the work required to do a fine job in meson spectroscopy will give an unprecedented and in-depth experimental view of the dependence of nucleon-antinucleon interactions and of hadronization on the discrete quantum numbers of the initial state. Once settled successfully in principle, questions concerning the internal couplings of undetectable fluons via glueball spectroscopy and the dynamics of the interactions between observable hadrons will still need to be understood in fundamental terms, and the available data bank may prove to be invaluable for this purpose.

REFERENCES

1. Oyanagi, Y., Takamatsu, K., and Tsuru, T., eds., Proc. Second Int.
   Conf. on Hadron Spectroscopy, KEK Report 87-7, Tsukuba (1987).
2. Gastaldi, U., Klapisch, R., and Close, F., eds., Spectroscopy of light
   and heavy quarks, Plenum Press, New York (1989).
3. Chung, S.U., ed., Glueballs, hybrids and exotic hadrons, American
   Institute of Physics Conf. Proc. 185, New York (1989).
4. Gastaldi, U., Detectors for glueballs, hybrids and exotic hadrons, in
   ref. 3:50(1989).
5. Gastaldi, U., Survey of the LEAR physics programme in ACOL time,
   Nucl. Phys., A478;813c(1988).
6. Gastaldi, U., Spectroscopy of non-exotic and exotic light mesons, in
   ref. 2:311(1989).
7. Gastaldi, U., The X-ray drift chamber, Nucl. Instrum. Methods, 157:
   441(1978).
8. May, B. et al., ASTERIX Collaboration, Observation of an isoscalar
   meson AX(1565) in annihilation of the $p\bar{p}$-atom from P states. Phys.
   Lett. B225:450(1989).
9. Duch, K. et al., ASTERIX Collaboration, Observation and analysis of E
   mesons in $p\bar{p}$ annihilation at rest in H_2 gas, Z. Phys., C 45:223(1989).
10. Gastaldi, U., The spiral projection chamber, Nucl. Instrum. Methods,
    188:459(1981).
11. Ahmad, S. et al., ASTERIX Collaboration, The ASTERIX spectrometer.
    Nucl. Instrum. Methods, A286:76(1990).
12. Gastaldi, U. et al., The X-ray drift chamber of the ASTERIX experi-
    ment at LEAR, to be submitted to Nucl. Instrum. Methods.
13, Dick, L. et al., The CERN polarized atomic hydrogen beam target,
    report CERN-EP/80-192(1980).
14. Armenteros, R. et al., 1963, in Proc. Int. Conf. on Elementary
    Particles, Sienna, G. Bernardini and G. Puppi, eds., SIF, Bologna,
    vol. 1:287(1963),

PARTICIPANTS OF THE FOURTH COURSE OF THE INTERNATIONAL SCHOOL OF PHYSICS
January 25-31, 1990
Erice, Sicily, Italy

PARTICIPANTS

M.J. Burchell   CERN - EP
                                   CH-1211 Geneva 23
                                   Switzerland

J.F. Donoghue   CERN - TH
                                   CH-1211 Geneva 23
                                   Switzerland

C.B. Dover   Brookhaven National Laboratory
                                   Upton, Long Island
                                   New York 11973
                                   USA

V. Filippini   Dipartimento di fisica nucleare
                                     e teorica
                                   Università di Pavia
                                   Via A. Bassi 6
                                   I-27100 Pavia
                                   Italie

U. Gastaldi   INFN - Sezione di Legnaro
                                   Lab. Naz. di Legnaro
                                   Via Romea 4
                                   I-35020 Legnaro PA
                                   Italy

N. Hamann   CERN - EP
                                   CH-1211 Geneva 23
                                   Switzerland

B.L. Ioffe   Institute for Theoretical and
                                     Experimental Physics (ITEP)
                                   B. Cheremushkinskaya ul. 25
                                   117 259 Moscow
                                   USSR

M. Karliner   School of Physics and Astronomy
                                   Tel Aviv University
                                   69 978 Tel Aviv
                                   Israel

R. Klapisch   CERN - DG
                                   CH-1211 Geneva 23
                                   Switzerland

R.A. Kunne

D P N C
Université de Genève
CH-1211 Geneva 4
Switzerland

Dr. M. Lamanna

INFN - Sezione di Trieste
Via Valerio, 2
I-34100 Trieste
Italy

R. Landua

CERN - EP
CH-1211 Geneva 23
Switzerland

A. Martin

CERN - TH
CH-1211 Geneva 23
Switzerland

S. Ohlsson

Department of Radiation Sciences
University of Uppsala
P.O. Box 535
S-75121 Uppsala
Sweden

M. Pachr

Nuclear Physics Institute
Czechoslovakian Academy
 of Sciences
CS-25068 Rez
Czechoslovakia

F. Pedersen

CERN - PS
CH-1211 Genev 23
Switzerland

J.-M. Richard

Institut des Sciences nucléaires
53, Avenue des Martyrs
F-38026 Grenoble
France

W. Roberts

High Energy Physics Laboratory
Havard University
41, Oxford Street
Cambridge
Massachusetts 02138
USA

T.N. Truong

Centre de Physique théorique
Ecole Polytechnique
F-91128 Palaiseau
France

P. Volkovitsky

INFN di Frascati
C.P. 13
I-00044 Frascati (Roma)
Italy

K. Zapfe

MPI for Nuclear Physics
Postfach 10 39 80
D-6900 Heidelberg 1
RFA

INDEX

Hyperon pair
    polarization, 161, 189
    production, 14, 114, 117, 135-138, 159-163
    spin correlation, 138, 159

Instantons, 4-6, 10

Kaon,
    electromagnetic form-factor, 71
    rare decays, 55, 59, 84-89

LEAR, 36, 97, 127, 129-130
    initial physics programme, 130
    internal jet target, 167
    polarized internal target (FILTEX), 209

Meson-exchange potential, 121-124
Meson-meson molecule, see Four-quark state

Optical model, 21
OZI rule violation, 4-6, 18, 93, 97, 175

Phase transition, 7
Pion,
    electromagnetic form-factor, 52-54
    hadronic form-factor, 68-72
P-wave annihilation, 17, 34-35, 133, 149 (see also
        Protonium)
Polarized antiproton beams, 154, 209-214, 224
        (see also LEAR)
Polarized target, 196, 224-226
Protonium,
    1s level shift, 142
    S- vs P-wave annihilation, 140-141, 204, 233
    cascade, 140
    and ρ parameter, 142
Proton spin puzzle, 93-101

QCD
    unsolved problems, 2-7
    coupling constant, 2, 4
    current quarks, 3, 93
    vacuum, 4, 43
    lattice theory, 7, 30
    sum rules, 30
    perturbative, 40

Quark-antiquark vertex
    (^3P_o vs $^3S_1$ model), 26, 107, 110, 136, 159
Quark model, 93
Quarks, 94, 98
    mass, 3, 41
Quasi-nuclear bound states, 15, 18, 23, 26 (see also
        Baryonium)
    decay, 29

Regge trajectory, see Trajectory
ρ parameter, 134, 153 (see also Protonium)

Sigma term (in πN scattering), 60, 94
Six-quark states, 7-8, 120 (see also Quasi-nuclear
        bound states)
Skyrme model, 46, 97
Soliton, 46, 97
Spin observables, 17, 19, 195, 223-227
Stark mixing, 141
Stochastic cooling, 94
SuperLEAR, 155
SU(3) symmetry, 5, 174 (see also Quark model)
S-wave annihilation, see Protonium

T-diquonium, see Diquonium
Tensor force, 17, 20-22
Trajectory, 23, 106, 110

Vector meson dominance, 72, 76-78

Zweig rule, see OZI rule